Lecture Notes in Physics

Springer
Berlin
Heidelberg
New York
Barcelona
Hong Kong
London
Milan
Paris
Singapore
Tokyo

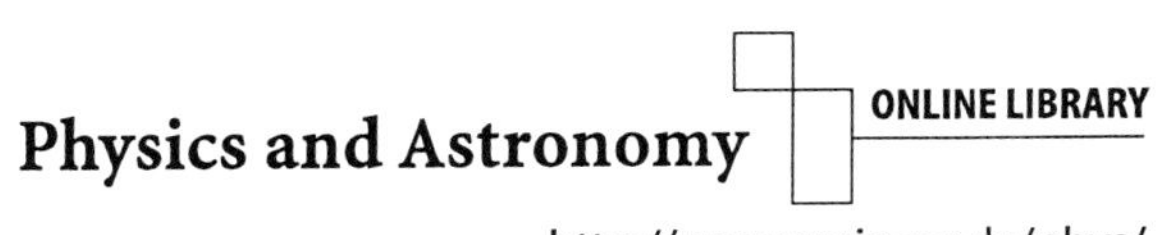

http://www.springer.de/phys/

The Editorial Policy for Proceedings

The series Lecture Notes in Physics reports new developments in physical research and teaching – quickly, informally, and at a high level. The proceedings to be considered for publication in this series should be limited to only a few areas of research, and these should be closely related to each other. The contributions should be of a high standard and should avoid lengthy redraftings of papers already published or about to be published elsewhere. As a whole, the proceedings should aim for a balanced presentation of the theme of the conference including a description of the techniques used and enough motivation for a broad readership. It should not be assumed that the published proceedings must reflect the conference in its entirety. (A listing or abstracts of papers presented at the meeting but not included in the proceedings could be added as an appendix.)
When applying for publication in the series Lecture Notes in Physics the volume's editor(s) should submit sufficient material to enable the series editors and their referees to make a fairly accurate evaluation (e.g. a complete list of speakers and titles of papers to be presented and abstracts). If, based on this information, the proceedings are (tentatively) accepted, the volume's editor(s), whose name(s) will appear on the title pages, should select the papers suitable for publication and have them refereed (as for a journal) when appropriate. As a rule discussions will not be accepted. The series editors and Springer-Verlag will normally not interfere with the detailed editing except in fairly obvious cases or on technical matters.
Final acceptance is expressed by the series editor in charge, in consultation with Springer-Verlag only after receiving the complete manuscript. It might help to send a copy of the authors' manuscripts in advance to the editor in charge to discuss possible revisions with him. As a general rule, the series editor will confirm his tentative acceptance if the final manuscript corresponds to the original concept discussed, if the quality of the contribution meets the requirements of the series, and if the final size of the manuscript does not greatly exceed the number of pages originally agreed upon. The manuscript should be forwarded to Springer-Verlag shortly after the meeting. In cases of extreme delay (more than six months after the conference) the series editors will check once more the timeliness of the papers. Therefore, the volume's editor(s) should establish strict deadlines, or collect the articles during the conference and have them revised on the spot. If a delay is unavoidable, one should encourage the authors to update their contributions if appropriate. The editors of proceedings are strongly advised to inform contributors about these points at an early stage.
The final manuscript should contain a table of contents and an informative introduction accessible also to readers not particularly familiar with the topic of the conference. The contributions should be in English. The volume's editor(s) should check the contributions for the correct use of language. At Springer-Verlag only the prefaces will be checked by a copy-editor for language and style. Grave linguistic or technical shortcomings may lead to the rejection of contributions by the series editors. A conference report should not exceed a total of 500 pages. Keeping the size within this bound should be achieved by a stricter selection of articles and not by imposing an upper limit to the length of the individual papers. Editors receive jointly 30 complimentary copies of their book. They are entitled to purchase further copies of their book at a reduced rate. As a rule no reprints of individual contributions can be supplied. No royalty is paid on Lecture Notes in Physics volumes. Commitment to publish is made by letter of interest rather than by signing a formal contract. Springer-Verlag secures the copyright for each volume.

The Production Process

The books are hardbound, and the publisher will select quality paper appropriate to the needs of the author(s). Publication time is about ten weeks. More than twenty years of experience guarantee authors the best possible service. To reach the goal of rapid publication at a low price the technique of photographic reproduction from a camera-ready manuscript was chosen. This process shifts the main responsibility for the technical quality considerably from the publisher to the authors. We therefore urge all authors and editors of proceedings to observe very carefully the essentials for the preparation of camera-ready manuscripts, which we will supply on request. This applies especially to the quality of figures and halftones submitted for publication. In addition, it might be useful to look at some of the volumes already published. As a special service, we offer free of charge LaTeX and TeX macro packages to format the text according to Springer-Verlag's quality requirements. We strongly recommend that you make use of this offer, since the result will be a book of considerably improved technical quality. To avoid mistakes and time-consuming correspondence during the production period the conference editors should request special instructions from the publisher well before the beginning of the conference. Manuscripts not meeting the technical standard of the series will have to be returned for improvement.
For further information please contact Springer-Verlag, Physics Editorial Department II, Tiergartenstrasse 17, D-69121 Heidelberg, Germany

Series homepage – http://www.springer.de/phys/books/lnpp

Agnès Maurel Philippe Petitjeans (Eds.)

Vortex Structure and Dynamics

Lectures of a Workshop Held in Rouen,
France, April 27-28, 1999

Springer

Editors

Agnès Maurel
Laboratioire Onde et Acoustique
Ecole Supérieure de Physique et de Chimie Industrielles
10 rue Vauquelin
75005 Paris, France

Philippe Petitjeans
Laboratoire de Physique et Mécanique des Millieux Hétérogènes
Ecole Supérieure de Physique et de Chimie Industrielles
10 rue Vanquelin
75005 Paris, France

Library of Congress Cataloging-in-Publication Data applied for.
Die Deutsche Bibliothek - CIP-Einheitsaufnahme

Vortex structure and dynamics : lectures of a workshop held in Rouen, France, April 27 - 28, 1999 / Agnés Maurel ; Philippe Petitjeans (ed.). - Berlin ; Heidelberg ; New York ; Barcelona ; Hong Kong ; London ; Milan ; Paris ; Singapore ; Tokyo : Springer, 2000
(Lecture notes in physics ; Vol. 555)
(Physics and astronomy online library)
ISBN 3-540-67920-0

ISSN 0075-8450
ISBN 3-540-67920-0 Springer-Verlag Berlin Heidelberg New York

Springer-Verlag Berlin Heidelberg New York
a member of BertelsmannSpringer Science+Business Media GmbH

Printed in Germany

Typesetting: Camera-ready by the authors/editor
Cover design: *design & production*, Heidelberg

Printed on acid-free paper
SPIN: 10779211 55/3141/du - 5 4 3 2 1 0

Preface

The workshop "Dynamics and Structure of vortices" was held in the Cloître des Pénitents in Rouen, France on the 27th and 28th April, 1999.

Our understanding of the structure and dynamics of vortices has improved considerably during the last few years, mainly thanks to progress in turbulence research, where these structures have been shown to play an important role. The aim of this French workshop was to gather theoreticians, computational researchers and experimentalists to illuminate various aspects of this subject. We wanted on the one hand to present the state of art, and on the other hand to collect the most recent contributions on the structure and dynamics of vortices.

This volume presents 22 articles corresponding to seminars and presentations given during this workshop. The first three articles correspond to general presentations: A. Babiano presents the two-dimensional aspects of vortices; S. Huberson and O. Daube give a review on numerical methods applied to vortical flow; and M. Rossi presents theories of vortex instability.

The following 19 papers correspond to presentations given by the participants on their research subjects related to experimental, numerical or theoretical aspects of vortices. Many of these studies are tied to related fields, such as turbulence, aerodynamics, wakes, geophysics, mixing, particles dynamics ...

The scientific committee of the workshop, A. Babiano, A. Maurel, P. Petitjeans and M. Rossi thank the CNRS for financial support through the Groupe de Recherche "Turbulence" and the Groupe de Recherche "Mécanique fondamentale des fluides géophysiques et astrophysiques", and also the Association Française de Mécanique.

We also take this opportunity to personally thank A. Babiano and M. Rossi for their collaboration in the scientific organization of the workshop; A. Babiano, S. Huberson and M. Rossi for the considerable work that went into preparing their overviews; and all the authors of this volume and participants who greatly contributed to the success of the workshop.

Paris,
June 2000

Agnès Maurel
Philippe Petitjeans

Contents

List of Contributors

Malek Abid
Institut de Recherche sur les Phénomènes Hors Equilibre
Service 252, Centre St-Jérôme
13397 Marseille Cedex 20, France
Malek.Abid@lrc.univ-mrs.fr

Bruno Andreotti
Laboratoire de Physique Statistique
Ecole Normale Supérieure
24 rue Lhomond
75005 Paris, France
Bruno.Andreotti@lps.ens.fr

Armando Babiano
Laboratoire de Meteorologie Dynamique
Ecole Normale Supérieure
24, rue Lhomond
75005 Paris, France
babiano@lmd.ens.fr

Christophe Baudet
Laboratoire des Ecoulements Géophysiques et Industriels
1025 rue de la Piscine
BP 53
38041, Grenoble cedex 9, France
Christophe.Baudet@hmg.inpg.fr

Eberhard Bodenschatz
Laboratory of Atomic and Solid State Physics
Cornell University, Ithaca
New York 14853-2501, USA

Frédéric Bottausci
Laboratoire de Physique et Mécanique de Milieux Hétérogènes
Ecole Supérieure de Physique et de Chimie Industrielles
10, rue Vauquelin
75005 Paris, France
freddy@pmmh.espci.fr

Olivier Cadot
Laboratoire de Mécanique
Université du Havre
25 rue Philippe Lebon
BP 540 76058 Le Havre cedex, France
cadot@univ-lehavre.fr

Francesca Chilla
Laboratoire de Physique
Ecole Normale Supérieure de Lyon
46, allée d'Italie
69007 Lyon, France
fchilla@physique.ens-lyon.fr

Jean-Marc Chomaz
LadHyX
Ecole Polytechnique
91128 Palaiseau Cedex, France

Alexandre Corjon
CERFACS
European Center for Research and Advanced Training in Scientific Computation
42, Av G. Coriolis
31057 Toulouse Cedex 1, France

Carlo Cossu
LadHyX
Ecole Polytechnique
91128 Palaiseau Cedex, France
carlo@LadHyX.polytechnique.fr

Olivier Daube
CEMI
Université d'Evry Val d'Essonne
Evry, France

Yvan Delbende
Université Paris-Sud, LIMSI
Bâtiment 508
91403 Orsay Cedex, France
delbende@limsi.fr

Stéphane Douady
Laboratoire de Physique Statistique
Ecole Normale Supérieure
24 rue Lhomond
75005 Paris, France
douady@peterpan.ens.fr

Uwe Ehrenstein
Laboratoire J.A.Dieudonne
Université de Nice Sophia-Antipolis
Parc Valrose
06108 Nice Cedex 2, France
uwe.ehrenstein@unice.fr

Afif Elcafsi
Université de Paris-Sud, LIMSI
Bâtiment 508
91403 Orsay Cedex, France
elcafsi@limsi.fr

Christophe Eloy
Centre Saint-Charles
Institut de Recherche sur les Phénomènes Hors Equilibre,
12, Avenue du Général Leclerc
13003 Marseille, France
eloy@marius.univ-mrs.fr

David Fabre
ONERA/DAFE
29, av. de la Division Leclerc
92322 Châtillon Cedex, France
fabred@onera.fr

Pierre Gougat
Université Paris-Sud, LIMSI
Bâtiment 508
91403 Orsay Cedex, France
gougat@limsi.fr

Rodrigo H. Hernández
LEAF-NL, Depto. Ingeniería Mecánica
Universidad de Chile
Casilla 2777, Santiago, Chile
rohernan@cec.uchile.cl

Serge Huberson
Laboratoire de Mécanique
Université du Havre
25 rue Philippe Lebon
BP 540 76058 Le Havre cedex, France
hu@cher.univ-lehavre.fr

Laurent Jacquin
ONERA
29, av. de la Division Leclerc
BP 72, 92322 Chatillon Cedex, France
jacquin@onera.fr

Satish Kumar
Department of Chemical Engineering
University of Michigan, 2300 Hayward,
3074 H.H. Dow Building, Ann Arbor,
MI 48105 USA
satishk@engin.umich.edu

Florent Laporte
CERFACS
European Center for Research and Advanced Training in Scientific Computation
42, Av G. Coriolis
31057 Toulouse Cedex 1, France
flaporte@cerfacs.fr

Arthur La Porta
Laboratory of Atomic and Solid State Physics
Cornell University, Ithaca
New York 14853-2501, USA

Stéphane Leblanc
Laboratoire d'Analyse Non Linéaire Appliquée
Université de Toulon et du Var
BP 132, F-83957 La Garde Cedex, France
sl@univ-tln.fr

Thomas Leweke
Institut de Recherche sur les Phénomènes Hors Equilibre
Université Aix-Marseille
12, avenue Général Leclerc
13003 Marseille, France
leweke@marius.univ-mrs.fr

Stéphane Le Dizès
Institut de Recherche sur les Phénomènes Hors Equilibre
Centre Saint-Charles
12, Avenue du Général Leclerc
13003 Marseille, France
legal@marius.univ-mrs.fr

Anne Le Duc
Lehrstuhl für Fluidmechanik,
Technische Universität München
Boltzmannstrasse 15
D-85748 Garching, Germany
leduc@flm.mw.tum.de

Patrice Le Gal
Institut de Recherche sur les Phénomènes Hors Equilibre
Centre Saint-Charles
12, Avenue du Général Leclerc
13003 Marseille, France
legal@marius.univ-mrs.fr

Lionel Le Penven
Laboratoire de Mécanique des Fluides et d'Acoustique
Ecole Centrale de Lyon
BP 163, F-69131 Ecully Cedex, France
lepenven@mecaflu.ec-lyon.fr

Sébastien Manneville
Laboratoire Ondes et Acoustique
Ecole Supérieure de Physique et Chimie Industrielles
10, rue Vauquelin
Paris 75005, France
Sebastien.Manneville@espci.fr

Agnès Maurel
Laboratoire Ondes et Acoustique
Ecole Supérieure de Physique et Chimie Industrielles
10, rue Vauquelin
Paris 75005, France
Agnes.Maurel@espci.fr

Patrice Meunier
Institut de Recherche sur les Phénomènes Hors Equilibre
Université Aix-Marseille
12, avenue Général Leclerc
13003 Marseille, France
meunier@marius.univ-mrs.fr

Frédéric Moisy
Laboratoire de Physique Statistique
Ecole Normale Supérieure
24, rue Lhomond
75005 Paris, France
moisy@physique.ens.fr

Caroline Nore
Université Paris-Sud, LIMSI
Bâtiment 508
91403 Orsay Cedex, France
nore@limsi.fr

Philippe Petitjeans
Laboratoire de Physique et Mécanique de Milieux Hétérogènes,
Ecole Supérieure de Physique et Chimie Industrielles
10, rue Vauquelin
75005 Paris, France
phil@pmmh.espci.fr

Jean-François Pinton
Laboratoire de Physique
Ecole Normale Supérieure de Lyon
46, allée d'Italie
69007 Lyon, France
pinton@physique.ens-lyon.fr

Alecsandra Rambert
Université de Paris-Sud, LIMSI
Bâtiment 508
91403 Orsay Cedex, France
alexan@limxi.fr

Maurice Rossi
Laboratoire de Modélisation en Mécanique
Université Pierre et Marie Curie
4, place Jussieu
75252 Paris cedex 05, France
maur@ccr.jussieu.fr

Catherine Simand
Laboratoire de Physique
Ecole Normale Supérieure de Lyon
46, allée d'Italie
69007 Lyon, France
csimand@physique.ens-lyon.fr

Denis Sipp
ONERA
29, av. de la Division Leclerc
BP 72, 92322 Chatillon Cedex, France
sipp@onera.fr

Jean-Hugues Titon
Laboratoire de Mécanique
Université du Havre
25 rue Philippe Lebon
BP 540 76058 Le Havre cedex, France

Greg Voth
Laboratory of Atomic and Solid State Physics
Cornell University, Ithaca
New York 14853-2501, USA

Charles H.K. Williamson
Mechanical & Aerospace Engineering
Cornell University
Ithaca, NY, USA

Non-homogeneous/Non-local Two-Dimensional Dynamics

Armando Babiano

Laboratoire de Météorologie Dynamique, Département de Géophysique de l'ENS de Paris, 24, rue Lhomond, 75005, Paris, France

Abstract. In this short review we examine two interesting aspects of the two-dimensional dynamics. On the one hand, we document the non-homogeneous/non-local description of the two-dimensional flows. On the other hand, we discuss the anomalous energy transfer properties of the non-linear term of the 2D Navier-Stokes equations induced by the nonlocal dynamics.

1 Introduction

In the past years a number of theoretical, numerical and experimental studies have been devoted to the understanding of the fluids dynamics which displays two-dimensional geometrical properties [1] (see [2] for extensive references). Recently, activity in this domain has intensified in connection to the development of the concept concerning the $non-homogeneous$ and *nonlocal* two-dimensional double-cascade scenario [3] [4] [5] and its relevance to describe the meso-scale transport and diffusion in dynamical oceanography as well as in other domains of atmospheric geophysical fluid dynamics [6] [7] [8]. This is the case, for example, of many large-scale geophysical flows when the synoptic circulation and their turbulent regimes can be represented in the framework of a quasi-two-dimensional system. The assumption is that three-dimensional effects basically enter as negligible correction compared with the robust constraint imposed by two-dimensional quadratic invariants of motion: kinetic energy and enstrophy.

It is well known from numerical simulations that two-dimensional dynamics distinguished by the presence of long-lived organized coherent shears and vortices which are characterized by high energy and vorticity concentrations. The contribution on the dynamics of coherent structures, including their time-evolution and interactions, plays an important role to the understanding of the spatial flows structure. Coherent vortices develop basic inhomogeneities and nonlocal dynamical properties which, in general, cannot be considered in phenomenological analyses or within the framework of statistical approachs.

As shown by Okubo [9] and later by Weiss [10], quasi- two-dimensional fields may be described in term of elementary partitioning following a simple criterion involving the distribution of vorticity and strain [3] [11] [12] [13] [14] [15] [9]. The relative dominance of one over the other allows us to define strongly elliptic domains (high vorticity concentrations which develop nonlocal dynamics), strongly hyperbolic domains (deformation cells characterized by high strain and turbulent energy) and moderate elliptic/hyperbolic domains (background field which, from dynamical point of view, may be considered as homogeneous and local turbulent medium). The basic issue consists in understanding whether different topological domains - which clearly represent the non-homogeneous character of the flow - may be associated with different features of the main non-linear processes characterizing the fluid motion: energy and enstrophy transfer, development of dynamical barriers, transport and turbulent diffusion, Lagrangian particle dispersion etc...

Such specificities of two-dimensional dynamics, including non-homogeneous and nonlocal features, are prominent not only within the framework of geophysical applications. Many laboratory observations indicate that often the fluid motion is "geometrically non-uniform" in the sense that the three-dimensional flow's structure is, in fact, strongly connected to quasi-two-dimensional regimes (see, for example, some flows of practical interest as rotating [9] and stratified flows, turbulent wake behing an obstacle [17] [18], etc...). To correctly describe such complex flows it is instructive to consider the main generic non-homogeneous/nonlocal two-dimensional features.

The purpose of the present review is to document the "non-homogeneous" two-dimensional description and to discuss the implication of the nonlocal dynamics to the energy transfer properties. We emphasize the 2D topology in the physical space following the Okubo-Weiss criterion - including further developments [15] - and the anomalous energy transfer properties of the non-linear term of the 2D Navier-Stokes equation generated by the non-homogeneous/nonlocal 2D dynamics in the presence of vortical coherent structures [19].

The paper is organized as follows. The basic properties of the two-dimensional turbulent dynamics as well as the numerical model used in reported experiments and the introduction to the analysis of the topology of strain and vorticity in physical space will be described in section 2. In section 3 we analyze the anomalous transfer properties of the nonlinear term of the 2D Navier-Stokes equation induced by the nonlocal dynamics. Finally, section 4 summarizes a brief general discussion.

2 Dynamics of 2D turbulence

The simplest model which contains the basic features of quasi-two-dimensional turbulent regimes is related to the barotropic equation for the relative vorticity in an incompressible fluid. Within the framework of the Taylor- Proudman theorem [20], we get

$$\partial_t \omega + J(\omega, \psi) = F - (D_\omega + D_E), \tag{1}$$

where the scalar $\omega = \boldsymbol{curl v}$ in two dimensions is the relative vorticity component orthogonal to plane of motion $z = const.$ defined as

$$\omega(t, x, y) = \nabla^2 \psi(t, x, y) = (\partial_x v - \partial_y u). \tag{2}$$

ψ refers to the stream function in the two-dimensional velocity field $\boldsymbol{v}(t, x, y) = \{u, v\}$, J is the two-dimensional Jacobian, $F(t, x, y)$ is an external energy and vorticity source; $D_\omega(t,x,y)$ and $D_E(t,x,y)$ are the vorticity sink associated with the dissipation at small scales and the large scale friction respectively.

2.1 Direct numerical simulations

The numerical experiments founded on the above approximation are performed integrating (1) on a doubly periodic square domain $(2\pi, 2\pi)$ by using a pseudo-spectral approximation [21]. The simplest form for the forcing F is defined by keeping the amplitude of the zonal mode $(0, k_I)$ constant in time. Dissipation of vorticity D_ω near the small cutoff scale is usually parametrized by the "iterated Laplacian method" [21]. A linear friction D_E dissipates energy at larger scales. We have in this case:

$$D_\omega + D_E = t_c^{-1}(-l_c^2\nabla^2)^\alpha \omega - t_d^{-1} l_d^{-2} \psi, \tag{3}$$

where α characterizes the hyperviscosity, l_c and l_d are the smallest and the largest resolved scales respectively; t_c and t_d are characteristic dissipation times. When the cutoff scale is of the order of the input scale, it is preferable to replace the hyperviscous dissipation at small scales by the "anticipated potential vorticity method" [22] and [23].

The equation (1) reduces to the two-dimensional Navier-Stokes equation for $\alpha = 1$ and $t_d \to \infty$. In the stationary conditions, when $F - (D_\omega + D_E) \approx 0$ in a statistical sense, $Z(t) \approx const.$ and $E(t) \approx const.$, where $Z = 1/2\langle\omega^2\rangle_x$ is the enstrophy, $E = 1/2\langle \boldsymbol{v}.\boldsymbol{v}\rangle_x$ is the turbulent kinetic energy and $\langle . \rangle_x$ refers to the average over all position vectors $\boldsymbol{x}$ in the flow domain. In the limit of zero viscosity (high Reynolds number) or when the forcing is compensated in a statistical sense by the dissipation, the turbulent kinetic energy E and the turbulent enstrophy Z are quadratic invariants of the motion. Two stationary cascade processes are present from injection scales in this case: a direct cascade of enstrophy toward small dissipative scales, and an inverse cascade of energy toward large scales. As was previously mentioned, the two-dimensional turbulent dynamics is also characterized by the existence of high concentrations of vorticity and energy into coherent structures which depend both on the localness of the flow dynamics in physical space. Moreover, it is well known that mechanism for direct two-dimensional enstrophy cascade is not universal and that energy spectra usually produced by the numerical simulations present typical shapes steeper than Kraichnan's theoretical k^{-3} prediction in the enstrophy range [1]. In the inverse energy cascade, linked with vortex merging and aggregation processes of the coherent structures, the universality is more robust. Energy spectra in $k^{-5/3}$ consistent with Kolmogorov's locally homogeneous energy cascade are indeed observed in numerical simulations when the two-dimensional turbulent system is maintained in a statistically steady state. In such situations - which contitute the simplest regimes for theory - the energy and the enstrophy transfer may be considered, in a first approximation, as locally homogeneous and isotropic at inertial length-scales even if in the physical space the existence of coherent vortices and shears induces a basic inhomogeneity of the velocity fields. However, locally homogeneous and isotropic character of the inverse cascade of energy is very precarious. In fact, the scale-uniformity of the energy transfer is not entirely guaranteed by 2D numerical simulations when the system is maintained in a equilibrium state using a large-scale dissipation [4]. The intermittency is surprisingly weak [24] [25] but the variance of the energy transfer per unit mass is scale-dependent in this case [26].

An interesting case is when the 2D dynamics deviates from the equilibrium state in the energy-cascading range. This is a typical situation when the inverse energy flux is not dissipated at largest scales in a confined flow domain. In this case, a non-stationary evolution of the inverse energy cascade characterized by a growth of the vortical structures and an accumulation of the turbulent kinetic energy at large scales is observed [5] [19]. In nature nonlocal properties of the 2D turbulent dynamics in the presence of vortical coherent structures intensified in the energy-cascading range. Then, the two-dimensional dynamics obeys the new transfer properties of the non-linear term of the Navier-Stokes equation which irreversibely develop at all scales the inhomogeneities induced by the nonlocal features [19].

Figure 1 show the vorticity levels in a two-dimensional direct numerical experiment with resolution 256 × 256 grid ($k_I = \pi/\ell_I = 30$). The experiment consisted of two stages. First, the equation (1) is integrated from random intitial conditions with a finite characteristic dissipation time t_d until a statistically steady state is reached (panel *a*). Then, we take this state as a new initial condition and we set to zero the large-scale dissipation ($t_d \to \infty$) but the forcing and the dissipation at small scales are retained. We integrated the system in the new conditions over $1600 t_R$ where t_R is the turnover time defined in terms of the enstrophy Z as $t_R = Z^{-1/2}$ (panel *b*). The initial flow, in which the kinetic energy production is balanced by the dissipation, is composed of a large number of opposite sign coherent vortices having approximately the same characteristic size $\ell_I = \pi/k_I$. For later times we observe that the dynamics is dominated by two vortical systems (cyclonic and anti-cyclonic) having a characteristic size 15 times the forcing scale ℓ_I. Both the cyclonic and anti-cyclonic system are generated via a non-stationary inverse cascade of energy and constitute a quasi-equilibrated dipole.

The energy spectra compensated for Kolmogorovs's scaling $k^{5/3}$ are displayed as a function of the wave number k in Figure 2. The energy spectrum at the initial quasi-equilibrium state ($t = 0$) is consistent with the $k^{-5/3}$ Kolmogorov-Kraichnan's [1] prediction in the range $0.1 < k/k_I < 1$, where k_I is the forcing wave number; the enstrophy range presents a typical shape steeper than the theoretical k^{-3}. The observed energy spectrum behavior in the enstrophy range $k > k_I$ is almost unchanged during the unbalanced evolution of the system. In contrast, the $k^{-5/3}$ spectrum is not maintained in the energy-cascading range. For $k < k_I$, we observe a continuous growth of the spectral slope from its initial steepness to being close to $k^{-2.4}$. This shows that the dynamics in the energy-cascading range tends to be nonlocal as time increases.

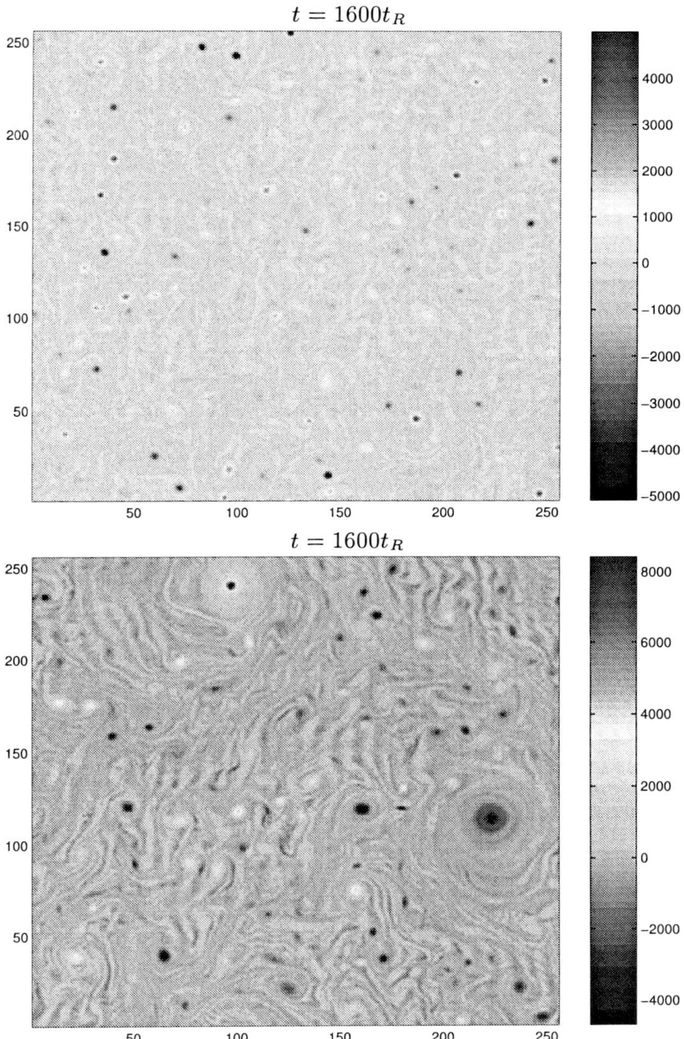

Fig. 1. Time evolution of the vorticity levels during the energy accumulation process.

Some parallelism may be established in the spectral signature between the *time evolution* in a confined domain of a 2D turbulent system not dissipated at large scale and the *spatial evolution* of the turbulent wake behind a cylinder [17] [18]. At the points near the cylinder, the Karman vortex street play a dominant role in the flow structure and a strong vorticity component aligned with the cylinder axis introduces some two-dimensional effect. In this case, the large-scale structures are confined by the turbulent wake. Figure 3a,b,c shows the energy spectra compensated for the Kolmogorov's scaling $k^{5/3}$ as a function of the

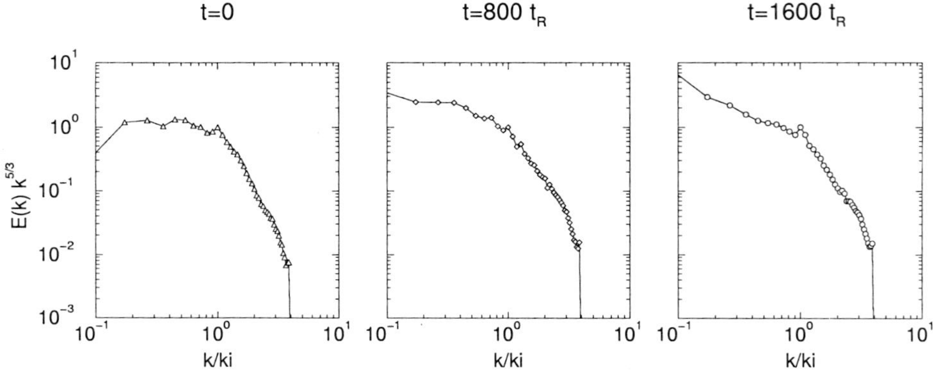

Fig. 2. Compensated energy spectra as a function of time $t = 0, 800t_R, 1600t_R$ corresponding to the 2D numerical simulation.

normalized wave number $k\eta$ at downstream distances $x/d = 3, 15, 80$ from the cylinder in the lateral position $y/d = -0.4$, where d is the cylinder diameter and η is the Kolmogorov's length scale. The spectra steepness is non universal in the turbulent wake and shows a close to k^{-3} behavior near the cylinder which clearly indicates that the dynamics in these regions is nonlocal. Far from the cylinder, the coherent structures become less and less pronounced, the dynamics remains local and the three-dimensional flow's structure tends to be more homogeneous and isotropic.

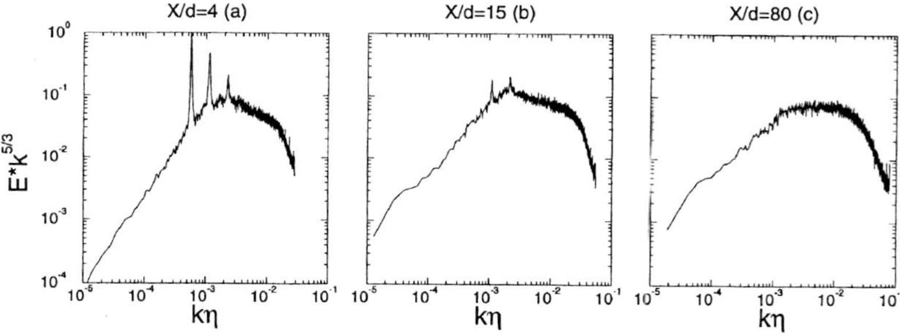

Fig. 3. Compensated energy spectra in the turbulent wake at different distances from the cylinder.

2.2 Elementary 2D-topology

To introduce this issue, we consider that the basic inhomogeneity characterizing the two-dimensional dynamics is induced by the carrying power of the long-lived stable coherent vortices and by their interactions. In other words, we consider in a simplified way a topology in physical space basically distinguishing the respective contribution of rotation and strain to the squared norm of the velocity gradients. The velocity gradient, in 2D case, is usually expressed as

$$||\nabla \boldsymbol{v}||^2 = \frac{1}{2}(\omega^2 + s^2), \tag{4}$$

where the scalar ω refers to the vorticity and s to the strain

$$\omega^2(\boldsymbol{x}) = (\partial_x v - \partial_y u)^2, s^2(\boldsymbol{x}) = (\partial_x u - \partial_y v)^2 + (\partial_x v + \partial_y u)^2. \tag{5}$$

Assuming adecuate boundary conditions as periodicity or vanishing velocity, we get

$$\langle ||\nabla \boldsymbol{v}||^2 \rangle_x = \langle \omega^2 \rangle_x = \langle s^2 \rangle_x. \tag{6}$$

In fact, the amplitude of the locally defined nondimensional quantity $\omega^2(\boldsymbol{x})/s^2(\boldsymbol{x})$ compared to 1 defines the local importance in the position point $\boldsymbol{x}$ in the flows domain of the rotation and the strain.

A more appropriate criterion has been proposed by Okubo [9] for the particle dispersion problem and later by Weiss [10]. The idea is based on the evolution equation for the two-dimensional vorticity gradient in the incompressible case. Assuming inviscid and unforced incompressible flow, from (1) we have

$$d_t \nabla \omega + \nabla \boldsymbol{v} . \nabla \omega \approx 0. \tag{7}$$

If we assume that the velocity gradients vary, in a statistical sense, slowly compared to the vorticity gradients, then (7) reduces to a linear constant coefficient equation which shows that the time evolution of the vorticity gradient is locally determined by the character of the velocity gradient eigenvalues:

$$\lambda^2(\boldsymbol{x}) = \frac{1}{4}[s^2(\boldsymbol{x}) - \omega^2(\boldsymbol{x})]. \tag{8}$$

Then, from the sign of $\lambda^2(\boldsymbol{x})$ the turbulent flow is basically divided into domains elliptic in character, where rotation locally dominates over strain (pure imaginaries eigenvalues), and domains hyperbolic in character where strain locally dominates over rotation (real eigenvalues). The relative dominance of elliptic and hyperbolic domains is measured by the sign and the amplitude of the quantity $\lambda^2(\boldsymbol{x})$.

A qualitative view of two-dimensional turbulence [3], [11], [12] shows that the distribution of λ^2 generates localized regions behaving as coherent entities. Then, the turbulent field may be further decomposed in (see [13])

(a) strongly elliptic domains where $\lambda^2 < -\lambda_s^2$ (cores of the coherent vortices), where λ_s^2 is a suitably defined threshold: $\lambda_s^2 \approx \langle s^2 \rangle_x$, for example (see relation (6));

(b) well structured hyperbolic domains where $\lambda^2 > 0$ and the locally defined kinetic energy is larger than the mean kinetic energy E; these domains, which we call deformation cells, typically surround the vortex cores;

(c) background elliptic domains where $-\lambda_s^2 < \lambda^2 < 0$ and the locally defined kinetic energy is smaller than E;

(d) background hyperbolic domains where $0 < \lambda^2 < \lambda_s^2$ and the locally defined kinetic energy is smaller than E.

The above parameterization, although heuristic, is physically motivated since it distinguishes coherent (elliptic vortex cores and surrounding hyperbolic deformation cells) from non-coherent dynamic structures (elliptic and hyperbolic background patches). Figure 4 shows an example of the distribution of $\lambda^2(\boldsymbol{x})$ produced by the two-dimensional numerical simulation at statistically steady state discussed above.

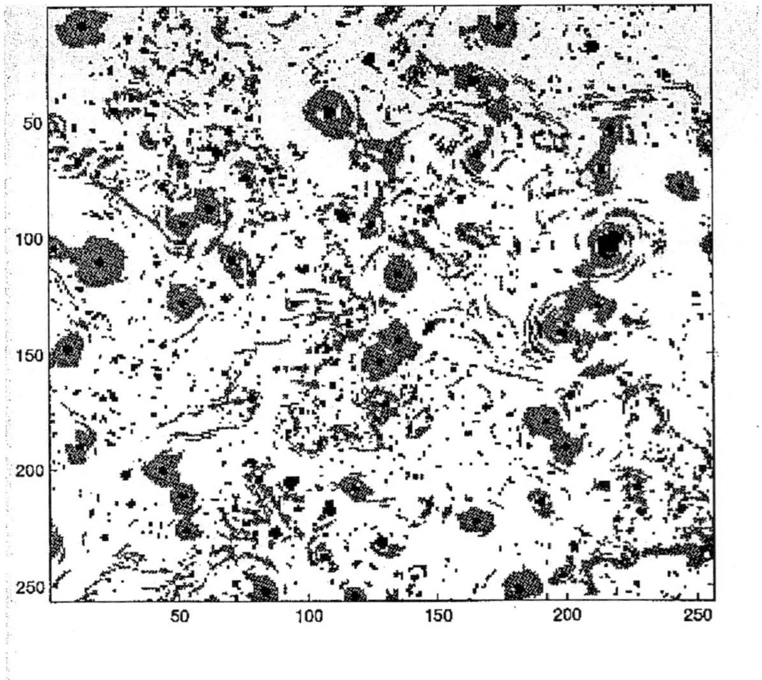

Fig. 4. The spatial distribution of λ_2; 2D numerical simulation at $t = 0$: cores of the vortices (black), deformation cells (grey), turbulent background (white).

Hyperbolic domains in the deformation cells and in the background turbulent field are associated with a large degree of Lagrangian chaoticity. In contrast, in the cores of existing coherent structures (strongly elliptic domains) the Lagrangian-Lyapunov exponents are close to zero [27]. The important observation is that the vorticity gradients provide a tendency for the alignment with the compressing eigenvector of the velocity gradient tensor [28]. The Lagrangian meaning is the following: since particle vorticity tends to be conserved in the inviscid 2D-case, then the most probable alignment of the fluid-particle trajectories is imposed by the stretching eigenvector.

The validity of the assumption which underlies the above simple criterion has been questioned by Basdevant and Philipovitch [14]. They have shown that its validity is restricted to a small fraction of the flow field, namely, the strongly elliptic centers of the vortex cores and the regions near saddle points between corotating vortices satisfying a constraint on the curvature of the pressure field. The main conclusion is that even if the criterion (8) seems pertinent to an approximative analysis of 2D dynamics the crucial topological properties are linked to the low and high values of the pressure field. This conclusion is consistent with the study performed by Larchevêque [29] on the diagnostic relation for the pressure in the simplifying, barotropic, statistically steady incompressible case. Taking the divergence operator of the incompressible Euler equation, this diagnostic relation is naturally related by

$$\nabla^2 P(\boldsymbol{x}) = -\frac{1}{4}[s^2(\boldsymbol{x}) - \omega^2(\boldsymbol{x})], \tag{9}$$

which shows that asymmetry characterizing the spatial behaviours of strain and vorticity is linked to the asymmetry of the pressure field P.

A more complete discussion of this problem was recently addressed by Hua and Klein [15] and [30] taking into account the equation which defines the particle Lagrangian accelerations $\boldsymbol{\Gamma} = \frac{d}{dt}\boldsymbol{V}$, where $\boldsymbol{V}$ is the Lagrangian velocity, in the incompressible inviscid case:

$$\boldsymbol{\Gamma} = -\boldsymbol{\nabla} P. \tag{10}$$

The Lagrangian acceleration eigenvalues which appears in the advection problem are, in this case [15],

$$\lambda_{\pm} = \frac{1}{4}(s^2 - \omega^2) \pm \frac{1}{2}\sqrt{(d_t s)^2 - (d_t \omega)^2}. \tag{11}$$

We see that the criterion (8) in the Lagrangian problem is obtained by neglecting the radicand in the right-hand side of (11). The second-order $HK-$ *criterion* (11) corresponds to the eigenvalues of the acceleration gradient tensor - the Hessian matrix of pressure that includes all nonlinear and nonlocal properties of the flow dynamics - while the criterion (8) corresponds to the eigenvalues of the velocity gradient tensor. This result is consistent with the well known property of the Navier-Stokes equation expressed in Lagrangian coordinates: nonlinearity is indeed transferred to the pressure field terms. The calculation of the second term in the right-hand side of (11), which is linked with the Hessian matrix of the pressure, requires to solve a Poisson equation over the whole flow domain [15]. This means that stirring properties of 2-D turbulence are indeed nonlocal. These arguments rectify the analysis performed in [28] and have led to show that the most probable alignment of the tracer gradient vectors (or the fluid particles trajectories) slightly differs from the stretching eigenvector [31] .

3 2D energy cascade

The purpose now is to show how the vortical structures and the nonlocal dynamics act on the energy transfer in two-dimensional flows. The interesting issue releases from the analysis of the energy cascade based on the energy flux sustained by the non-linear term of the 2D Navier-Stokes equation [19]. In general, the function of the nonlinear term is to transfer the energy among the various scales of motion following a cascade process, locally homogeneous or not, which tends to smooth the inhomogeneities generated by the external constraints acting on the flow. By definition, fully developed homogeneous state is quasi-isotropic in the sense that the scale-by-scale distribution of both the kinetic energy and energy transfer are quasi-independent of the azimuthal direction. In a non-homogeneous situation, the ratio between "*transversal*" and "*longitudinal*" to given direction $\boldsymbol{\ell}$ structures of the velocity field deviates from the universal prediction established in the framework of Karman's relation valid for homogeneous flow [21]. In this case, it is advisable to decompose the turbulent energy in two basic components with respect to the vector $\boldsymbol{\ell}$: "*tranversal/rotating*" and "*longitudinal*" parts. It appears that, when the dynamics is nonlocal, then only the longitudinal component of the total kinetic energy participates in such a *cascade* process: the *transversal/rotating* part of the energy anomalously accumulates at the scale where it is generated.

3.1 Energy flux

For a given length-scale ℓ characterizing a control surface Σ_ℓ, the flux per unit mass induced by the non-linear term in the equation of two-dimensional motion of incompressible flow is given by the integral over the contour S_ℓ containing the surface Σ_ℓ. We get

$$\sigma_\ell(\boldsymbol{x}, t) = \frac{1}{\Sigma_\ell} \oint_{S_\ell} (v^2 + P)\boldsymbol{v}.\boldsymbol{n} d\boldsymbol{x}_c, \tag{12}$$

where σ_ℓ denotes the energy flux, $\boldsymbol{v}$ is the velocity, $\boldsymbol{n}$ is the unit vector normal to the boundary element dx_c and P denotes the pressure.

Let Σ_ℓ be a circle of radius $|\boldsymbol{\ell}|/2$ centered on $\boldsymbol{x}$; in every position $\boldsymbol{x}_c$ on the boundary S_ℓ of the control surface Σ_ℓ we can define $v_{||}(\boldsymbol{x}_c) = \boldsymbol{v}(\boldsymbol{x}_c).\boldsymbol{\ell}/\ell$ and

$v_\perp(\boldsymbol{x}_c) = \boldsymbol{v}(\boldsymbol{x}_c) \times \boldsymbol{\ell}/\ell$ where $\boldsymbol{\ell}$ is applied to x_c. Note that, $v^2(\boldsymbol{x}_c) = v_{||}^2(\boldsymbol{x}_c) + v_\perp^2(\boldsymbol{x}_c)$, where $v_\perp^2$ is the *transversal/rotating* component of the kinetic energy. The expansion in the first term of the integrand of (12) yields

$$\oint_{S_\ell} v^2 \boldsymbol{v}.\boldsymbol{n} dx_c = \frac{1}{2} \oint_{S_\ell} \{[v_{||}^3(\underline{\boldsymbol{x}}_c) - v_{||}^3(\boldsymbol{x}_c)] - B\} dx_c, \tag{13}$$

where

$$B = v_\perp^2(\boldsymbol{x}_c) v_{||}(\boldsymbol{x}_c) - v_\perp^2(\underline{\boldsymbol{x}}_c) v_{||}(\underline{\boldsymbol{x}}_c) \tag{14}$$

and positions $\boldsymbol{x}_c$ and $\underline{\boldsymbol{x}}_c = \boldsymbol{x}_c + \boldsymbol{\ell}$ correspond to the opposite sides of the contour S_ℓ. Then, the energy flux can be decomposed as

$$\sigma_\ell = \frac{1}{2\Sigma_\ell} \oint_{S_\ell} \delta v_{||}^3(\boldsymbol{\ell}, \boldsymbol{x}_c)[1 - \frac{(B-3A)}{\delta v_{||}^3(\boldsymbol{\ell}, \boldsymbol{x}_c)} - \frac{C}{\delta v_{||}^3(\boldsymbol{\ell}, \boldsymbol{x}_c)}] dx_c, \tag{15}$$

where

$$A = v_{||}^2(\underline{\boldsymbol{x}}_c) v_{||}(\boldsymbol{x}_c) - v_{||}^2(\boldsymbol{x}_c) v_{||}(\underline{\boldsymbol{x}}_c) = v_{||}(\underline{\boldsymbol{x}}_c) v_{||}(\boldsymbol{x}_c) \delta v_{||}(\ell), \tag{16}$$

$$C = P(\boldsymbol{x}_c) v_{||}(\boldsymbol{x}_c) - P(\underline{\boldsymbol{x}}_c) v_{||}(\underline{\boldsymbol{x}}_c), \tag{17}$$

$$\delta v_{||}(\ell) = v_{||}(\underline{\boldsymbol{x}}) - v_{||}(\boldsymbol{x}). \tag{18}$$

The "*transversal*" velocity $v_\perp$, which refers to the rotation in the orthogonal direction to $\boldsymbol{\ell}$, does not participate in the transfer and the transversal energy $v_\perp^2$ is only transported by the "*longitudinal*" velocity $v_{||}$ which, from a physical standpoint, refers to the elongation in the direction $\boldsymbol{\ell}$. Under the *homogeneity hypothesis*, usually assumed in the framework of statistical theories of turbulence, the correlation terms (14),(16) and (17) can be quantitatively estimated as negligible statistical quantities, i.e. $A = B = C = 0$ in a statistical sense. In this case, the energy flux σ_ℓ is related only to the so-called third-order longitudinal velocity structure function [32], [33]. In a more general context, $A \neq B \neq C \neq 0$ and the energy flux per unit mass σ_ℓ, in addition, is also related to both the *transversal/rotating* part of the kinetic energy $v_\perp^2$ and the work of the pressure forces.

3.2 Energy transfer and nonlocal dynamics

From the physical standpoint, longitudinal velocity increment (18) refers to the typical turbulent fluctuation of the longitudinal to $\boldsymbol{\ell}$ component of the velocity field. It is well known from theory of incompressible and homogeneous turbulence that the general relationship between the second-order moments of (18) and its orthogonal counterpart $\delta v_\perp(\ell) = v_\perp(\underline{\boldsymbol{x}}) - v_\perp(\boldsymbol{x})$ is given by [21]

$$\langle \delta v_\perp^2 \rangle = (1 + q\ell \frac{d}{d\ell}) \langle \delta v_{||}^2 \rangle \tag{19}$$

where $q = 1$ in 2D turbulence and $q = 1/2$ in the 3D case; $\langle . \rangle$ refers to averaging over all position vectors $\boldsymbol{x}$.

Figure 5 shows a comparative study on the experimental and the theoretical $\langle \delta v_\perp^2 \rangle$ as a function of normalized length scale ℓ/ℓ_I and time in homogeneous and non-homogeneous/nonlocal cases considered in Figure 1. Theoretical $\langle \delta v_\perp^2 \rangle$ is computed from the experimental $\langle \delta v_\parallel^2 \rangle$ using relation (19). We observe that relation (19) is satisfied at quasi-equilibrium state ($t = 0$) and becomes non valid during the unbalanced evolution of the 2D inverse cascade of energy. Deviation from the homogeneous indicator (19) is due to an anomalous distribution of the transverse energy among largest scales of the motion (in this case, the scales of the energy-cascading range).

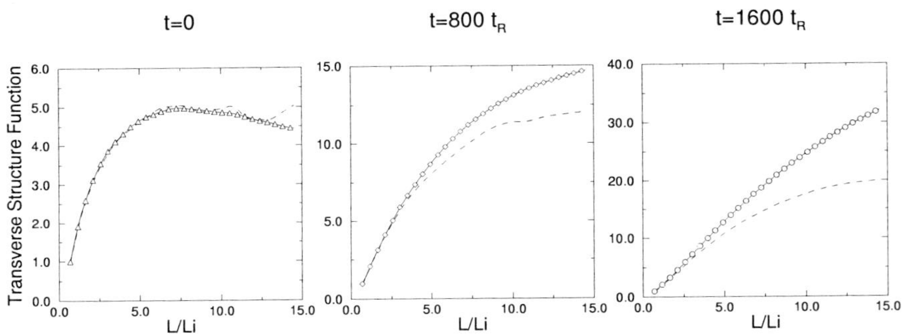

Fig. 5. The 2D experimental transverse velocity structure function and the theoretical estimate using (19) (long-dashed curves).

Relation (19) can be rewritten as:

$$\frac{\langle \delta v_\perp^2 \rangle}{\langle \delta v_\parallel^2 \rangle} = 1 + q \frac{d(\mathrm{Log}\langle \delta v_\parallel^2 \rangle)}{d(\mathrm{Log}\ell)} = 1 + q\xi_2^\parallel, \tag{20}$$

where $\xi_2^\parallel$ is the local scaling exponent of the second-order longitudinal velocity increments, $\langle \delta v_\parallel^2 \rangle \sim \ell^{\xi_2^\parallel}$. Relaxing the restriction of homogeneity and denoting by $Q = \langle \delta v_\perp^2 \rangle / \langle \delta v_\parallel^2 \rangle$, we have the following approximation in the non-homogeneous case:

$$Q \approx 1 + q\xi_2^\parallel. \tag{21}$$

It is found that $\xi_2^\parallel \approx 2/3$ and $Q \approx 5/3$ in locally homogeneous 2D inertial energy-cascading range ($q = 1$). In terms of non-averaged quantities, definition of Q can rewritten

$$[v_\perp(\underline{\boldsymbol{x}}) - v_\perp(\boldsymbol{x})]\delta v_\perp \approx Q[v_\parallel(\underline{\boldsymbol{x}}) - v_\parallel(\boldsymbol{x})]\delta v_\parallel. \tag{22}$$

Multiplying (22) by factors $v_{||}(\boldsymbol{x})v_{\perp}(\boldsymbol{x})$ and $v_{||}(\underline{\boldsymbol{x}})v_{\perp}(\underline{\boldsymbol{x}})$ and combining the result of such an operation using definitions (16,14), we get

$$B \approx QA[1 - \frac{v_{\perp}(\underline{\boldsymbol{x}})v_{\perp}(\boldsymbol{x})}{QA}\delta v_{||} - \frac{v_{||}^2(\underline{\boldsymbol{x}}) - v_{||}^2(\boldsymbol{x})}{A}\delta v_{||}]. \tag{23}$$

Using definition (16) again, the approximate relationship between A and B in the definition of the energy flux (15) is given by

$$B \approx Q[1 - T]A, \tag{24}$$

where, in terms of averaged quantities, T is given by

$$T = \frac{1}{Q}\frac{\langle v_{\perp}^2(\boldsymbol{x})\rangle}{\langle v_{\perp}^2(\underline{\boldsymbol{x}})\rangle}\frac{R_{\perp}}{\underline{R}_{||}} + (1 - \frac{\langle v_{||}^2(\boldsymbol{x})\rangle}{\langle v_{||}^2(\underline{\boldsymbol{x}})\rangle})\frac{1}{\underline{R}_{||}}, \tag{25}$$

where

$$R_{\perp} = \frac{\langle v_{\perp}(\underline{\boldsymbol{x}})v_{\perp}(\boldsymbol{x})\rangle}{\langle v_{\perp}^2(\boldsymbol{x})\rangle}, \underline{R}_{||} = \frac{\langle v_{||}(\underline{\boldsymbol{x}})v_{||}(\boldsymbol{x})\rangle}{\langle v_{||}^2(\underline{\boldsymbol{x}})\rangle}. \tag{26}$$

Terms $R_{\perp}$ and $\underline{R}_{||}$ refer to the transverse and longitudinal velocity correlation coefficients respectively. They usually define the transverse and the longitudinal integral length scales.

Relation (24) shows that the link between terms A and B as a function of the lengh-escale in (15) is insured by the ratio Q and the function T which is related to the transverse and longitudinal velocity correlation coefficients. By definition, Q is the ratio between transverse and longitudinal second-order velocity structure functions. According to (21), on the other hand, Q is roughly related to the local exponent $\xi_2^{||}$ balanced by the factor q which characterizes flow geometry.

3.3 Transfer-collapse

Let $E(k) \sim k^{-n}$ be the energy spectrum and m be the scaling exponent of the second-order velocity structure function. It is well known from Fourier analysis that the classical correspondance $m = n - 1$ is only valid for $1 < n < 3$. For nonlocal energy spectrum behavior – close to or steeper than k^{-3} – m saturates to 2 (see [35] for a more detailed discussion in 2D case) and, therefore, Q saturates to 3. According to (24), (21) and (15), when Q deviates from 5/3 in non-homogeneous dynamics, we have to consider two cases of interest:

(i) $Q(1 - T) < 3$; this implies that transverse part of the energy is transferred by the non-linear term of the Navier-Stokes equation and, consequently, transversal to $\boldsymbol{\ell}$ velocity increments contribute to the properties of the energy transfer.

(ii) $Q(1-T) \approx 3$; in this case the contribution of the second term of the integrand of (15) tends to be a negligible quantity; then, the scale-by-scale transfer of the *transversal/rotating* part of energy tends to be inhibited.

The first implication of (ii) is that, in this case, the *transversal/rotating* part of the turbulent kinetic energy is weakly scale-by-scale transferred by the non-linear term of the Navier-Stokes equation and accumulates at the scales where it is generated. This property, called *transfer-collapse*, is discussed in [19]. The second implication of (ii) is that the global energy flux basically becomes governed by the third-order longitudinal velocity increments only, as in homogeneous case, and by the work of the pressure forces. Obviously, the reality of such a situation depends of the statistical properties of Q and T.

Figure 6a shows the behavior of Q as a function of normalized length scale ℓ/ℓ_I in homogeneous and non-homogeneous/nonlocal two-dimensional cases considered in Figure 1. We observe that the self-similarity shows a very weak consistency at equilibrium state at low numerical Reynolds number considered in this experiment: the value 5/3 for Q is roughly satisfied in the energy-cascading range. In contrast, we find that a more robust self-similarity is steadily recovered before the development of the nonlocal dynamics: Q remains steadily close to 2.5 in this case. This behavior and the increase of the spectral slope displayed in Figure 2 ($t = 1600t_R$) are in agreement: $\xi_{\|}^2$ is of the order of 1.5 ($q = 1$) and the energy spectrum behavior in $k^{-2.4}$ is entirely consistent with this value. This consolidates the assumption (21) in non-homogeneos/nonlocal dynamics.

Figure 6b shows a experimental study on T computed using (25) and (26). T is negative at largest-scales at the initial equilibrium state and increases monotoneously toward a positive saturation value weaker, in mean sense, than 0.2. At largest-scales the saturation is reached when T is of the order of 0.1 According to (24) and (15), negative values of T amplify the accumulation of the *transversal/rotating* part of the turbulent kinetic energy even for moderate values of Q. Conversely, positive values slow down such a process. The interesting property of T in the energy-cascading range is that it increases the factor $Q[1-T]$ amplifying the transfer-collapse even when Q is small and close to the "*homogeneous value*" 5/3 and slows down one when Q is large and close to the "*nonlocal value*" 3.

4 Discussion

In this short review we have examined two interesting aspects of the two-dimensional dynamics in the incompressible approximation. We emphasized the link between, on the one hand, the tendency to generate localized concentrations of vorticity and energy behaving as coherent entities in the physical space and, on the other hand, the changing of the energy transfer properties of the non-linear term of the two-dimensional equation of motion when the dynamics

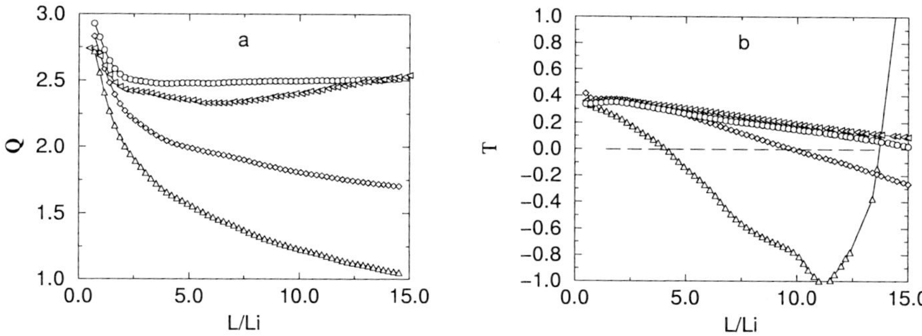

Fig. 6. The behaviors of Q and T at $t = 0$, (triangles up), $t = 800\ t_R$ (diamond), $t = 1400\ t_R$ (triangles left), $t = 1600\ t_R$ (circles).

is nonlocal. We find that only the longitudinal component of the total kinetic energy participates in the *energycascade* process: the *transversal/rotating* part of the energy anomalously accumulates at the scale where it is generated. The *transfer − collapse* have an irreversible character in the sense that it generate nonlocal properties or, in other words, they are favorable for the accumulation of the rotating antisymmetric component of the velocity gradient tensor: the vorticity. We see that, on the one hand, the *transfer − collapse* as a non-linear property of the equation of motion and, on the other hand, the two-dimensional topology described within the framework of Okubo-Weiss's or Hua-Klein's criteria are closely connected.

Above problem have been numerically analyzed in the 2D case. However, some features discussed here are also prominent to the understanding of 3D flows. The energy spectra displayed in Figure 3 inditate that the second-order longitudinal structure function changes in the turbulent wake behing the cylinder from the nonlocal anomalous behavior ℓ^2 near the cylinder to $\ell^{2/3}$ far from the cylinder where a $k^{-5/3}$ spectrum is observed. Above discussion have been demostrated that the energy flux in the turbulent wake is basically related to the third-order logitudinal velocity increments. This found the consistency of several studies performed using the mesurement of the longitudinal velocity structure only. However, we see that more extensive and correct undertanding of such a non-homogeneous/nonlocal flow as a function of the distance from the cylinder requires a substantial experimental progress in the meassurement of the transverse velocity structure of the velocity field in order to explore how the *geometrical* parameter q evolves in such a "*geometrically non − uniform*" flow.

References

1. Kraichnan, R., 1967 Inertial ranges in two-dimensional turbulence. Phys. Fluids 10, 1417.
2. Lesieur, M. 1990. Turbulence in Fluids. Second revised edition. Kluwer Academic Publishers.
3. McWillians, J.C., 1984. Fluid Mech. 146, 21-43.
4. Borue, V., 1994. Phys.Rev. Lett. 72, 1475
5. Smith, L.,M. & Yakhot, V., 1994 Finite-size effect in forced two-dimensional turbulence. J. Fluid Mech. 274, 115-138.
6. Charney, J. (1971). Geostrophic turbulence, J. Atmos. Sci. 28, 1071-1095.
7. Rhines, P. B., 1979. Ann. Rev. Fluid Mech. 11, 401-411.
8. Nastrom, G.D., Gage, K.S. & Jasperson, W.H. 1984 Kinetic energy spectrum of large and mesoscale atmospheric process. Nature 310, 5 July, 36-38.
9. Okubo, A. 1970 Horizontal dispersion of floatable particles in the vicinity of velocity singularities such as convergence. Deep-Sea Res. 17, 445-454.
10. Weiss, J. 1991 The dynamics of enstrophy transfer in two-dimensional hydrodynamics. Physica D 48, 273-294.
11. Brachet, M., Meneguzzi, M., Politano, H. & Sulem, P. 1988 The dynamics of freely decaying two-dimensional turbulence. J. Fluid Mech. 194, 333-349
12. Ohkitani, K. (1991) Wave number space dynamics of enstrophy cascade in a forced two-dimensional turbulence. Phys. Fluids A 3, 1598-1611.
13. Elhmaidi, D., Provenzale, A. & Babiano, A. 1993 Elementary topology of two-dimensional turbulence from a Lagrangian viewpoint and single-particle dispersion. J. Fluid Mech. 257, 533-558.
14. Basdevant, C. & Philipovich, T. (1994) On the validity of the "Weiss criterion" in two-dimensional turbulence, Physica D 73, 17-30.
15. Hua, B.L. & Klein P. 1998 An exact criterion for the stirring properties of nearly two-dimensional turbulence. Physica D 113, 98-110.
16. Andreotti, B., Douady, S. & Couder, Y. 1997. About the interaction between vorticity and stretching. In Turbulent modeloing and vortex dynamics. Edired by O. Boratv, A. Eden & A. Erzan, Lectures Notes in Physics, 92-107
17. Meneveau, C. & K. R. Sreenivasan. J. Fluid Mech. 224, 429 (1991).
18. Gaudin, E., B. Protas, S. Goujon-Durand, J. Wojciechowski, J. E. Wesfreid. Phys. Rev. E. 57, 9-12 (1998).
19. Babiano, A. 2000 On non-homogeneous two-dimensional inverse energy cascade. submited to Phys. of Fluids.
20. Pedlosky, J. 1987 Geophysical Fluid Dynamics. Springer, Berlin.
21. Basdevant, C. Legras, B., Sadourny, R. & Beland, M. 1981 A study of barotopic model flows: Intermittency waves and predictability. J. Atmos. Sci 38 2305-2326.
22. Sadourny, R. & Basdevant, C. 1985 Parameterization of subgrid scale barotropic eddies in quasi-geostrophic models: anticipated potential vorticity method. J. Atmos. Sci. 42, 1353-1363.
23. Vallis, G.K. & Hua, B.L. 1988 Eddy diffusivity of the Anticipated Potential vorticity method, J. Atmos. Sci., 45, 617-627.
24. Paret, J. & Tabeling, P. 1998 Intermittency in the 2D inverse cascade of energy: experimental observations. Phys. Fluids, vol 10, 12, 3126-3136.
25. Babiano, A., Dubrulle, B., Frick, P. 1995 Scaling properties of numerical two-dimensional turbulence. Physical Review E, 52, 4, 3719-3729.

26. Babiano, A., Dubrulle, B., Frick, P. 1997 Some properties of two-dimensional inverse energy cascade dynamics. Physical Review E, 55, 3, 2693-2706.
27. Babiano, A., Boffetta, G., Provenzale, A. & Vulpiani, A. 1994 Chaotic advection in point vortex models and two-dimensional turbulence. Phys. Fluids A, 6 (7),2465-2474.
28. Protas, B., Babiano, A. & Kevlahan K.-R. 1999. On Geometrical Alignment Properties of Two-dimensional Forced Turbulence. Physica D, 128, 169-79.
29. Larchevêque, M. 1993 Pressure field, vorticity field, and coherent structures in two-dimensional incompressible turbulent flows. Theoret. Comput. Fluid Dynamics 5, 215-222.
30. Hua, B.L., McWilliams J. & Klein, P. 1998 Lagrangian accelerations in geostrophic turbulence. J. Fluid Mech., 366, 87-108.
31. Lapeyre, G. Klein, P. & Hua, B.L. 1999. Does the tracer gradient vector align with the strain eigenvectors in 2D turbulence? Phys. Fluids, 11, 12, 3729-3737.
32. Monin, A.S. & Yaglom, A.M. 1975 Statistical Fluid Mechanics. The MIT Press.
33. Frisch, U. Turbulence. Cambridge University Press, (1995).
34. Landau, L.D. & E.M. Lifschitz, *Fluid Mechanics*, 2nd ed. (Pergamon, Oxford, 1987).
35. Babiano, A., Basdevant, C. & Sadourny, R. 1985 Structure functions and dispersion laws in two-dimensional turbulence. J. Atmos. Sci. 42, 942-949.

Numerical Simulation of Vortex Flows

Serge Huberson[1] and Olivier Daube[2]

[1] Laboratoire de Mécanique, Université du Havre, France
[2] CEMI, Université d'Evry Val d'Essonne, France

Abstract. This paper presents a selection of numerical methods among the most currently used to solve the Navier Stokes equation in a vorticity formulation. The paper focuses on the connection that exists between vorticity boundary condition and the divergence free condition. The different methods has been cast in two classes : the methods which are based on divergence free approximations of the vorticity fields, and those which use a correction procedure in order to satisfy this constraint. For both classes, the respective advantages of Eulerian grid methods and lagrangian methods are briefly discussed.

1 Introduction

The numerical simulation of incompressible vortex flows was probably one of the very first attempts to model a physical phenomena with the help of a computer. The huge number and variety of methods which have been proposed makes it almost impossible to gather them in a short paper and many books have not been enough to exhaustively treat the subject. This statement will be invoked hereafter as a motivation to limit this text to a very partial point of view. It will be partial since it will not include some very important works. It will also be partial in the choice of the material which will be directly derived from the personnal knowledge, and sometimes from the personnal taste of the authors.

From a general point of view, it can be said that the vorticity forms of the incompressible Navier-Stokes equation have been developed as alternative forms of the classical primitive variables formulation in an attempt to overcome the difficulties which are linked to the treatment of the incompressibility constraint. Beside this, it was also considered of some interest to compute directly the vorticity in order to have some kind of direct glance at the object under study.

2 The basic vorticity formulations

2.1 Two dimensional flows

In addition to the two previously mentioned points, the early 2D vortex methods took advantage from the vorticity- stream function formulation which has only two unknown functions whereas the usual velocity- pressure involves three. In these pioneering works, the vorticity transport equation and the stream function

Poisson equation were solved by means of finite differences :

$$\begin{aligned} &\frac{\partial \omega}{\partial t} + \operatorname{div}(\mathbf{U}\omega) = \nu \Delta \omega \\ &\Delta \psi = -\omega \\ &\mathbf{U} \ = \nabla \times (\psi \mathbf{e}_z) \end{aligned} \tag{1}$$

This approach was probably one of the most widely used in the early sixties, when computers started their invasion of modern fluid mechanic. Although the method was rather sucessfull, it was not really the first and older computational results by Rosenhead [33] and Westwater [38] are mentioned in the well known Hydrodynamic book by Batchelor[2]. These calculations were performed in the early thirties. They were based on the integro-differential formulation of the vorticity transport equation for inviscid flows :

$$\begin{aligned} &\frac{\partial \omega}{\partial t} + \operatorname{div}(\mathbf{U}\,\omega) = 0. \\ &\mathbf{U} \ = \frac{1}{2\pi} \iint_{\mathrm{Supp}(\omega)} \omega(\mathbf{x}') \frac{\mathbf{e}_z \times (\mathbf{x} - \mathbf{x}')}{|\, \mathbf{x} - \mathbf{x}' \,|^2} dx' \end{aligned} \tag{2}$$

The first equation means that the vorticity attached to any material particle remains constant so that Rosenhead and Westwater were able to compute flows in which the initially concentrated vorticity was discretized into a dozen of vortex carrying particles. This was probably the oldest version of the now well known vortex method. It must be mentioned that an other way to solve these equations for a viscous flow was proposed in 1974 by Wu and Thomson [39] who reduced the two previous equations into one single integro-differential equation :

$$\frac{\partial \omega}{\partial t} + \operatorname{div}\left(\frac{1}{2\pi} \iint_{\mathrm{Supp}(\omega)} \omega(\mathbf{x}') \frac{\mathbf{e}_z \times (\mathbf{x} - \mathbf{x}')}{|\, \mathbf{x} - \mathbf{x}' \,|^2} dx' \quad \omega \right) = \nu \Delta \omega \tag{3}$$

which was solved on a finite differences grid with adaptive size.

2.2 Three dimensional flows

The improvement of discretization methods, together with that of the computers performances led scientists to turn their interest to new and larger problems and the first three dimensionnal calculations soon appeared. The extension of the previous methods to three dimensionnal flows was not straightforward for many reasons.

- The first one was that the main advantage of the method that was to reduce the number of unknown functions turned out to be a severe drawback since the three dimensionnal vorticity- stream function formulation involves six unknown functions - the three components of both vorticity and stream function - whereas the velocity pressure formulation involves only four :

$$\begin{aligned}
\frac{\partial \omega}{\partial t} &+ (\mathbf{U} \cdot \nabla)\,\omega = (\omega \cdot \nabla)\,\mathbf{U} + \nu \Delta \omega \\
\omega &= \nabla \times \mathbf{U} \\
\mathrm{div}\mathbf{U} &= 0
\end{aligned} \tag{4}$$

- The second problem came from the divergence free condition on the vorticity field for the discrete problem. This condition which was automatically satisfied for plane flows has to be added to the three dimensional equations as an additional constraint. This problem is comparable to that of the incompressibility condition for the velocity- pressure formulation, yielding similar difficulties.
- The last problem was already present in the two dimensional formulation : there are no explicit boundary conditions for vorticity, except in very special cases such as some free surface flows. Once again, the problem was similar to that of setting the pressure value at the boundary in a velocity- pressure formulation. It is clear that these two difficulties could be avoided by solving directly the fully coupled set of equations (4).

2.3 Velocity–Vorticity formulations

An additional difficulty comes from the fact that the stream function components cannot be directly connected to the velocity boundary values : the velocity components are thus sums of its partial derivatives. Consequently, velocity–vorticity formulations were found more tractable by some authors [16], [28]:

$$\begin{aligned}
\frac{\partial \omega}{\partial t} &+ (\mathbf{U} \cdot \nabla)\,\omega = (\omega \cdot \nabla)\,\mathbf{U} + \nu \nabla^2 \omega \\
\nabla^2 \mathbf{U} &= -\omega
\end{aligned}$$

An integral solution for the last equation can be easily obtained in the form of the Biot- Savart relation. However, very few attempts have been made to use this formulation in the context of grid methods, probably because of a prohibitive CPU time requirement. As a result, this formulation has been mainly used within the framework of vortex methods, the velocity being then expressed by means of the Biot–Savart law :

$$\begin{aligned}
\frac{\partial \omega}{\partial t} &+ (\mathbf{U} \cdot \nabla)\,\omega = (\omega \cdot \nabla)\,\mathbf{U} + \nu \nabla^2 \omega \\
\mathbf{U} &= \frac{1}{4\pi} \iint_{\mathrm{Supp}(\omega)} \frac{\omega(\mathbf{x}') \times (\mathbf{x} - \mathbf{x}')}{\mid \mathbf{x} - \mathbf{x}' \mid^3} dx'
\end{aligned}$$

It resulted from these problems that the "vorticity and something" formulation was hampered by the number of unknown functions whereas the divergence free problem and the boundary conditions were basically the same for the two formulations.

Although the only remaining advantage of vorticity formulation for three dimensionnal flows reduces to the questionable satisfaction of having the vorticity as an explicit unknown of the formulation, a lot of works have been done in order to overcome the two main difficulties arising from the solenoidal conditions and from the boundary conditions. Actually, these two problems where closely linked as it is the case for the divergence free condition of the velocity field and the pressure at the wall. Similarly, two main classes of solution procedures were proposed : correction methods, and direct discretization. In the first kind of method, a first guess is corrected with such method as artificial compressibility [7], projection methods [19] or influence matrices [15]. In the second class of methods, the discretized vorticity belongs to a convenient subspace of divergence free functions. To build such functions on a grid can be very tricky [21] and this kind of methods have been mainely developped as grid free methods such as the vortex filament method by Leonard [24]. Here are the two classes of methods which will be discussed in this paper.

2.4 Divergence free vorticity versus vorticity boundary conditions

The connection between the boundary conditions and the divergence free condition for the vorticity can be heuristically explained by considering the vortex filaments which are material lines for inviscid flows. Helmholtz theorem states that in those cases the circulation must be constant along such filaments so that they can not end in the fluid domain. Actually, vortex filaments can only be closed lines like vortex rings or lines attached to the solid walls. In this last case, the vorticity at the wall also satisfies the previous criteria in some way since the total vorticity must be kept constant. Therefore, the vortex filaments can be looked at as closed lines lying for a part in the free stream, and for an other part on the wall. It must be pointed out that this analysis has been widely used in numerical simulation of the flow around lifting surfaces yielding the well known horse-shoe pattern by Belotserkovskii [4] and others, after Prandtl lifting line model. A more recent work on the same subject is that of Summers and Chorin [35] who introduced the notion of small divergence free structures which were named magnets.

An important remark on this problem is that the horse-shoe scheme also preserves the divergence free condition. This is a straightforward way to relate the discretization scheme within the flow field and the discrete boundary conditions. It will be seen now that the discretization which is used at the wall must satisfy some compatibility conditions in order to ensure that this relation can be correctly written. This problem was solved ten years ago in a set of papers by Quartapelle and colleagues [30,17]. Although the three dimensional case is rather different since the divergence free condition for the vorticity field is automatically satisfied for two dimensional flows, the basic formulations to be used for the boundary values of the vorticity components are similar. It is only the vectorial nature of the three dimensional vorticity field that requires the additional constraint of zero divergence.

3 A Velocity–Vorticity formulation

In this section, a velocity–vorticity formulation which adresses the main problems is described. Let us consider an open bounded *p–multiconnected* domain Ω of R^n (n=2 or 3). Its boundary Γ is the union of Γ_0, the outer boundary and of p interior connected component Γ_i.

$$\Gamma = \Gamma_0 \bigcup_{i=1,\ldots,p} \Gamma_i$$

For each of the interior components, we define an arbitrary simple loop γ_i which surrounds it (see figure 1). It is important to quote that the choice of these loops is completly arbitrary in the sense that the results will not depend on it. As usual, $\mathbf{n}$ denotes the outwards unit vector normal to the boundary Γ, dS denotes an elementary surface element on Γ and $\boldsymbol{dl}$ denotes the elementary displacement vector along the path γ_i. For the sake of simplicity, only cartesian grids will be considered here. The general case of curvilinear coordinates is adressed in [6].

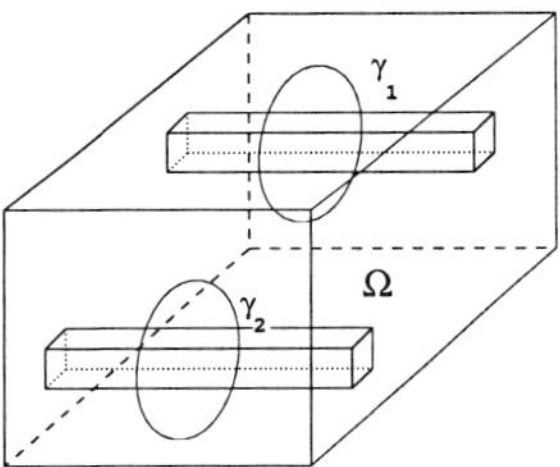

Fig. 1. sketch of a 2–multiconnected domain Ω of R^2

Let us consider the flow in Ω of an incompressible viscous fluid, possibly subject to body forces $\boldsymbol{f}$. The viscous effects are characterized by the Reynolds number Re. Using the velocity $\mathbf{U}$ and the dynamic pressure $P = p + 1/2\mathbf{U}^2$, the Navier–Stokes equations for this flow read:

$$\begin{aligned}
&\frac{\partial \mathbf{U}}{\partial t} + \omega \times \mathbf{U} = -\nabla P + \frac{1}{\mathrm{Re}}\nabla^2\mathbf{U} + \boldsymbol{f} && \text{in} \quad \Omega \\
&\nabla \cdot \boldsymbol{U} \qquad\qquad\qquad\qquad = 0 && \text{in} \quad \Omega \\
&\boldsymbol{U} \qquad\qquad\qquad\qquad\quad = \boldsymbol{b} && \text{on} \quad \Gamma
\end{aligned}$$

These equations have to be supplemented by initial conditions:

$$\begin{aligned}
\boldsymbol{U}\big|_{t=0} &= \boldsymbol{U}_0 && \text{in} \quad \Omega \\
\nabla \cdot \boldsymbol{U}_0 &= 0 && \text{in} \quad \Omega \\
\boldsymbol{U}_0 \cdot \mathbf{n} &= \boldsymbol{b} \cdot \mathbf{n} && \text{on} \quad \Gamma
\end{aligned}$$

Here and in the whole text, $\boldsymbol{b}$ stands for a vector field given on Γ and which is assumed to be in a suitable trace space. As we are dealing with incompressible flows, it is also assumed to satisfy the constraint:

$$\iint_{\Gamma} \mathbf{b} \cdot \mathbf{n} \, dS = 0 \tag{5}$$

Taking the curl of the momentum equation (5) and taking into account the continuity equation yields the so–called velocity $\mathbf{U}$–vorticity $\boldsymbol{\omega}$ formulation:

$$\begin{array}{lrll}
\dfrac{\partial \omega}{\partial t} + \nabla \times (\omega \times \mathbf{U}) = & \dfrac{1}{\text{Re}} \nabla^2 \omega + \nabla \times \boldsymbol{f} & \text{in} & \Omega \\
\nabla \times \boldsymbol{U} & = \omega & \text{in} & \Omega \\
\nabla \cdot \boldsymbol{U} & = 0 & \text{in} & \Omega \\
\boldsymbol{U} \cdot \boldsymbol{n} & = \boldsymbol{b} \cdot \boldsymbol{n} & \text{on} & \Gamma \\
\boldsymbol{U} \times \mathbf{n} & = \boldsymbol{b} \times \mathbf{n} & \text{on} & \Gamma
\end{array}$$

Daube *et al* [14] have established the conditions that ensure the equivalence between the two formulations . In particular, they have shown that they are equivalent if the domain Ω is simply connected. Otherwise, let us denote

$$\boldsymbol{W} = \frac{\partial \mathbf{U}}{\partial t} + \omega \times \mathbf{U} - \frac{1}{\text{Re}} \nabla^2 \mathbf{U}$$

The vorticity transport equation (6) reads then:

$$\nabla \times (\boldsymbol{W} - \boldsymbol{f}) = 0$$

To ensure that the vector $\boldsymbol{W} - \boldsymbol{f}$ is a true gradient (of the pressure P), the p additional following conditions along each loop γ_i have to be satisfied:

$$\oint_{\gamma_i} \boldsymbol{W} \cdot dl = \oint_{\gamma_i} \left[\frac{\partial \mathbf{U}}{\partial t} + \omega \times \mathbf{U} + \frac{1}{\text{Re}} \nabla \times \omega \right] \cdot dl = \oint_{\gamma_i} \boldsymbol{f} \cdot dl \quad \forall\, i = 1, \ldots, p \tag{6}$$

Here we made use of the vector identity:

$$\nabla^2 \mathbf{U} = \nabla(\nabla \cdot \mathbf{U}) - \nabla \times (\nabla \times \mathbf{U})$$

Remark :

- If the body forces field $\boldsymbol{f}$ is such that $\nabla \times \boldsymbol{f} = 0$, it does no longer play any role in equation (6). If its circulation is also zero along each loop γ_i, its only role is to modify the pressure by adding a kind of hydrostatic term. However, if these circulations are not all equal to zero, the body forces have an influence which is taken into account through relation (6).

- The choice of the loops γ_i is arbitrary and will be made for numerical convenience as it will be explained later.
- The non satisfaction of this condition along a loop γ_i may be responsible for a physically meaningless gap of the pressure at either the leading or the trailing edge of Γ_i

4 Direct methods

4.1 Solving the coupled set of equations

The main problem for grid methods is to define discrete approximations for both velocity and vorticity such that the discrete divergence vanishes. This problem was first adressed for the velocity- pressure formulation, leading to different technics, some of which will be shortly described in this section. We will concentrate on the discretization procedure since the resulting set of non-linear algebraic equations is not specific of the present problem. Therefore, any solution procedure suitable for some similar problems can be used. Many of these methods have been reviewed and discussed in the classical CFD books, for example [22,29].

The first works on this problem were based on finite differences and led to the well known staggered grid method. The main idea is that the divergence of a given vector field within a given volume is directly connected to the flux through the boundary according to the Gauss theorem :

$$\iiint_{\mathcal{M}_o} \text{div}(\mathbf{V}) \quad dx = \iint_{\partial\mathcal{M}_o} \mathbf{V}\cdot\mathbf{n} \quad d\sigma = 0$$

The simplest discretization scheme for this equality consists in approximating div($\mathbf{V}$) at the center of the cell whereas the flux is approximated at the center of the faces constituting the cell boundary.

For a velocity- pressure formulation, there exists a close relation between the pressure and the divergence of the velocity field so that the pressure and the divergence free condition have to be writen at the same point, that is the center of the cell. The simultaneous solution of the previous discrete equation and the linearized momentum equations yield a regular linear system. The corresponding method is known as the MAC method (Marker and Cell) the name of which has also been retained for the staggered grid itself.

The connection with the vorticity- velocity formulation is not straightforward. The solution of this problem can be looked for on a staggered grid with the vorticity defined at the center of the grid edges and the velocity defined on the cell faces. This procedure was shown to provide a well-posed discrete problem by Daube [5] for the case of two dimensionnal flows.

4.2 Divergence free approximations

Building a divergence free basis

An other way to enforce the divergence free condition consists in using a Hilbert basis of divergence free functions. This procedure can be applied locally within the framework of finite elements. It seems that this last method applied to vortical flows has been first introduced by Leonard *et al* and has been used by Virk, Melander and Hussain [36] in order to define what they called polarized vortex rings. A class of eingenfunctions $\mathbf{W}_j$ for the curl operator is first defined :

$$\nabla \times \mathbf{W}_j = \lambda \mathbf{W}_j$$

Taking then the curl of this equation yields the elliptic equation :

$$-\Delta \mathbf{W}_j = \lambda^2 \mathbf{W}_j$$

for which a complete system of eignefunctions in a convenient Hilbert space does exist. In the particular case of an axisymmetric flow with axial periodicity and looking for separate variables solutions, Virk *et al* were able to express the eigenfunctions in a closed form. It is then possible to express any solenoïdal vector field as a sum of these eigenfunctions and therefore to derive for the vorticity a similar expression which naturally satisfies the divergence free condition. Virk and is co-workers pointed out that each eigenfunction corresponds to a vortex ring, the support of which is a torus with its section radius defined by the zeros of the Bessel functions. These vortex rings are said to be polarized since each elementary solution corresponds to a vortex ring with swirl.

4.3 The vortex filament method

Vortex filaments stretching

By direct methods, we mean methods in which a divergence free form of dicrete vorticity is used. This is the case of the vortex filament method for inviscid incompressible flows. The discrete elements considered in this method are discrete approximation of the actual vortex filament which are known to constitute an elementary solution of the problem. Therefore, the flow can be approximated by a set of interacting vortex filament for which evolution equations have to be written. We start from the vorticity transport equation for inviscid three dimensionnal flows :

$$\frac{\partial \boldsymbol{\omega}}{\partial t} + (\mathbf{U} \cdot \nabla)\, \boldsymbol{\omega} = (\boldsymbol{\omega} \cdot \nabla)\, \mathbf{U} \tag{7}$$

The right hand side is the stretching term which vanishes for plane flows. The discretization scheme presented hereafter is that of Knio and Ghoniem which is a compound of particle methods and vortex filament methods [23]. In this method, the filament is discretized into a set of small segments $\mathcal{S}_i$ for $i \in I$. Each segment represents a small amount of vorticity which is related to the circulation γ_n. Since this circulation is constant according to Helmhotz and Kelvin theorems, we have :

$$\boldsymbol{\Omega}_i = \int_{\mathcal{S}_i} \boldsymbol{\omega} dx = \gamma_n (\mathbf{X}_{i+1} - \mathbf{X}_i)$$

where $\mathbf{X}_i$ and $\mathbf{X}_{i+1}$ are the extremities of $\mathcal{S}_i$. A transport equation for the filament $\mathcal{L}_n$ can be readily obtained :

$$\frac{d\mathbf{X}_i}{dt} = \mathbf{U}\mathbf{X}_i$$

and the velocity can be easily computed from the Biot Savart Law :

$$\mathbf{U}(\mathbf{X}_i) = \sum_n \int_{\mathcal{L}_n} \gamma_n \frac{\boldsymbol{\tau} \times (\mathbf{X}_i - \mathbf{x})}{2\pi \mid \mathbf{X}_i - \mathbf{x} \mid^3} dl(\mathbf{x})$$

where $\boldsymbol{\tau}$ is the unit vector tangent to the line $\mathcal{L}_n$. Using the segments $\mathcal{S}_i$ yields :

$$\mathbf{U}(\mathbf{X}_i) = \sum_n \gamma_n \sum_{j \in I} \frac{(\mathbf{X}_{j+1} - \mathbf{X}_j) \times (\mathbf{X}_i - \mathbf{x})}{2\pi \mid \mathbf{X}_i - \mathbf{x} \mid^3} dl(\mathbf{x})$$

The Knio & Ghoniem method differs from that of Leonard and others [4], [13], by the fact that the integral over $\mathcal{S}_j$ is approximated once more rather than analytically computed.

$$\mathbf{U}(\mathbf{X}_i) = \sum_n \sum_{j \in I} \frac{\boldsymbol{\Omega}_j^n \times (\mathbf{X}_i - \bar{\mathbf{X}}_j)}{2\pi \mid \mathbf{X}_i - \bar{\mathbf{X}}_j \mid^3} dl(\mathbf{x})$$

where $\bar{\mathbf{X}}_j = 1/2(\mathbf{X}_{j+1} + \mathbf{X}_j)$. It must be noted that the circulation γ_n is constant, so the discrete vorticity $\boldsymbol{\Omega}^n$ has to be computed at each time step. It is worth to point out that this provides a discrete evaluation for the stretching term :

$$\begin{aligned} \frac{d\boldsymbol{\Omega}_j^n}{dt} &= \gamma_n \frac{d(\mathbf{X}_{j+1} - \mathbf{X}_j)}{dt} \\ &= \gamma_n(\mathbf{U}_{j+1} - \mathbf{U}_j) \\ &= (\gamma_n(\mathbf{X}_{j+1} - \mathbf{X}_j)) \cdot \frac{(\mathbf{X}_{j+1} - \mathbf{X}_j) \times (\mathbf{U}_{j+1} - \mathbf{U}_j)}{\mid \mathbf{X}_{j+1} - \mathbf{X}_j \mid^2} \\ &= (\boldsymbol{\Omega}_j^n) \cdot \mathrm{grad}(\mathbf{U})_j \end{aligned} \tag{8}$$

One of the main interest of the method lies in the satisfaction of the divergence free condition through the discretization of the flow since this condition is automatically satisfied by a set of closed vortex filament. However, a particular problem was found to appear due to the necessity to use closed line. If the flow pattern is somewhat complicated, the numerical representation of the vortex filament by a set of connected vortex carrying segments can become very tricky. The discretization must be compatible with the local curvature. This property can not be ensured with a fixed numer of segments so that authors where confronted to the necessity of splitting some of the filaments during the calculation. This problem was quite close to that of the representation of turbulent flows and this was the next step for the method.

Simulating the viscous effects

Two effects of the viscous diffusion will be discussed in this section : the local diffusion within the filament cross section, and the reconnection of vortex filament. The first point has been adressed by many authors due to the importance of the decay of aircraft trailing vortices in airport management. An asymptotic model proposed by Guiraud [20] provided a support to estimate the viscous effects within the section and a collection of these results can be found in a recent work on vortex rings by Margerit [26]. It is clear from these works that the modification of the section under diffusion at high Reynolds number is weak and that the structure of the vortex filament is not really affected by viscosity. However, even in the case of very high Reynolds number, viscosity can play an important role since it is necessary to allow for reconnection which is an essential mechanism in the dynamic of a set of vortex filament. These considerations led Chorin [8] to suggest some kind of vortex filament surgery in which new segments are introduced in the region of high curvature, whereas reconnection is enforced when the situation appears to be relevant of this phenomena.

It is important to point out that the use of stochastic models to simulate diffusion is not straightforward. This model has been used by Chorin in his work on plane flows and have been widely used since with several improvements concerning particularly the interpretation of the results. The extension of the model to three dimensional flows was far from being evident since the brownian motion not only affects the vortex filament motion, but also the vorticity. The simple argument that the circulation is constant along the filament and that the brownian motion only concern the material line itself does not hold since the resulting model does not satisfy the vorticity diffusion equation. Besides this, the conservation of vorticity on a set of points undergoing a Brownian motion does not preserve the divergence free condition for the vorticity. Despite this, numerical results by Gharakani and Ghoniem seems to indicate that this last method can represent an essential part of the actual viscous flow [18]. The only way to use stochastic diffusion models for three dimensionnal flows seems to consist in using a Brownian motion on a set of particles which are eventually agregated in one macro-particle for the transport step.

5 Predictor-Corrector methods

5.1 Fractional Step methods

Time Discretization

In this section, only the time discretization of the incompressible Navier–Stokes equations in velocity–vorticity form is considered. The time derivative is approximated at time level $(n+1)\Delta t$ by a second order Euler Backward scheme, the viscous terms are implicitely written at he same time level whereas the non linear terms are explicitely evaluated by means of an Adams–Bashforth time extrapolation from the previous time steps. With this time discretization, the vorticity transport equation reads:

$$\frac{3\omega^{(n+1)} - 4\omega^{(n)} + \omega^{(n-1)}}{2\Delta t} + 2\nabla \times (\omega \times \mathbf{U})^{(n)} - \nabla \times (\omega \times \mathbf{U})^{(n-1)}$$
$$= \frac{1}{\mathrm{Re}}\nabla^2 \omega^{(\mathbf{n+1})} + \nabla \times \boldsymbol{f}^{(\mathbf{n+1})} \quad (9)$$

and the relations (6) read:

$$\oint_{\gamma_i} \left[\frac{3\mathbf{U}}{2\Delta t}^{(n+1)} + \frac{1}{\mathrm{Re}} \nabla \times \omega^{(n+1)} \right] \cdot dl$$
$$= \oint_{\gamma_i} \left[\frac{4\mathbf{U}^{(n+1)} - \mathbf{U}^{(n)}}{2\Delta t} - \left(2(\omega \times \mathbf{U})^{(n)} - (\omega \times \mathbf{U})^{(n-1)} - \boldsymbol{f}^{(n+1)} \right) \right] \cdot dl \quad (10)$$

Setting $\sigma = 3\mathrm{Re}/2\Delta t$, we get an Helmholtz type vector equation for ω^{n+1}:

$$(\sigma \boldsymbol{I} - \nabla^2)\omega^{(\mathbf{n+1})} = \mathbf{S}^{\mathbf{n},\mathbf{n-1}}$$

Where $\mathbf{S}^{n,n-1}$ is a vector source term which contains all the terms that are evaluated at time n and $n-1$ and also the curl of the body forces. It is straightforward to check that $\mathbf{S}^{n,n-1}$ reads as the curl of a vector $\boldsymbol{g}^{n,n-1}$, $\mathbf{S}^{n,n-1} = \nabla \times \boldsymbol{g}^{n,n-1}$ and is therefore divergence free. Let us consider now the condition (6). By using the same notations we get:

$$\oint_{\gamma_i} (\sigma \mathbf{U}^{(n+1)} + \nabla \times \omega^{(n+1)}) \cdot dl = \oint_{\gamma_i} \boldsymbol{g}^{n,n-1} \cdot dl = g_i^{(n,n-1)} \quad (11)$$

This result has been established in the continuous case. In order to have the same property in discrete form, it is necessary to write the non linear term in the form of the discrete curl of some vector. Also noteworthy is the fact that any time discretization with an implicit treatment of the linear terms and an explicit treatment of the non liniear terms leads to a similar form of the semi-discretized equations.

The Generalized Stokes Problem

For the sake of simplicity, the time indexes will be dropped from now on, and the previously described time discretization of the Navier–Stokes equations leads to a generalized Stokes problem $\mathcal{P}(\boldsymbol{g}, \boldsymbol{b}, g_1, \ldots, g_p)$ defined by:

$$(\sigma \mathbf{Id} - \nabla^2)\,\omega = \nabla \times \boldsymbol{g} \quad \text{in } \Omega \quad (12)$$
$$\nabla \times \boldsymbol{U} = \omega \quad \text{in } \Omega \quad (13)$$
$$\nabla \cdot \boldsymbol{U} = 0 \quad \text{in } \Omega \quad (14)$$
$$\boldsymbol{U} \cdot \boldsymbol{n} = \boldsymbol{b} \cdot \boldsymbol{n} \quad \text{on } \Gamma \quad (15)$$
$$\boldsymbol{U} \times \mathbf{n} = \boldsymbol{b} \times \mathbf{n} \quad \text{on } \Gamma \quad (16)$$
$$\oint_{\gamma_i} (\sigma \mathbf{U} + \nabla \times \omega) \cdot dl = \oint_{\gamma_i} \boldsymbol{g} \quad i = 1, \ldots, p \quad (17)$$

We will assume that this fourth order system has a unique solution $(\boldsymbol{\omega}, \mathbf{U})$. Note also that, as it is usual in vorticity formulations of the Navier Stokes equation, the boundary conditions concern only the velocity components, a fact which prevents an easy decoupling of the vorticity transport equation from the other equations.

Solving directly this coupled system directly is a formidable task. Thus, it is generally preferred to compute, by means of an iterative process, a consistant (when the time step $\Delta t \longrightarrow 0$) approximation of the solution of the full GSP. Each iteration of this process consists in the resolution of two decoupled second order elliptic problems for which user or package subroutines are generally available:

- The first one will solve a Helmholtz type problem (the vorticity transport equation) *supplemented with appropriate boundary conditions* to obtain an approximate vorticity field $\widetilde{\boldsymbol{\omega}}$.
- the second computes an approximate velocity field $\widetilde{\mathbf{U}}$ through the resolution of a so–called div–curl problem using $\widetilde{\boldsymbol{\omega}}$ as a source term.

These two subproblems and their combination into a so-called trial problem are discussed in the next sections.

The Helmholtz Problem

Focusing on equation (12) two points have to be emphasized:

- It is a standard elliptic equation which can be solved provided appropriate boundary conditions are given.
- Since its purpose is to obtain an approximate vorticity field $\boldsymbol{\omega}$ which has to be the curl of the velocity field $\mathbf{U}$, the divergence $\nabla \cdot \boldsymbol{\omega}$ has necessarily to be equal to 0.

The last point is a crucial one and is of interest only for 3D flows. It is addressed in the definition of the following elliptic Helmholtz type problem $\mathcal{H}(\boldsymbol{g}, \boldsymbol{t})$ where $\boldsymbol{t}$ is an arbitrary vector field belonging to a suitable tangential trace space on Γ and $\boldsymbol{g}$ is the vector which was defined in the previous sections.

$$\begin{aligned} (\sigma \mathbf{Id} - \nabla^2)\, \boldsymbol{\omega} &= \nabla \times \mathbf{g} && \text{in } \Omega \\ \boldsymbol{\omega} \times \mathbf{n} &= \boldsymbol{t} && \text{on } \Gamma \\ \nabla \cdot \boldsymbol{\omega} &= 0 && \text{on } \Gamma \end{aligned}$$

This problem has a unique solution $\boldsymbol{\omega}$ which is obviously divergence free since $\nabla \cdot \boldsymbol{\omega}$ satisfies an homogeneous Helmholtz equation with homogeneous Dirichlet boundary conditions.

The div–curl problem

In this section we adress the problem of finding a solenoïdal vector field $\boldsymbol{U}$ when its curl $\boldsymbol{\omega}$ is known. Roughly speaking, this problem has a solution when the normal component of $\boldsymbol{U}$ is given on the boundary. Actually, in the case of a multiconnected domain, one has also to impose the circulation $\{c_i\}_{i=1,\dots,p}$ on p arbitrary independant which, for convenience, may be the loops γ_i defined in §3 but this not mandatory. Let us define a *div–curl* problem $\mathcal{DC}(\boldsymbol{\omega}, b, c_1, \dots, c_p)$ in the following way :

Let $\boldsymbol{\omega}$ be a solenoïdal vector field defined in Ω and b be a scalar field defined on Γ which satisfies the constraint $\oint_\Gamma b\,dS = 0$. Let c_i , $i = 1, \dots, p$ be p arbitrary real numbers.

Then the problem

$$\begin{aligned} \nabla \times \boldsymbol{U} &= \boldsymbol{\omega} && \text{in } \Omega \\ \nabla \cdot \boldsymbol{U} &= 0 && \text{in } \Omega \\ \boldsymbol{U} \cdot \boldsymbol{n} &= b && \text{on } \Gamma \\ \int_{\gamma_i} \boldsymbol{U} \cdot \boldsymbol{dl} &= c_i && i = 1, \dots, p \end{aligned}$$

has a unique solution $\boldsymbol{U}$ in Ω.

To solve this problem, a Helmholtz decomposition is performed, i.e. the vector field $\mathbf{U}$ is computed by a two steps process:

- Find an intermediate vector field $\boldsymbol{w}$ which satisfies:

$$\nabla \times \boldsymbol{w} = \boldsymbol{\omega} \quad , \quad \int_{\gamma_i} \boldsymbol{w} \cdot \boldsymbol{dl} = c_i \quad \text{and } \boldsymbol{w} \cdot \mathbf{n} = b \quad \text{on } \Gamma$$

  It will be shown in the next section that there is a very simple way to get such a vector.
- Project this vector field $\boldsymbol{w}$ onto the space of divergence free vectors. This is achieved by means of the usual procedure which consists in solving a scalar Poisson equation for an auxiliary function ϕ with Neuman boundary conditions:

$$\begin{aligned} -\nabla^2 \phi &= \nabla \cdot \boldsymbol{w} && \text{in } \Omega \\ \frac{\partial \phi}{\partial n} &= 0 && \text{on } \Gamma \end{aligned}$$

- set $\mathbf{U} = \boldsymbol{w} + \nabla \phi$. This vector is *the* solution of the div–curl problem.

We have now all the ingredients to define a procedure which will allow us to compute an approximate solution $(\widetilde{\boldsymbol{U}}, \widetilde{\boldsymbol{\omega}})$ of the generalized Stokes problem $\mathcal{P}$. For this purpose, a trial problem which combines the two previous problems and which constitutes the basis of the resolution of the Navier–Stokes equations in velocity–vorticity form, is defined.

The trial problem

Let $\mathbf{g}$, $\boldsymbol{b}$ be data defined as in §5.1 and $\widetilde{\mathbf{t}}$ defined as in §5.1. It is now possible to define a trial problem $\widetilde{\mathcal{P}} = \widetilde{\mathcal{P}}(\mathbf{g}, \boldsymbol{b}, \widetilde{\mathbf{t}})$ consisting of two successive steps :

1. solve $\mathcal{H}(\mathbf{g}, \widetilde{\mathbf{t}})$ in order to get a trial "vorticity" field $\widetilde{\boldsymbol{\omega}}$. Compute then a vector field $\mathbf{U}^*$ which is defined by:

$$\begin{aligned} \sigma \mathbf{U}^* &= \mathbf{g} - \nabla \times \widetilde{\boldsymbol{\omega}} && \text{in the interior of } \Omega \\ \mathbf{U}^* \cdot \mathbf{n} &= \mathbf{b} \cdot \mathbf{n} && \text{on } \Gamma \end{aligned}$$

   It is straightforward to check that $\nabla \times \mathbf{U}^* = \widetilde{\boldsymbol{\omega}}$ and that $\mathbf{U}^*$ satisfies relation (17). Noteworthy is the fact that this determination of $\mathbf{U}^*$ is explicit since it does not require the resolution of any additional system.
2. Project $\mathbf{U}^*$ onto the space of divergence free vectors by solving a Poisson equation with Neumann boundary conditions for a scalar function ϕ:

$$\begin{aligned} -\nabla^2 \phi &= \nabla \cdot \mathbf{U}^* && \text{in } \Omega \\ \frac{\partial \phi}{\partial n} &= 0 && \text{on } \Gamma \end{aligned}$$

   and finally set:

$$\widetilde{\mathbf{U}} = \mathbf{U}^* + \nabla \phi$$

It is a simple matter to see that, whatever the distribution $\widetilde{\boldsymbol{t}}$ is, the vector fields $\widetilde{\boldsymbol{\omega}}$ and $\widetilde{\mathbf{U}}$ satisfy equations (12) through (15) and (17) but not the remaining boundary condition (16) of the generalized Stokes problem of §5.1, i.e. the tangential components of $\mathbf{U}$ do not have the values which are prescribed by the no–slip condition.

On the vorticity boundary distribution

Therefore, solving the problem $\mathcal{P}$ amounts now to find the "correct" distribution of $\widetilde{\mathbf{t}}$ on the boundary, i.e. the one which will ensure that the tangential components of the velocity have the prescribed values on Γ. We stress the fact that it would be extremely lucky to find this correct distribution $\widetilde{\mathbf{t}}$ at the first shot. Once again, this reminds us that the behaviour of the boundary vorticity distribution is of integral nature and that speaking of vorticity boundary conditions is rather misleading. For deeper insight into this problem, the reader is referred to Quartapelle [31].

The determination of this vorticity distribution can be achieved through an iterative process, each iteration of which consists in solving a trial problem as previously defined and in consequently updating the distribution of $\widetilde{\mathbf{t}}$ on Γ in order to improve the satisfaction of relation (16). An iterative process of this kind requires at each iteration the resolution of a trial problem and can be therefore very much time consuming. From a practical point of view, we will have to restrict ourselves to a small number of iterations: 1,2 or at most 3. The influence matrix techniques fall into these methods and corresponds to finding the exact solution of the discrete fully coupled system in two iterations. Such a technique, similar to the one described in Daube [15] in the case of 2D problems, is defined in the next section.

Coupled Solution of the GSP Problem

Starting from an arbitrary distribution $\widetilde{\mathbf{t}}$ of tangential components for $\boldsymbol{\omega}$ as described in the previous sections, a solution $(\widetilde{\mathbf{t}}, \widetilde{\boldsymbol{\omega}})$ of the associated trial problem $\widetilde{\mathcal{P}}(\mathbf{S}, \boldsymbol{b}, \widetilde{\mathbf{t}})$ is first computed. Let us now consider the differences $\widehat{\boldsymbol{\omega}} = \boldsymbol{\omega} - \widetilde{\boldsymbol{\omega}}$ and $\widehat{\boldsymbol{U}} = \mathbf{U} - \widetilde{\boldsymbol{U}}$ between the solutions of the original problem $\mathcal{P}$ and those of this last problem $\widetilde{\mathcal{P}}$. These differences are solution of the homogeneous problem $\widehat{\mathcal{P}}$

$$
\begin{aligned}
(\sigma \mathbf{Id} - \nabla^2)\, \widehat{\boldsymbol{\omega}} &= 0 && \text{in } \Omega \\
\nabla \times \widehat{\boldsymbol{U}} &= \widehat{\boldsymbol{\omega}} && \text{in } \Omega \\
\nabla \cdot \widehat{\boldsymbol{U}} &= 0 && \text{in } \Omega \\
\widehat{\boldsymbol{U}} \cdot \boldsymbol{n} &= \mathbf{0} && \text{on } \Gamma \\
\oint_{\gamma_i} \widehat{\boldsymbol{U}}\, dl &= 0 && i = 1, \ldots, p
\end{aligned}
$$

where γ_i are the loops defined in §3

The solution of this system constitutes obviously a vector space which is in one to one correspondance with the distribution of $\widehat{\boldsymbol{\omega}}$ on the boundary Γ. Therefore, the mapping $\mathcal{H}$ defined by :

$$\mathcal{H}(\widehat{\boldsymbol{\omega}}) = \widehat{\boldsymbol{U}} \times \mathbf{n}$$

is a one to one linear mapping, owing to the linearity of the problems $\mathcal{P}$,$\widetilde{\mathcal{P}}$ and $\widehat{\mathcal{P}}$. Therefore, solving the problem $\mathcal{P}$ is equivalent to find the distribution $\widehat{\boldsymbol{t}}$ on Γ which satisfies:

$$\mathcal{H}(\widehat{\boldsymbol{t}}) = \mathbf{b} \times \mathbf{n} - (\widetilde{\boldsymbol{U}} \times \mathbf{n})_{\Gamma} \tag{18}$$

The second step of the calculations consists then in computing the correct values of $\boldsymbol{\omega} \times \mathbf{n}$ on Γ by setting $\mathbf{t} = \widetilde{\boldsymbol{t}} + \widehat{\boldsymbol{t}}$. These final boundary values are then used as boundary conditions to solve once more a problem $\widetilde{\mathcal{P}}$ yielding the solution of the original GSP problem.

From a discrete point of view, solving the system (18) amounts to solve a linear sytem the rank of which is N_{Γ} for 2D flows and $2 \times N_{\Gamma}$ for 3D flows, where N_{Γ} is the number of discretization points on the boundary. Therefore, this cannot be afforded in 3D configuration, and real iterative techniques would have to be envisioned.

5.2 The vortex particle methods

The vorticity transport equation

We consider first the case of an unbounded inviscid flow induced by an initially bounded vorticity field. The vortex particle method consists in discretizing the vorticity into a set of vortex carrying particles. These particles are also material particles which move with the fluid and therefore, the carried vorticity must satisfy the vorticity transport equations. The first problem to be considered is to generate an initial set of particles. Since the initial vorticity support is bounded, it can be covered with a finite cartesian grid evenly spaced in each direction. A vortex particle $\mathcal{P}_i$ is then associated to each cell of the mesh and defined by its location $\mathbf{X}_i$ and the affected vorticity $\boldsymbol{\Omega}_i$:

$$\begin{aligned}
\mathbf{X}_i &= \frac{\iiint_{\mathcal{P}_i} \mathbf{x} dx}{\iiint_{\mathcal{P}_i} dx} \\
\boldsymbol{\Omega}_i &= \iiint_{\mathcal{P}_i} \boldsymbol{\omega} dx
\end{aligned} \tag{19}$$

In order to compute the time evolution of $\mathbf{X}_i$ and $\boldsymbol{\Omega}_i$, the Navier Stokes equations are written in Lagrangian coordinates :

$$\begin{aligned}
\frac{d\mathbf{X}(\mathbf{x}_o, t)}{dt} &= \mathbf{U}(\mathbf{x}, t) \\
\frac{d\boldsymbol{\omega}(\tilde{\mathbf{X}}(\mathbf{x}_o, t))}{dt} &= (\boldsymbol{\omega} \cdot \nabla)\ \mathbf{U}\ |_{\mathbf{X}(\mathbf{x}_o, t)}
\end{aligned}$$

A discrete approximation of this set of ordinary differential equations can be easily obtained as :

$$\begin{aligned}
\frac{d\mathbf{X}_i}{dt} &= \mathbf{U}(\mathbf{X}_i) \\
\frac{d\boldsymbol{\Omega}_i}{dt} &= (\boldsymbol{\omega} \cdot \nabla)\ \mathbf{U}\ |_{\mathbf{X}_i}
\end{aligned}$$

The velocity field is obtained as an approximation of the Biot Savart integral relation

$$\begin{aligned}
\mathbf{U}(\mathbf{x}) &= \frac{1}{4\pi} \iiint_{\mathcal{D}(t)} \frac{\boldsymbol{\omega}(\mathbf{x}') \times \mid \mathbf{x}' - \mathbf{x} \mid}{\mid \mathbf{x}' - \mathbf{x} \mid^3} d\mathbf{x}' \\
&= \frac{1}{4\pi} \sum_j \iiint_{\mathcal{P}_j} \frac{\boldsymbol{\omega}(\mathbf{x}') \times \mid \mathbf{x}' - \mathbf{x} \mid}{\mid \mathbf{x}' - \mathbf{x} \mid^3} d\mathbf{x}' \\
&\simeq \frac{1}{4\pi} \sum_j \iiint_{\mathcal{P}_j} \frac{\boldsymbol{\omega}(\mathbf{x}') \times \mid \mathbf{X}_j - \mathbf{x} \mid}{\mid \mathbf{X}_j - \mathbf{x} \mid^3} d\mathbf{x}' \\
&\simeq \frac{1}{4\pi} \sum_j \frac{\boldsymbol{\Omega}_j \times \mid \mathbf{X}_j - \mathbf{x} \mid}{\mid \mathbf{X}_j - \mathbf{x} \mid^3}
\end{aligned} \tag{20}$$

The most important problem in the use of this approximation is that it is singular for $\mathbf{x} = \mathbf{X}_i$ and $j = i$. This singularity which has been artificially introduced through the discretization has been extensively discussed in the seventies. It was eventually elucidated by the mathematical analysis of the method which enlighted the fact that the Biot Savart integral is actually a convolution product of which a regular approximation can be obtained. The simplest regular approximation is probably the so-called desingularisation method :

$$\mathbf{U}(\mathbf{X}_i) \simeq \frac{1}{4\pi} \sum_j \frac{\boldsymbol{\Omega}_j \times \mid \mathbf{X}_j - \mathbf{x} \mid}{\mid \mathbf{X}_j - \mathbf{x} \mid^3 + \epsilon^3} \tag{21}$$

More elaborate method has been derived [3], [12] by using the following approximation for the vorticty :

$$\begin{aligned} \boldsymbol{\omega}_\epsilon(\mathbf{x}) &= \boldsymbol{\omega} \star \zeta_\epsilon \\ &= \iiint \boldsymbol{\omega}(\mathbf{x}')\zeta_\epsilon(\mathbf{x} - \mathbf{x}')dx' \end{aligned} \tag{22}$$

where ζ_ϵ is a smooth approximation of the dirac measure. It is clear that replacing ζ_ϵ with this measure yield the exact formulae :

$$\begin{aligned} \boldsymbol{\omega}_\epsilon(\mathbf{x}) &= \boldsymbol{\omega} \star \delta \\ &= \iiint \boldsymbol{\omega}(\mathbf{x}')\delta(\mathbf{x} - \mathbf{x}') \end{aligned} \tag{23}$$

The computation of the stretching term in the equation for $\boldsymbol{\Omega}_j$ can be readily derived from this approximated velocity field. Different forms have been proposed, first by Rehbach [32], and also by Cottet and Choquin [9]. In this paper, we use the form proposed by Rehbach in 1977 in the first published work on the three dimensional particle method :

$$\begin{aligned} (\boldsymbol{\Omega} \cdot \nabla)\mathbf{U}) \mid_{\mathbf{X}_i} = &-\frac{1}{4\pi} \sum_j \frac{3}{\mid \mathbf{X}_j - \mathbf{X}_i \mid^5 + \epsilon^5} (\mathbf{X}_j - \mathbf{X}_i)(\boldsymbol{\Omega}_i \cdot (\boldsymbol{\Omega}_j \times (\mathbf{X}_j - \mathbf{X}_i)) \\ &+ \frac{1}{\mid \mathbf{X}_j - \mathbf{X}_i \mid^3 \epsilon^3} (\boldsymbol{\Omega}_j \times \boldsymbol{\Omega}_i) \end{aligned} \tag{24}$$

Using these approximation for the velocity field and the stretching term, the three dimensionnal particles method results in a set of discrete ordinary differential equations :

$$
\begin{aligned}
\frac{d\mathbf{X}_i}{dt} &= \frac{1}{4\pi}\sum_j \frac{\boldsymbol{\Omega}_j \times \mid \mathbf{X}_j - \mathbf{x} \mid}{\mid \mathbf{X}_j - \mathbf{x} \mid^3 + \epsilon^3} \\
\frac{d\boldsymbol{\Omega}_j}{dt} &= -\frac{1}{4\pi}\sum_j \frac{3}{\mid \mathbf{X}_j - \mathbf{X}_i \mid^5 + \epsilon^5}(\mathbf{X}_j - \mathbf{X}_i)\,(\boldsymbol{\Omega}_i \cdot (\boldsymbol{\Omega}_j \times (\mathbf{X}_j - \mathbf{X}_i)) \\
&\quad + \frac{1}{\mid \mathbf{X}_j - \mathbf{X}_i \mid^3 \epsilon^3}(\boldsymbol{\Omega}_j \times \boldsymbol{\Omega}_i)
\end{aligned}
\tag{25}
$$

It is clear that the solution of these equation does not satisfy the divergence free condition and a correction procedure has to be used. However, despite many attempts, no one of the many solution proposed have yield a reliable solution. This is probably due to the geometrical character of any correction procedure aiming at connecting explicitely the vortex segments represented by the particles. Indeed, the dependency of the previous equation on the particles location is highly non linear. The many numerical experiments wich have been carried out with the vortex particle method have also demonstrated that the satisfaction of the exact divergence free constraint is not really important except close to solid wall where the possible interference with the vorticity generation makes it crucial.

Viscous flows

Many methods have been proposed in order to solve the Navier Stokes equations in unbounded domains. We restrict this presentation to the Particle Strength Exchange model which is an extension of the method proposed by Choquin, Cottet, Huberson and Mas-Gallic [9] [11]. We start from the Green function for the unbounded diffusion problem, that is the solution of the following problem :

$$
\begin{aligned}
\frac{\partial \omega}{\partial t} &= \nu \Delta \omega \\
\omega \mid_{t=0} &= \delta
\end{aligned}
\tag{26}
$$

where the initial condition is the Dirac measure at point 0. The solution of this problem is :

$$
\omega(\mathbf{x}, t) = \frac{1}{(2\sqrt{\pi \nu t})^d} \exp -\frac{(\mathbf{x})^2}{4\nu t}
$$

Using this Green function, the solution of the problem :

$$
\begin{aligned}
\frac{\partial \omega}{\partial t} &= \nu \Delta \omega \\
\omega \mid_{t=0} &= \omega_o(\mathbf{x})
\end{aligned}
\tag{27}
$$

can be expressed as :

$$
\omega(\mathbf{x}, t) = \frac{1}{(2\sqrt{\pi \nu t})^d} \iiint_{R^3} \omega_o(\mathbf{x}') \exp -\frac{(\mathbf{x} - \mathbf{x}')^2}{4\nu t} d\mathbf{x}'
$$

Substituting $\boldsymbol{\omega}(\mathbf{x},t)$ to ω_o the solution at time $t+\delta t$ can be approximated by :

$$\boldsymbol{\Omega}_i(t+\delta t) = \boldsymbol{\Omega}_i(t)$$
$$+\frac{1}{(2\sqrt{\pi\nu\delta t})^d}\sum_j\left[\boldsymbol{\Omega}_j(t)\iiint_{\mathcal{P}_i}d\mathbf{x}-\boldsymbol{\Omega}_i\iiint_{\mathcal{P}_j}d\mathbf{x}\right]\exp-\frac{(\mathbf{X}_j-\mathbf{X}_i)^2}{4\nu\delta t} \quad (28)$$

The divergence free condition

Up to now, nothing has been made in this particle method to satisfy the divergence free condition for the velocity and vorticity fields. Moreover, the approximate velocity field can have a non zero divergence so that the divergence free condition is only satisfied within the limit $\epsilon \to 0$. A lot of attempts has been made in order to satisfy ecplicitely this condition. An obvious way is to derive an iterative correction procedure from the locally computed divergence of the vorticity. However, this quantity, as well as the velocity divergence, is strongly related to the smoothing function that is used to compute the velocity field. We have :

$$\begin{aligned}\operatorname{div}(\boldsymbol{\omega}_\epsilon) &= \operatorname{div}(\iiint(\boldsymbol{\omega}(\mathbf{x}')\zeta_\epsilon(\mathbf{x}-\mathbf{x}'))dx' \\ &= (\iiint(\boldsymbol{\omega}(\mathbf{x}')\operatorname{div}\zeta_\epsilon(\mathbf{x}-\mathbf{x}'))dx' \end{aligned} \quad (29)$$

yielding the discrete approximation :

$$\operatorname{div}(\boldsymbol{\omega}_\epsilon)(\mathbf{x}) = \sum_i \boldsymbol{\omega}_i\operatorname{div}\zeta_\epsilon(\mathbf{x}-\mathbf{x}_i)) = 0. \quad (30)$$

Using a collocation method at points $\mathbf{X}_j$,the following set of equation can be obtained :

$$\operatorname{div}(\boldsymbol{\omega}_\epsilon)(\mathbf{X}_j) = \sum_i \boldsymbol{\omega}_i\operatorname{div}(\mathbf{X}_j-\mathbf{x}_i)) = 0. \quad (31)$$

This could be interpreted has a set of linear algebraic equations for $\boldsymbol{\Omega}_i$ which has a non trivial solution if and only if it is singular. This condition cannot be fullfilled unless the location of the discretisation points, that is the particles, moves. Since the function ζ_ϵ is generally non linear, the problem is very difficult to solve and not too much attempts, at least sucessfull attempts have been made in that direction.

The second condition to be imposed is that of the zero divergence for the velocity field. This one is obtained in the form of the integral of a cross product through the Biot Savart law :

$$\mathbf{U}_\epsilon = \frac{1}{4\pi} \iiint \boldsymbol{\omega}_\epsilon(\mathbf{x}') \times \frac{\mathbf{x} - \mathbf{x}'}{\mid \mathbf{x} - \mathbf{x}' \mid^3} dx' \tag{32}$$

yielding :

$$\begin{aligned} \text{div}\mathbf{U}_\epsilon &= \frac{1}{4\pi} \iiint \text{div}(\boldsymbol{\omega}_\epsilon(\mathbf{x}') \times \frac{\mathbf{x} - \mathbf{x}'}{\mid \mathbf{x} - \mathbf{x}' \mid^3}) dx' \\ &= \frac{1}{4\pi} \iiint \left(\nabla \times (\boldsymbol{\omega}_\epsilon(\mathbf{x}')) \cdot \frac{\mathbf{x} - \mathbf{x}'}{\mid \mathbf{x} - \mathbf{x}' \mid^3} + \boldsymbol{\omega}_\epsilon(\mathbf{x}') \cdot \nabla \times \left(\frac{\mathbf{x} - \mathbf{x}'}{\mid \mathbf{x} - \mathbf{x}' \mid^3} \right) \right) \end{aligned} \tag{33}$$

It can be pointed out that :

$$\nabla \times \left(\frac{\mathbf{x} - \mathbf{x}'}{\mid \mathbf{x} - \mathbf{x}' \mid^3} \right) = 0.$$

whereas

$$\nabla \times (\boldsymbol{\omega}_\epsilon(\mathbf{x})) = \sum_i \mathbf{grad}(\zeta_\epsilon(\mathbf{x} - \mathbf{X}_i) \times \boldsymbol{\Omega}_i)$$

Therefore, the expression for the velocity field divergence is :

$$\text{div}\mathbf{U}_\epsilon(\mathbf{x}) = \left(\sum_i \mathbf{grad}(\zeta_\epsilon(\mathbf{x} - \mathbf{X}_i) \times \boldsymbol{\Omega}_i) \right) \cdot \left(\frac{\mathbf{x} - \mathbf{x}'}{\mid \mathbf{x} - \mathbf{x}' \mid^3} \right) = 0. \tag{34}$$

Although this equation can be interpreted as a regular set of linear equation for $\boldsymbol{\Omega}_i$, no more success has been achieved. The more elaborated treatments are those proposed by Winkelman [37] although no one of the proposed algorithms as been found reliable up to now.

6 Conclusions

Although this paper was restricted to a small part of the effort made to compute vortical flows, a few conclusive, if not definitive ideas can be pointed out.

It seems that recent work on the projection methods have really enlighted the crucial aspect of building divergence free approximation [1]. Therefore, it should be of no use to deal with a vorticity formulation which show very few advantages for the resolution procedure, exepted perhaps for the particular case of two dimensional flows. A correct approximation for the vorticity field can be reasonably expected from a method which resolve the divergence free condition up to the computer accuracy.

Besides this, the numerical solution of the Navier Stokes equation does not reduce to a theoretical exercise and a large part of the design of a method is still based on the personal skill of the designer. The difficulty encountered in combining the different ingredients of a Navier Stokes solver is so large that there remains an open space which cannot be filled by anything else but experiment and eventually personal taste for a given method. It is clear that its convective nature is a strong advantage of vorticity from this last point of view.

The last tendancy for the computation of compressible flows [34], [27] is to include an additional vorticity transport equation which can be used as a correction for the computed solution. One probable advantage of this method is that it allows for some kind of visual control of the quality of the solution. However, it must be kept in mind that a correct evaluation of the vorticity field is also a necessary condition to the correct evaluation of the performances of most aerodynamical and hydrodynamical device.

References

1. Y. Achdou, J. L. Guermond : SIAM J. Numer. Anal., **37** 3, L799 (2000)
2. G.K. Batchelor, *Introduction to Fluid Dynamics*, (Cambridge University Press, 1967)
3. J. T. Beale, A. Majda : Maths of Comp., **39**, 1 (1982)
4. S.M Belotserkovskii, 'Calculation of the flow about wings of arbitrary plan form at a wide range of angles of attack', RAE library Translation, Mekhanika Zhidkosti I Gaza, **4**, 32, 1963.
5. F. Bertagnolio, O. Daube : J. Comp. Phys., **138**, L121 (1997).
6. F. Bertagnolio, O. Daube : Intern. J. Numer. Meth. Fluids, **28**, L917 (1998)
7. A.J. Chorin : Math. of Comp. **22**, L745 (1968)
8. A.J. Chorin : J. Comp. Phys., **91**, L1 (1990)
9. J.P. Choquin, G.H.Cottet : C. R. Acad. Sci. **306**, Série 1, L739 (1990)
10. J.P. Choquin, S. Huberson : Computers and Fluids, **17**, L397 (1989)
11. G.H. Cottet, S. Mas-Gallic : Numer. Math., **57**, L805 (1990)
12. G.H. Cottet : Ann. Inst. H. Poincaré, **5**, L227 (1988)
13. B. Couet, O. Buneman, A. Leonard : J. Comp. Phys., **39**, L305 (1981)
14. O. Daube, J.L. Guermon, A. Sellier, C. R. Acad. Sci. **313**, Série II, L377 (1991)
15. O. Daube, J. Comp. Phys., **138**, L121 (1997)
16. S.C.R. Denis, D.B. Ingham, R.N Cook: J. Comp. Phys., **33**, L325 (1979)
17. S.C.R. Denis, L. Quartapelle : Int. J. Num. Meth. Fluids , **9**, L871 (1989)
18. A. Gharakani, A. Ghoniem : ASME J. Fluid Eng., **120**, 2 , L319 (1999)
19. R. Glowinski, O. Pironneau : SIAM Rev., **12**, L167 (1979)
20. J.P. Guiraud, R.K. Zeytounian : J. Fluid Mech., **79**, L93 (1977)
21. F. Hecht, Construction d'une base d'un élément fini P_1 non conforme à divergence nulle dans R^3, Thèse Universit Paris VI (1980)
22. C. Hirsch : *Numerical computation of internal and external flows*, (Wiley 1990)
23. O. Knio, A. Ghoniem : J. Comp. Phys., **86**, L75 (1992)
24. A. Leonard : Lecture Notes in Physics, **35**, L245 (1974)
25. A. Leonard : J. Comp. Phys., **37**, L289 (1980)
26. D. Margerit : Mouvement dynamique des filaments et des anneaux tourbillons de faible épaisseur, Thèse de l'Institut Polytechnique de Loraine (1997)
27. V. Nastasie, Etude numérique du tourbillon d'extrémité de pale du rotor d'hélicoptère en régime compressible, Thèse Paris VI, (1998)
28. P. Orlandi : Comput. and Fluids, **15**, L137 (1987)

29. R. Peyret, T. Taylor : *Computational methods for fluid flows*, (Springer Verlag New York 1983)
30. L. Quartapelle : 'Factorized Formulations of the Incompressible Navier–Stokes equations 'in *Int. Conf. on Computational Methods in Flow Analysis at Okayama, Japan, 5-8 september, 1988,* ed. by H. Niki and M. Kawahara (Okayama University of Sciences) pp 337-348
31. L. Quartapelle, *Numerical Solutions of the Incompressible Navier–Stokes Equations*, (Birkhäuser Verlag, Basel 1993)
32. C. Rehbach, 'A numerical calculation of three dimensional unsteady flows with vortex sheets', in : *AIAA* 16th *Aerospace Sciences meeting, Huntsville, USA, 1978*
33. L. Rosenhead : Proc Roy. Soc. London, Serie A, **127**, L170 (1931)
34. J. Steinhof, R. Ramachandranm : AIAA J., **28**, 3, L426 (1990)
35. D. Summers, A. Chorin : *Vortex Flows and Related Numerical Methods II*, Esaim Proc. Vol **1**, www.emath.fr/Maths/Vol.1/summers.html
36. D. Virk, M. Mellander, F. Hussain : J. Fluid Mech. **260**, L23 (1994)
37. G. Winkelmans, A. Leonard : J. Comp. Phys. , **109** , 2, L247 (1993)
38. Westwater, *Rolling up of the surface of discontinuity behind an airfoil of finite span*, Aeronautical research commitee report and memoranda # 1962, 1935.
39. J.C. Wu, J.F. Thomson, Comput. Fluids, **1**, L197 (1973)

Of Vortices and Vortical Layers: An Overview

Maurice Rossi

Laboratoire de Modélisation en Mécanique
URA CNRS Université Pierre et Marie Curie,
F-75256 Paris - France.

Abstract. A theoretical overview of local flow models such as hyperbolic point flows or localized vorticity structures is presented. Vortex layers and tubes are particularly emphazised. Various exact Navier-Stokes or Euler solutions are introduced to analyse generic features of vorticity dynamics: vorticity gradients, vorticity stretching, interplay between axial and azimuthal vorticity, effect of a large scale strain rate or the existence of a helical symmetry. The linear stability of some of these basic flows is considered.

1 Introduction

Vorticity is a central concept in Fluid Mechanics [1] [2] [3] [17] as evidenced by the prominent role played by Helmholtz' laws of vortex motion. Experimentally, localized vorticity structures are pervasive throughout all fluid flows, e.g. in the fine scales of turbulence, and it is thus tempting to study many features of generic velocity fields through the analysis of local vorticity models. Two classes of models may be distinguished [5]: on the one hand, *vortex sheets or layers*, on the other hand, *filaments or vortex tubes*. From a theoretical standpoint, such patterns appear to be universal since they are geometrically the simplest topological structures that are consistent with the Helmholtz laws. This is confirmed by laboratory [12] [8] and numerical [8] [9] [3] experiments of three-dimensional turbulence where both types of concentrated vorticity structures have been identified. Such elementary flows may therefore be used to build an effective model of turbulence in physical space rather than in spectral space as in most studies : one of the few analytical connections between the Navier-Stokes equations and the Kolmogorov spectrum of turbulence has in fact been established *via* a statistical approach based on such elementary solutions [11] [12]. More recently, a subgrid-scale model has been proposed for Large Eddy Simulations [13] which relies on stretched vortices. Aside from turbulence, laminar or transitional flows are largely a world populated by vortices or shear layers [14] [10]. The study of such vortex patterns and of their stability is therefore a necessary prerequisite, though not the only one to be sure, if one is to achieve a deeper understanding of fluid motion in general and of turbulence in particular.

In the first part of this review, various elementary flows which are for the most part exact solutions of the Navier-Stokes or Euler equations are presented. After introducing in section 2.1 basic concepts pertaining to vortex layers and vortex tubes, section 2.2 is concerned with the simplest cases of two-dimensional steady configurations : a pure rotational flow i.e. a *solid body rotation*, a pure straining flow i.e. a *hyperbolic stagnation point*, a flow with equal amounts of vorticity and rate of strain i.e. a *simple shear flow* and finally configurations which "contain" more vorticity than rate of strain i.e. *elliptic flows.* All these *unbounded linear* models are subsequently generalized to three-dimensions. Starting from such primitive solutions, the flow complexity is gradually increased. The finite spatial extent of the vorticity distribution is first taken into consideration in section 3. Each of the above primitive models has a bounded vorticity counterpart. Hyperbolic stagnation point or simple shear flow are respectively the limiting case of *two-dimensional vortex layers subjected to strain* and *parallel free shear layers* (section 3.1). A solid body rotation is the limiting case of *axisymmetric vortices* whereas elliptic flows describe the inner core of *non-axisymmetric vortices* (section 3.2). A large scale strain produces such an elliptic deformation of streamlines which may be computed exactly (inviscid dynamics of vortex patches) or perturbatively (inviscid or viscous smooth velocity profiles). The flow topology is then further complexified to three-dimensions by adding an *axial velocity field.* When this axial component only depends on the radial and azimuthal coordinates, *swirling jets* are obtained (section 3.3). When the axial component also depends on the axial coordinate, the effect of *axial stretching* may be examined (section 3.4). Different cases can be analyzed : vortices embedded in an axisymmetric and uniform, steady or unsteady, strain [16] [17] or vortices with a more intricate *radial cell* structure [18] [19]. As a final step, one considers a general class of three-dimensional vortices which may break the rotational symmetry but still preserve *helical symmetry* (section 3.5) as encountered in vortex chamber experiments [20].

The second part of the paper is devoted to the generic instabilities which the above basic flows may sustain. Parallel free shear layers are associated to a *shear layer instability*, two-dimensional hyperbolic stagnation points to a *hyperbolic instability*, two-dimensional axisymmetric vortices to a *centrifugal instability*, vortices distorted by a large scale strain to an *elliptic instability. Inflexion point instabilities* of parallel shear layers [14] [10] are not considered but the effect of a strain field on the *instability of vortex layers* is examined, in particular, the hyperbolic point instability [21](section 5.1) or the Kelvin-Helmholtz instability in the presence of stretching [22] [23] [24] [25] (section 5.2). This second part however is mostly concerned with various *instabilities of vortex tubes.* The zeroth order model of unbounded solid body rotation expounded in section 4, is simple but not too simple since it can, by itself, support as a result of Coriolis forces, neutrally stable *Kelvin waves* which are largely responsible for the dynamical behaviour of more realistic vortices. If vorticity remains purely axial but is assumed to be *localized* in the radial direction, a new phenomenon may appear: the centrifugal instability (section 6.1) which can disrupt any vortex with an absolute circulation that decreases away from the axis. In analogous fashion, the famous inflexion point theorem in two-dimensional flows [14] [10] can be generalized to rotating flows (section 6.2) : according to this criterion, the radial gradient of absolute vorticity should change sign in order for an instability to arise. Concentrated vortices observed experimentally are generally stable to the centrifugal instability (section 6.3) : these vortices then act as *waveguides* [26] for the so-called *inertial waves* which are extensions, for concentrated vortices, of Kelvin waves (note that the terminology adopted in this paper is sometimes used in the reverse way by some authors !). Various types of inertial waves do exist: axisymmetric or *varicose waves* for which the vortex core undergoes a periodic bulging along its axis; *helical waves* which cause a displacement of the vortex position along its axis without modifying – at least to leading order– the internal vortical structure. Inertial waves can display more complex deformations in which the vortex axis is not changed but the internal structure is altered. All these classical linear concepts have found a new life in the nonlinear regime. For instance, inertial waves in tubes have been shown analytically to be exact nonlinear solutions (section 6.4).

The loss of axisymmetry due to the presence of a large scale strain, is responsible for a significant change of the filament dynamics : it induces the celebrated *elliptic instability* (section 7) which has been studied extensively in various experiments [12] [28] [6] [30] (and references therein). It may be described, in the context of a *local* theory, as a *parametric instability* of Kelvin modes for the unbounded vorticity configuration [31]. An alternative interpretation for concentrated vortices is associated to a *global resonance* mechanism that couples inertial waves and large scale strain.

The presence of an added axial velocity profile, as in the *Batchelor vortex*, considerably alters the dynamics of stable axisymmetric vortices (section 8) : unstable waves typical of *swirling jets* may appear due to a *generalized centrifugal* [32] or *Kelvin-Helmholtz instability. Stretching* (section 9) is another factor that can be determinant on vortex tube dynamics. In particular, it may eliminate the elliptic instability altogether [33] [7] but, in general, its influence is far from being elucidated.

This review solely addresses dynamical phenomena that are directly linked to the inner core structure of a single vortex. This implies that multiple vortex basic states (vortex pair instability, secondary instability of vortex arrays) will not be covered. However it should be emphasized that many features of single vortex dynamics remain relevant in multiple vortex configurations. For instance, the small scale instability of a vortex pair which appears superimposed on the large-scale Crow instability is directly related to the elliptic instability [35] [36] of a single vortex. Similarly secondary instabilities in shear layers [37] and wakes [38] may, in certain cases, be interpreted in terms of the elliptic instability.

For general references on vortex dynamics, the reader is advised to consult treatises on general hydrodynamics [2] [3] or hydrodynamic stability theory [14], specific monographs on vortex dynamics [1] [17] [39] or else conference proceedings [40] [41]. Except when explicitly mentioned, the following notations are consistently used : bold letters refer to vectors, ω denotes the complex frequency and $\boldsymbol{\Omega}$ or Ω the vorticity. The spatial position vector reads $\mathbf{x} = (x, y, z)$ or $\mathbf{x} = (x_1, x_2, x_3)$, x_j with $j = 1, 2, 3$, in Cartesian coordinates and $\mathbf{x} = (r, \theta, z))$ in cylindrical coordinates. The partial derivatives ∂_i, $\partial/\partial x_i$ are interchangeably used. Quantities ϵ_{ijk} and δ_{ij} respectively denote the antisymmetric and

Kronecker tensors. The symbol $\mathbf{e}_i$ (resp. $\mathbf{e}_y$) stands for a unit vector along the x_i-axis(resp. y-axis). Finally the Einstein repeated index summation convention is used and *c.c.* always stands for the complex conjugate.

2 General concepts on tubes and layers

2.1 Vorticity and Strain

In the neighbourhood of a given point $\mathbf{x} = \mathbf{x}^0$, the velocity field $\mathbf{U}(\mathbf{x}, t)$ of an incompressible fluid of density ρ, is, aside from an overall uniform velocity $\mathbf{U}(\mathbf{x}^0, t)$, represented by the combination of two effects : a deformation and a rotation. The deformation – i.e. an extension or a compression – is expressed by the rate of strain or deformation tensor $S_{ij} = \frac{1}{2}(\partial_i U_j + \partial_j U_i)$, while the solid body rotation is associated with the vorticity $\boldsymbol{\Omega}$, a polar vector with components $\Omega_k = \epsilon_{kmn}\partial_m U_n$. This combination can be mathematically formulated by using the decomposition into symmetric and antisymmetric tensors

$$\partial_i U_j = S_{ij} + \frac{1}{2}\epsilon_{ijk}\Omega_k. \tag{1}$$

The vorticity and the rate of strain tensor are dynamically coupled as can be seen by differentiating the Navier-Stokes equations

$$[\partial_t + U_m\partial_m]\partial_i U_j + (\partial_i U_m)(\partial_m U_j) = \nu\partial_m\partial_m\partial_i U_j - \frac{1}{\rho}\partial_i\partial_j p \tag{2}$$

and then separating the symmetric and antisymmetric parts to obtain

$$[\partial_t + U_m\partial_m]\Omega_k = \nu\partial_m\partial_m\Omega_k + S_{kp}\Omega_p, \tag{3}$$

$$[\partial_t + U_m\partial_m]S_{ij} = \nu\partial_m\partial_m S_{ij} + \frac{1}{4}(\Omega_m\Omega_m\delta_{ij} - \Omega_i\Omega_j) - S_{im}S_{mj} - \frac{1}{\rho}\partial_i\partial_j p. \tag{4}$$

Most of fluid dynamics is actually contained in these equations[1]! The l.h.s. of the vorticity equation (3) clearly corresponds to vorticity transport, while the first term on the r.h.s. represents diffusion and the second term on the r.h.s. the stretching action of the rate of strain tensor. This equation is extensively considered in this review. In the rate of strain evolution equation (4), one again recognizes diffusion (first term on the r.h.s.), while the second and third terms respectively describe the local action of vorticity and rate of strain tensor on the rate of strain tensor. The last term – i.e. the Hessian of pressure divided by

[1] The first equation is well-known whereas the second one is not generally considered. See [42] for a derivation.

density– represents the non local action of pressure on the rate of strain. It is non local since pressure satisfies the elliptic equation

$$\frac{1}{\rho}\partial_m \partial_m p = \frac{1}{2}\Omega_n \Omega_n - S_{ij} S_{ij}. \tag{5}$$

This feature necessitates to take into account the entire flow: including these effects seems at the present time only possible numerically and it is difficult to deal with in local models which, by definition, are restricted to a localized flow region. In general, the rate of strain tensor and the vorticity are fully coupled through equations (3)-(5). In much of what is exposed here, equation (4) is not considered: two-dimensional or three-dimensional strain is assumed to be imposed by the experimentalist ! This assumption is rather stringent but it has proven to be useful. Fluid mechanics is mainly expressed in terms of vorticity dynamics : man likes local mechanisms ! Furthermore local vorticity models have shown their interest not only to identify various dynamical processes but also to disentangle the structural complexity of flows in nature. The fields S_{ij} and $\boldsymbol{\Omega}$ are used to categorize "coherent" or structureless regions in turbulent flows [43] [44]. These identification methods have demonstrated that regions of concentrated vorticity seem to be structured in *layers* and *vortices* [5]. Both configurations are of basic interest, as shown by examining the fate of an initial vorticity field of finite extent embedded in a uniform strain $\mathbf{U} = (a_1 x_1, a_2 x_2, a_3 x_3)$. Because of incompressibility, the condition $a_1 + a_2 + a_3 = 0$ is satisfied with $a_1 \geq 0$, $a_3 \leq 0$ and $a_3 \leq a_2 \leq a_1$. For an axial strain ($a_2 < 0$), vortex tubes aligned with the x_1-axis are generated. For a biaxial strain ($a_2 > 0$) vortex layers are obtained. In the latter instance, vortex layers are subjected to a Kelvin-Helmholtz instability modified by stretching [24] which may lead to the roll-up of the vorticity field into an array of vortex tubes. Finally, it has been numerically determined [45] that initially compact vorticity packets rapidly connect to form tube-like structures. In turbulent flows, strong vortices are widely observed that connect large coherent eddies: they have been referred to as the sinews of turbulence [15].

2.2 Unbounded linear models of vortex tubes and vortex layers

One of the simplest example of velocity field is a two-dimensional flow (here in the x_1-x_2 plane) characterized by the superposition of a uniform unbounded axial vorticity $\boldsymbol{\Omega} = \Omega \mathbf{e}_3$ and a stagnation flow field

$$\mathbf{U}_{strain} = -\gamma x_2 \mathbf{e}_1 - \gamma x_1 \mathbf{e}_2 \tag{6}$$

of given rate of strain γ (coordinate axes are defined so that $\Omega \geq 0$ and $\gamma \geq 0$). These Navier-Stokes and Euler solutions are given by the streamfunction[2] and pressure fields

$$\Psi(x_1, x_2) = -\frac{\Omega}{4}({x_1}^2 + {x_2}^2) + \frac{\gamma}{2}({x_1}^2 - {x_2}^2), \tag{7}$$

[2] The convention $U_1 = \frac{\partial \Psi}{\partial x_2}$ and $U_2 = -\frac{\partial \Psi}{\partial x_1}$ is used.

$$P(x_1, x_2) = \frac{\rho}{2}((\frac{\Omega}{2})^2 - \gamma^2)({x_1}^2 + {x_2}^2). \tag{8}$$

Such unbounded linear two-dimensional velocity fields of uniform vorticity and rate of strain are rather primitive but already contain a lot of physics. When $\gamma = 0$ (figure 1 (a)), equations (7)-(8) describe a *solid body rotation* which approximates the flow field near an axisymmetric vortex core. The case $\Omega = 0$ (figure 1 (b)) pertains to a pure strain field with stretching (resp. compression) axis at an angle of $-\frac{\pi}{4}$ (resp. $\frac{\pi}{4}$) to the x_1-axis. It may be regarded as the leading order Taylor expansion of a potential flow near a *stagnation point* at the origin. Note that, for such a case, pressure reaches a local maximum at the origin.

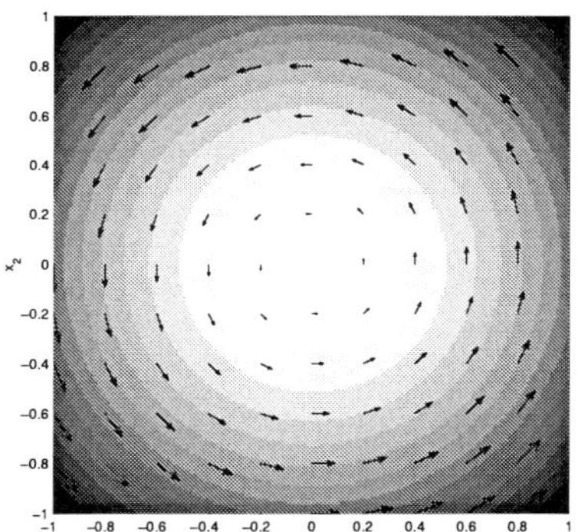

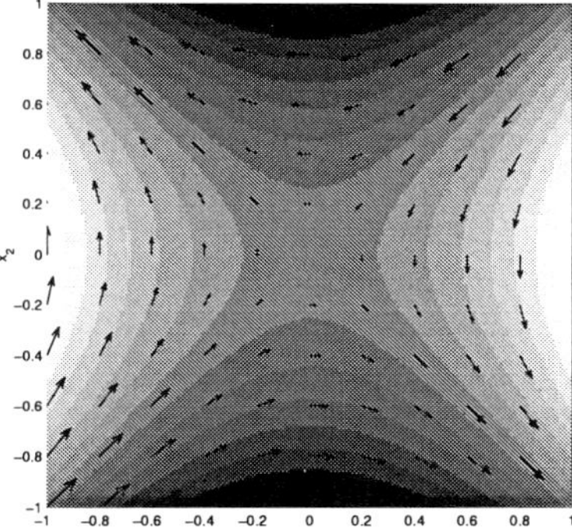

Fig. 1. Various unbounded linear velocity fields (7) : (a) Solid body rotation $\Omega > 0$ and $\gamma = 0$; (b) Stagnation point $\Omega = 0$ and $\gamma > 0$. The grey shading is proportional to the magnitude of the streamfunction Ψ and helps to visualize the streamlines. Arrows indicate the direction and magnitude of velocity field.

When $0 < \gamma < \frac{\Omega}{2}$ (figure 2 (a)), streamlines are similar ellipses with major and minor semi-axes a and b along the x_1- and x_2- directions. The aspect ratio $E = \frac{a}{b}$ and ellipticity $\mu = \frac{a-b}{a+b}$ are given by

$$E = \sqrt{\frac{\frac{\Omega}{2} + \gamma}{\frac{\Omega}{2} - \gamma}}, \; \mu = \frac{E-1}{E+1}. \tag{9}$$

Such an *elliptic flow* approximates the core of a viscous localized vortex embedded in a potential strain field [15] [47]. When $0 < \frac{\Omega}{2} < \gamma$ (figure 2 (c)), streamlines become *hyperbolic* as encountered locally in stretched layers. Finally, the case $\gamma = \frac{\Omega}{2}$ (figure 2 (b)) describes *unbounded Couette flow* which is a special case of free shear flow. As in more general parallel free shear flows, rate of strain and vorticity contributions counterbalance and the pressure field becomes uniform (see equation (8)).

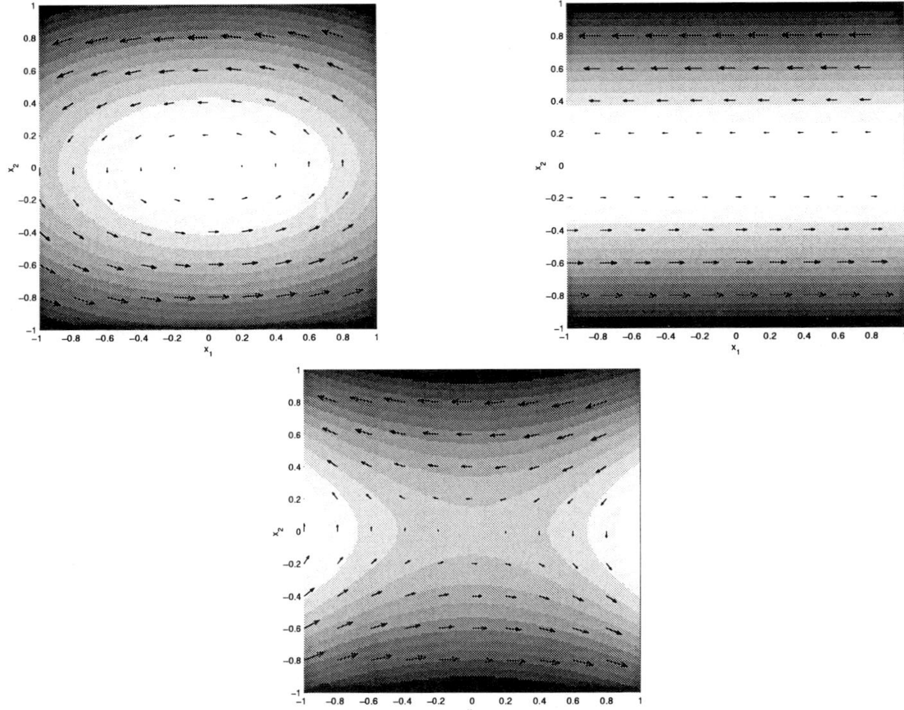

Fig. 2. Various unbounded linear velocity fields (7) : (a) Elliptic flow $0 < \gamma < \frac{\Omega}{2}$; (b) Unbounded Couette flow $0 < \gamma = \frac{\Omega}{2}$; (c) Hyperbolic flow $0 < \frac{\Omega}{2} < \gamma$. The grey shading is proportional to the magnitude of the streamfunction Ψ and helps to visualize the streamlines. Arrows indicate the direction and magnitude of velocity field.

The above unbounded family of two-dimensional flows displays a *linear dependence* with respect to the space coordinates. As shown in [48], this property can be generalized to three-dimensional unsteady flows

$$U_i = T_{ij}(t)x_j \ , \tag{10}$$

whenever the tensor T_{ij} is *traceless* and the tensor

$$\Pi_{ij} = -\rho(\frac{dT_{ij}}{dt} + T_{ik}T_{kj}) \tag{11}$$

is *symmetric*[3]. The latter condition can be justified by a simple introduction of the anzatz (10) in equation (2) : the quantity $\partial_i U_j = T_{ji}$ is independent of the space coordinates and the advection and diffusion terms are hence zero. Thus,

[3] In reference [48], a more general field $U_i(\mathbf{x},t) = U_i^{inst}(t) + T_{ij}(t)x_j$ is actually considered which includes a spatially uniform velocity $\mathbf{U}^{inst}(t)$. However this term brings no new physics and it is omitted here for the sake of clarity.

such a non-parallel velocity field is governed by equation (2) if and only if Π given by (11) is the Hessian of the pressure $P(\mathbf{x}, t)$, i.e.

$$\Pi_{ij} = \frac{\partial^2 P}{\partial x_i \partial x_j}. \tag{12}$$

The time-dependent tensor Π is therefore symmetric. In that case, equation (12) determines the pressure field $P(\mathbf{x}, t)$ for a given Π. Finally, in order to satisfy the continuity equation, the tensor T_{lm} should be traceless since

$$T_{ii} = \partial_i U_i = 0. \tag{13}$$

Aside from the two-dimensional steady flows (7)-(8), a second family of steady solutions satisfies both conditions (12) and (13): three-dimensional irrotational fields of the form

$$U_1 = \alpha x_1 \ , U_2 = \beta x_2 \ , U_3 = -(\alpha + \beta) x_3. \tag{14}$$

These are typical *three-dimensional stagnation point flows* written here in the appropriate principal axes. A third family of solutions consists in the superposition of a time-dependent uniform vorticity field

$$\Psi(x_1, x_2) = -\frac{\Omega(t)}{4}({x_1}^2 + {x_2}^2) \tag{15}$$

and of a three-dimensional uniform strain (14). This is an exact Navier-Stokes solution provided that the vorticity evolves according to the law :

$$\Omega(t) = \Omega(0) \exp[-(\alpha + \beta)t]. \tag{16}$$

It illustrates the amplification (resp. attenuation) of vorticity by axial stretching $\alpha + \beta < 0$ (resp. compression $\alpha + \beta > 0$). This model is easily generalized to an arbitrary uniform vorticity with time-dependent stretching $\alpha(t)$ and $\beta(t)$. In that case, the three-dimensional vorticity field reads after a straightforward integration of (3) :

$$\begin{aligned} &\frac{\Omega_1(t)}{\Omega_1(0)} = \exp[\int_0^t \alpha(t)dt] \ , \frac{\Omega_2(t)}{\Omega_2(0)} = \exp[\int_0^t \beta(t)dt] \ , \\ &\frac{\Omega_3(t)}{\Omega_3(0)} = \exp[-\int_0^t (\alpha(t) + \beta(t))dt]. \end{aligned} \tag{17}$$

Such fields have been used in particular to study time-periodic stretching effects [49][50]. Note that the unbounded linear three-dimensional flows (10) satisfy both the Euler and Navier-Stokes equations ! For such velocity fields, viscosity does not play a role : the viscous diffusion Laplacian identically vanishes because of the linear dependence on the space coordinates and the no-slip condition does not need to be enforced since the fields are assumed to be unbounded.

Finally, let us mention another easy extension of the two-dimensional flows (7)-(8) : the vorticity is still uniform and axial but the linear basic strain is replaced by a multipolar irrotational field

$$\Psi(r,\theta) = -\frac{\Omega}{4}r^2 + \gamma\frac{r^n}{n}\cos n\theta \tag{18}$$

in the usual (r, θ) cylindrical coordinate system[4]. For $n = 2$, the steady state family (7)-(8) is recovered. For $n \neq 2$, vortex lines are no longer ellipses or hyperbolas : close to the center they are almost circles while they display a triangular (n=3)(figure 3 (a)), square (n=4) (figure 3 (b)), ..., shape with increasing distance from the center until a separatrix is reached.

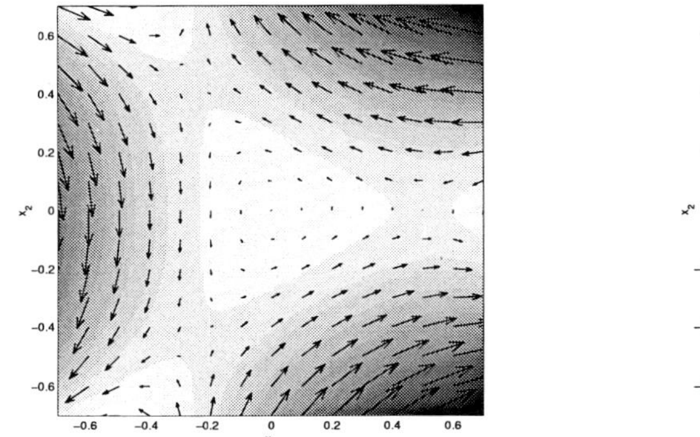

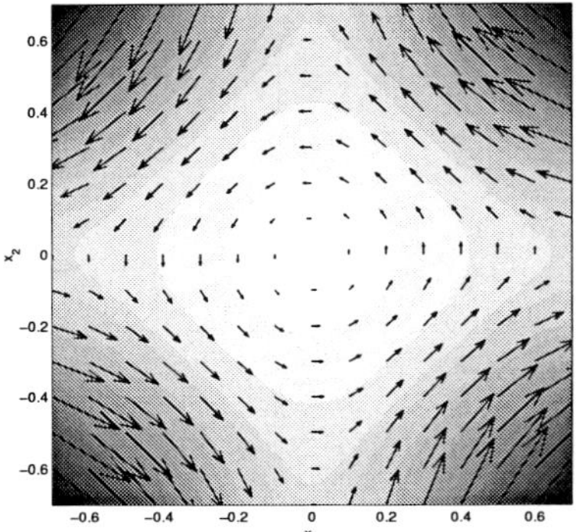

Fig. 3. Various multipolar velocity fields (18) : (a) $n = 3$; (b) $n = 4$. The grey shading is proportional to the magnitude of the streamfunction Ψ and helps to visualize the streamlines. Arrows indicate the direction and magnitude of velocity field.

3 Bounded vorticity flows

Until now, we have discussed flows with a uniform and unbounded vorticity. The next structural feature that should be introduced to get a more realistic flow is the finite spatial extent of the vorticity distribution. The easiest way is to assume that vorticity is still unidirectional along the z-axis but concentrated either in the y-direction only (figure 4 (a)) which gives *vortex layers*, the desingularized version of vortex sheets, or in both the x- and y-directions (figure 4 (b)) which gives *vortex tubes*, the desingularized version of vorticity filaments.

3.1 Two and three-dimensional vortex layers

In parallel free shear flows $\mathbf{U} = U(y)\mathbf{e}_x$, vorticity is concentrated in a layer or, in the singular case of the vortex sheet, in a plane. When viscosity is introduced,

[4] In cylindrical coordinates, the convention $U_r = \frac{1}{r}\frac{\partial\Psi}{\partial\theta}$ and $U_\theta = -\frac{\partial\Psi}{\partial r}$ is used.

Fig. 4. (a) Vortex layers; (b) Vortex tubes. Shaded area approximately covers the vortical region in the x-y plane. Arrows depict the velocity field $\mathbf{U}$ and the corresponding vorticity $\boldsymbol{\Omega}$.

these solutions of the steady Euler equations become time-dependent since vorticity $\boldsymbol{\Omega} = -\frac{\partial U(y,t)}{\partial y}\mathbf{e}_z$ is subjected to diffusion. As an example, consider the mixing layer between two parallel streams along the x-axis of velocity $U_{+\infty}$ and $U_{-\infty}$ at $y = \pm\infty$ respectively, described by the self-similar Navier-Stokes solution [2]

$$U(y,t) = \frac{U_{+\infty} + U_{-\infty}}{2} + \frac{U_{+\infty} - U_{-\infty}}{2} F(\chi). \qquad (19)$$

The similarity variable is defined by $\chi = \frac{y}{2\delta(t)}$ where the shear layer thickness is $\delta(t) = \sqrt{\nu t}$ and the function $F(\chi)$ reads

$$F(\chi) = \frac{2}{\sqrt{\pi}} \int_0^{\chi} \exp(-s^2) ds \ . \qquad (20)$$

Under such conditions, the vorticity

$$\boldsymbol{\Omega}(y,t) = -\frac{U_{+\infty} - U_{-\infty}}{2\sqrt{\pi\nu t}} \exp(-[\frac{y}{2\delta(t)}]^2)\mathbf{e}_z \qquad (21)$$

is clearly confined to a region of characteristic thickness $\delta(t)$. In the presence of *stretching*, this diffusion process may be stopped as rigorously shown by examining the effect of a uniform unsteady stretching (figure 5)

$$\mathbf{U} = U(y,t)\mathbf{e}_x - \gamma(t)y\mathbf{e}_y + \gamma(t)z\mathbf{e}_z \ . \qquad (22)$$

In this case, vorticity is aligned with the stretching z-direction of this bi-axial strain and it satisfies

$$\frac{\partial \Omega}{\partial t} - \gamma(t)y\frac{\partial \Omega}{\partial y} - \gamma(t)\Omega = -\nu\Delta\Omega \qquad (23)$$

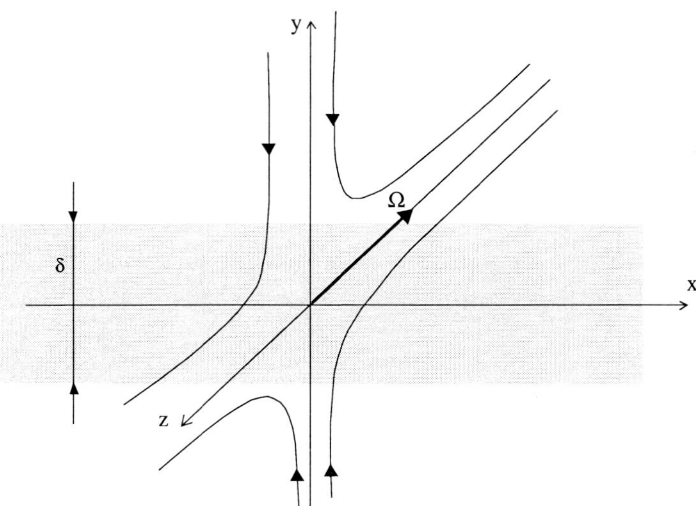

Fig. 5. Stretched vortex layer subjected to a uniform bi-axial strain. Shaded area approximately covers the vortical region in the x-y plane. The parameter δ denotes the characteristic layer thickness and $\boldsymbol{\Omega}$ the vorticity vector.

instead of a purely diffusive equation.
Let us introduce the time variable $\tau = \int_0^t s^2(u)\, du$ and the space variable $\eta = s(t)y$, where the dimensionless quantity $s(t) \equiv \exp(\int_0^t \gamma(u)du)$ represents stretching. It is then easily found that

$$\Omega(y,t) = s(t)\Omega_{unst}(\eta,\tau), \tag{24}$$

where the rescaled quantity $\Omega_{unst}(\eta,\tau)$ now satisfies a pure diffusion equation without stretching. To any unstretched solution Ω_{unst} is thus associated a stretched one Ω ! In the context of inviscid dynamics, $\Omega_{unst}(\eta)$ becomes arbitrary and does not depend on τ: the solution (24) embodies the essence of stretching ! Returning to the viscous case, the velocity field (22) with profile (19) remains a valid self-similar solution of equation (23) provided that $\delta(t)$ in equations (19) and (21) now becomes

$$\delta^2(t) = \nu \frac{\int_0^t s^2(u)du}{s^2(t)}. \tag{25}$$

For a steady stretching $\gamma = \gamma_0$ and large time, one recovers, from equations (19)-(25), the famous *Burgers vortex layer*

$$\Omega(y) = -\frac{U_1 - U_2}{\sqrt{2\pi}} \sqrt{\frac{\gamma_0}{\nu}} \exp(-\frac{\gamma_0}{2\nu}y^2) \ , \tag{26}$$

of constant thickness $\delta_0^2 = \frac{\nu}{2\gamma_0}$. In such a case, the vorticity is enhanced by strain and damped by diffusion. Note that the strength $U_{+\infty} - U_{-\infty}$ of the layer remains, contrary to its thickness, arbitrary. From the intrinsic length scale δ_0 and this velocity scale, one can build the Reynolds number

$$Re \equiv \frac{(U_{+\infty} - U_{-\infty})}{2\sqrt{\nu\gamma_0}}. \tag{27}$$

3.2 Two-dimensional vortex tubes

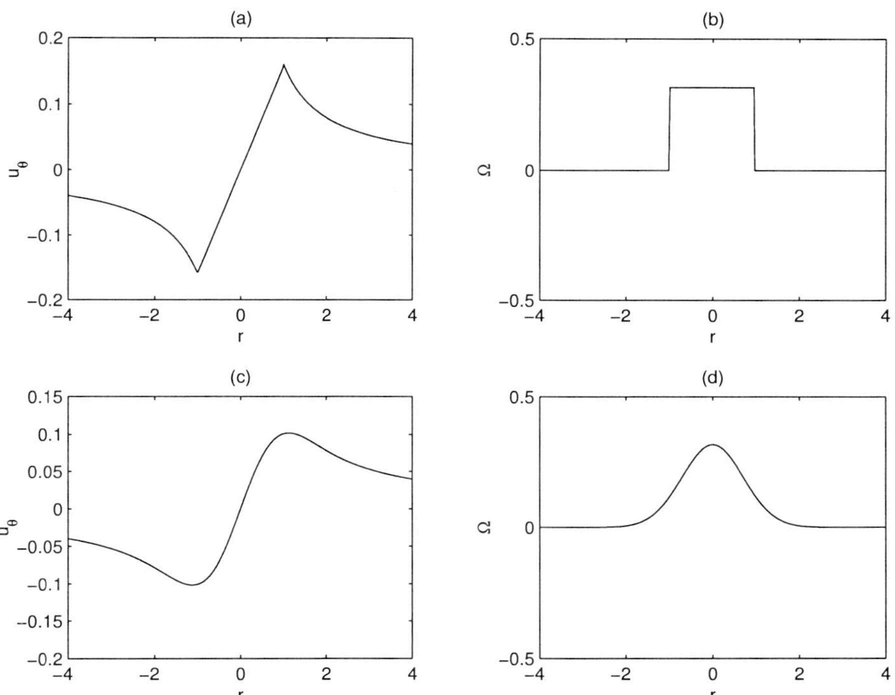

Fig. 6. Various azimuthal velocity fields $U_\theta(r)$ and corresponding axial vorticity $\Omega(r)$ made non-dimensional with respect to the velocity scale Γ/a and length scale a. (a) and (b) Rankine vortex; (c) and (d) Lamb vortex.

An azimuthal velocity field $U_\theta(r,t)$ is associated to a unidirectional and concentrated vorticity along the z-axis given by

$$\boldsymbol{\Omega} = \Omega \mathbf{e}_z = \frac{1}{r}\frac{d}{dr}(rU_\theta)\mathbf{e}_z. \tag{28}$$

In the framework of inviscid dynamics, this profile remains steady and arbitrary: it can be any smooth or discontinuous function of the radial coordinate r. A *rigidly rotating fluid column* located in a pipe of radius a constitutes the simplest example of such an inviscid flow. Its velocity profile reads:

$$\mathbf{U} = \frac{\Omega}{2} r \mathbf{e}_\theta \ , r < a. \tag{29}$$

Note that such a velocity field is also a solution of the Navier-Stokes equations provided that the pipe be rotating about its axis with angular velocity $\Omega/2$. Another solution, more appropriate for unbounded flows, is the *Rankine vortex* of total circulation Γ :

$$U_\theta(r,t) = \frac{\Gamma}{2\pi a^2} r \ , \Omega(r,t) = \frac{\Gamma}{\pi a^2} \ , r < a \tag{30}$$

$$U_\theta(r,t) = \frac{\Gamma}{2\pi r} \ , \Omega(r,t) = 0 \ , r > a. \tag{31}$$

Inside the core of size a, vorticity is assumed uniform and the fluid flows in solid body rotation. Outside this zone, vorticity jumps to zero: the fluid moves as in a potential point vortex flow. The Rankine vortex (figure 6 (a) and figure 6 (b)) is only relevant in the inviscid limit since it displays a jump in vorticity. Note that rigidly rotating pipe flow and the Rankine vortex are dynamically different: oscillations in the fluid column are such that the normal velocity vanishes on the boundary while the Rankine vortex may deform. This is confirmed by a direct comparison of the dynamical waves they can support (section 6.3). For smooth vorticity profiles, a typical solution is the famous *Lamb vortex* (figure 6 (c) and figure 6 (d)) of core size $a(t)$ and constant circulation Γ given by

$$U_\theta(r,t) = \frac{\Gamma}{2\pi r}[1 - \exp(-\frac{r^2}{a^2})] \ , \tag{32}$$

$$\Omega(r,t) = \frac{\Gamma}{\pi a^2} \exp(-\frac{r^2}{a^2}) \ . \tag{33}$$

While the core size remains constant in the inviscid case, it diffuses if viscosity is included. In the latter instance, the profile $U_\theta(r,t)$ satisfies a diffusion equation and a is time-dependent such that

$$a(t) = \sqrt{4\nu t + a_0^2}, \tag{34}$$

where a_0 is the initial core radius at $t = 0$. Moreover vorticity exponentially decreases with increasing distance from the axis and its maximum varies as $1/t$. A Reynolds number can be defined based on the characteristic velocity Γ/a and length a :

$$Re_\Gamma = \frac{\Gamma}{\nu}. \tag{35}$$

It has recently been demonstrated that this is the only possible similarity solution in viscous two-dimensional flows [51]. Introducing into the Navier-Stokes equations a velocity profile with a unique azimuthal component $U_\theta(r,t)$, it is easily found that the diffusion term and the time derivative term cancel each other : under the action of viscous diffusion all smooth vorticity profiles converge for large time to the self-similar *Lamb vortex solution.* Furthermore an inward pressure gradient counterbalances, in the Navier-Stokes equations, the centrifugal force caused by the inertial term so that

$$P(r,t) = \rho \int \frac{U_\theta^2}{r} dr. \tag{36}$$

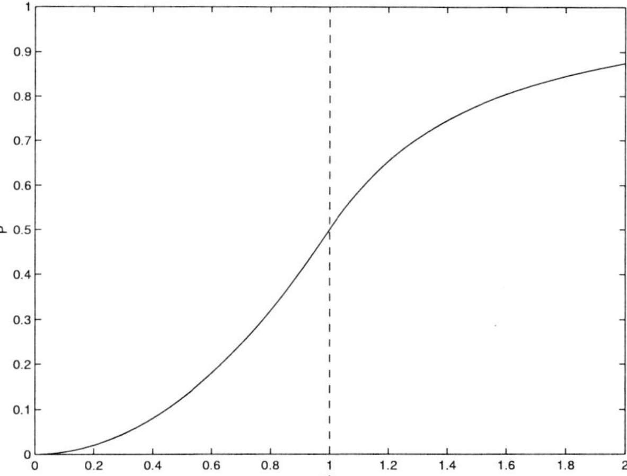

Fig. 7. Rankine vortex pressure field $P(r)$ made non-dimensional so that $P(0) = 0$ and $P_\infty = 1$. The area on the left of the dashed line corresponds to the Rankine vortex core.

For the *Rankine vortex*, the pressure field (figure 7) thus reads:

$$P(r,t) = P_\infty - \rho(\frac{\Gamma}{2\pi})^2 \frac{1}{2r^2} \; , r > a, \tag{37}$$

$$P(r,t) = [P_\infty - \rho(\frac{\Gamma}{2\pi a})^2] + \frac{\rho}{2}(\frac{\Gamma}{2\pi a^2})^2 r^2 \; , r < a, \tag{38}$$

where P_∞ stands for the pressure at infinity. Invoking equations (37)-(38), the vortex core is seen to be a low-pressure region (see also equation (5) in section 2.1) : half of the pressure loss is contained within the vortex. Such a property is used to identify vortices in experiments in which bubbles migrate towards low-pressure regions within vortex cores [12]. As previously shown, axisymmetric solutions are always steady in the inviscid framework. However, more general isolated unsteady solutions can be found for two-dimensional inviscid flows. Let us mention a generalization of the Rankine vortex called the *Kirchhoff vortex* : it is an elliptic vortex patch of uniform vorticity Ω and major and minor semi-axes a and b steadily rotating with an angular velocity $\Omega \frac{ab}{(a+b)^2}$ [17]. For smooth vortices, it is believed that there is a tendency for isolated nonaxisymmetric vortices to become axisymmetrical though Dritschel [52] has recently shown that an isolated nonaxisymmetric vortex with sufficiently sharp edges can persist.

When a two-dimensional vortex is not isolated, some asymmetry may be present due to boundaries or to the local strain generated by other vortices, for instance in a counter-rotating vortex pair [35]. In the remainder of this section, we analyze the structure of a viscous vortex deformed by boundaries, or various inviscid or viscous vortices distorted by the local strain due to the presence of other vortices. A simple theoretical extension of the rotating column (29) in a rotating pipe is

$$\mathbf{U} = [(-\frac{\Omega}{2} - \gamma)y, (\frac{\Omega}{2} - \gamma)x, 0]. \tag{39}$$

It exactly satisfies the Navier-Stokes and Euler equations and represents a *confined vortex* where the asymmetry is here due to to the presence of elliptic boundaries. In laboratory experiments, this solution almost corresponds, beyond a spin-up phase, to the elliptic flow observed by Malkus [12] inside a rotating elastic cylinder which is compressed by rollers and rotates at a constant velocity[5]. Malkus' experimental flow is not represented by (39) in a thin boundary layer near the cylinder boundary since the velocity on the cylinder wall in equation (39) is not a constant as in the experimental set-up.

In the case of an unbounded domain, several exact or approximate Euler or Navier-Stokes solutions are known for a vortex subjected to a given uniform steady strain field : an *inviscid vortex patch*, an *inviscid smooth profile* and finally a *viscous vortex*. The first velocity field to be examined extends the Rankine vortex (30)-(31) to include the presence of strain [17]: it is a patch of uniform vorticity Ω delineated by an ellipse of major and minor semi-axes a and b (figure 8). Here the strain counterbalances the propension of the elliptic vortex to rotate as its Kirchhoff's counterpart. When the $x-$ (resp. $y-$) direction coincides with the major (resp. minor) axis, the steady streamfunction Ψ within the vortex reads

$$\Psi(x,y) = -\frac{\Omega a^2 b^2}{2(a^2+b^2)}[(\frac{x}{a})^2 + (\frac{y}{b})^2 - 1]. \tag{40}$$

The vorticity is clearly uniform inside the ellipse and the strain field $-\gamma y \mathbf{e}_x - \gamma x \mathbf{e}_y$ leads to a compression along the first diagonal $x = y$ as in the linear

[5] Other experiments of a transient nature [28] [6] have been performed using a rotating elliptic solid cylinder which is suddenly brought to rest.

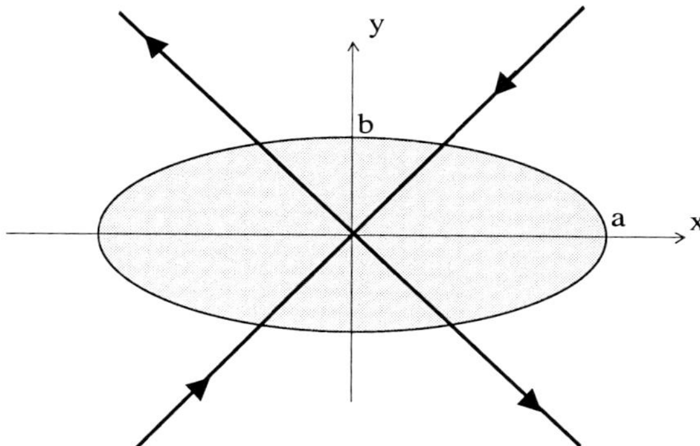

Fig. 8. Steady elliptic uniform vorticity patch embedded in irrotational strain field. Associated streamfunction inside the patch is given by (40). Shaded area exactly covers the vortical region in the x-y plane. Compression and stretching axes are also shown.

unbounded case (7)-(8). The exact relationship

$$\frac{\gamma}{\Omega} = \frac{E(E-1)}{(E^2+1)(E+1)} , \tag{41}$$

between the aspect ratio $E = \frac{a}{b}$ of the ellipse and the dynamical ratio $\frac{\gamma}{\Omega}$ is readily obtained by matching solution (40) with the outer potential flow. Relation (41) shows that the steady elliptic vortex only exists if $\frac{\gamma}{\Omega} \leq 0.15$, which corresponds to the maximum aspect ratio $E \sim 2.9$. In the regime $0 \leq \frac{\gamma}{\Omega} \leq 0.15$, other unsteady solutions (e.g. a rotating or nutating vortex patch) may be obtained [53]. Above this critical rate of strain, the steady solution no longer exists and the vortex is elongated infinitely in the direction of the strain [53].

Smooth inviscid vortices subjected to an *arbitrary potential field*, can be constructed by exploiting the vorticity equation (3) with $\nu = 0$: in two-dimensional steady inviscid incompressible flows, the vorticity $\Omega(x,y)$ is conserved along each streamline $\Psi(x,y) = \text{const}$. As a result there exists a function F relating the vorticity $\Omega(x,y) = -\Delta\Psi$ and the streamfunction $\Psi(x,y)$ of the form $\Omega = F(\Psi)$. The velocity field is therefore obtained by solving the two-dimensional non-linear equation

$$\Delta\Psi = -F(\Psi). \tag{42}$$

The condition at infinity is essential in specifying the potential flow field which has to be superimposed on the above vorticity field. This formulation has been used by Moore and Saffman [15] to compute a family of vortices subjected to the uniform steady strain

$$\mathbf{U}_{strain} = -\beta x \mathbf{e}_x + \beta y \mathbf{e}_y \tag{43}$$

in the weak strain limit (figure 9). Note that, for future convenience, the external strain takes a different form than the one adopted previously: the coordinate axes

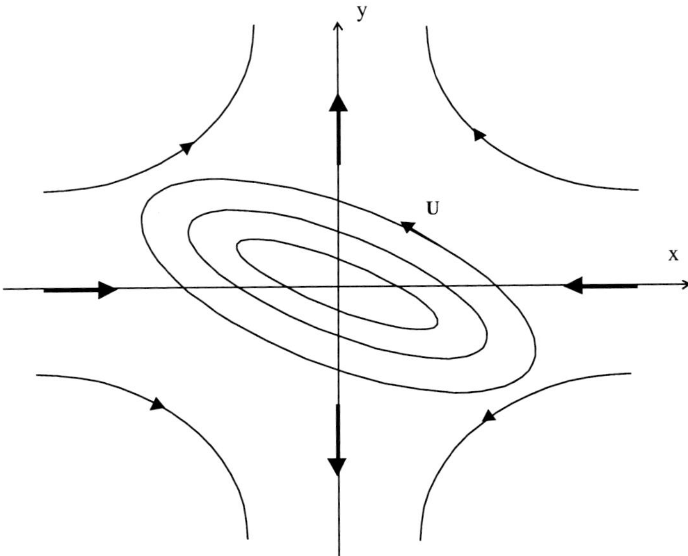

Fig. 9. Steady inviscid smooth elliptic vortex in weak irrotational straining field (43). Compression and stretching axes are also shown. The vorticity is assumed positive along the z-axis.

have been simply rotated so that, for $\beta > 0$, the stretching axis is now along the y-axis instead of the diagonal $x = -y$. In the weak strain approximation, the circulation $\Gamma > 0$ and characteristic core size a_c are, by definition, such that the external strain in the vortex region βa_c has a much smaller effect than the velocity $\frac{\Gamma}{a_c}$ induced by the vortex :

$$\beta a_c << \frac{\Gamma}{a_c}. \tag{44}$$

This assumption ensures the presence of a small parameter $\epsilon_1 \equiv \frac{\beta {a_c}^2}{\Gamma}$. In the ensuing development down to equation (51), the streamfunction Ψ and radial coordinate r are made non-dimensional with respect to Γ and a_c. The condition at infinity consists in the superposition of the uniform strain and the far field induced by the vortex itself:

$$\Psi(r,\theta) \sim -\frac{\epsilon_1}{2} r^2 \sin 2\theta - \frac{1}{2\pi} \ln r + ... \tag{45}$$

For $\epsilon_1 = 0$, a given axisymmetric smooth profile $U_\theta(r)$ prevails which corresponds to a precise relation $\Omega_0 = F(\Psi_0)$ between its vorticity $\Omega_0(r)$ and streamfunction $\Psi_0(r)$. If the latter relation is assumed to be unchanged for $\epsilon_1 \neq 0$, the function F appearing in equation (42), defines a unique branch of solutions. In order to determine this branch in the weak strain limit, the streamfunction is expanded in powers of the small parameter ϵ_1 according to

$$\Psi \sim \Psi_0(r) + \epsilon_1 \Psi_1(r) + ... \tag{46}$$

Equation (42) is automatically satisfied at zeroth order and, at first order, it reads

$$\Delta \Psi_1 = q(r)\Psi_1, \tag{47}$$

where

$$q(r) \equiv -\frac{dF}{d\Psi}(\Psi_0) = -\frac{\partial \Omega_0}{\partial r} / \frac{\partial \Psi_0}{\partial r} \tag{48}$$

is specified by the axisymmetric zeroth-order solution. According to the condition at infinity (45), one may look for perturbative solutions of the form

$$\Psi_1(r) = 2f(r)\sin 2\theta \tag{49}$$

with the condition at infinity

$$f(r) \to -\frac{1}{4}r^2. \tag{50}$$

Note that the logarithmic term in (45) has already been taken into account in the zeroth order solution $\Psi_0(r)$. From (47)-(49), it is straightforward to derive the equation

$$\frac{d^2 f}{dr^2} + \frac{1}{r}\frac{df}{dr} - (q(r) + \frac{4}{r^2})f = 0. \tag{51}$$

Up to a multiplicative constant, equation (51) possesses only one solution that remains regular near $r = 0$. Moreover the function $q(r)$ goes very rapidly to zero as r increases since vorticity is concentrated in the vortex core: regular solutions $f(r)$ at $r = 0$ thus go to infinity like r^2 times a constant. This behaviour implies that the far field condition (50) uniquely defines the function $f(r)$ which must be computed numerically. According to (49), it can be seen that streamlines are ellipses, the major axis of which is making an angle of $\pi/4$ with the stretching y- axis (figure 9). This is the result previously obtained for the vortex patch and here extended to smooth profiles. The aspect ratio of streamlines is now non uniform and varies with r. This perturbation analysis of concentrated vortices can be generalized to a weak external multipolar strain field $\Psi_{strain} \sim \frac{r^n}{n}\sin(n\theta)$. Figure 10(a) (resp.(b)) displays the case of the inviscid Rankine vortex when embedded in a multipolar strain field $n = 3$ (resp. $n = 4$).

The same analysis has also been performed in the *viscous case, for unbounded vortices* [47] [55] in the weak steady strain limit and at high circulation Reynolds numbers $Re_\Gamma = \frac{\Gamma}{\nu} >> 1$. As in the inviscid case, a relation of the type $\Omega = F(\Psi)$ is verified to ensure inviscid equilibrium. This state is reached within a time scale $\frac{a_c{}^2}{\Gamma}$ characteristic of the circulation motion around the vortex of characteristic

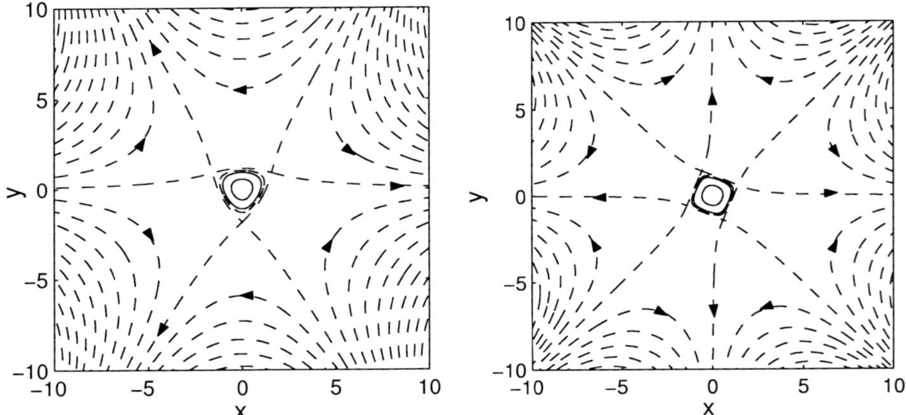

Fig. 10. Steady Rankine vortex subjected to multipolar strain field (a) $\frac{r^3}{3}\sin 3\theta$ and (b) $\frac{1}{4}\frac{r^4}{4}\sin 4\theta$. Solid (resp. dashed) streamlines are located inside (resp. outside) the core. Courtesy of C.Eloy.

core radius a_c. Thereafter, on the slow viscous time scale $\frac{a_c{}^2}{\nu} = Re_\Gamma \frac{a_c{}^2}{\Gamma} >> \frac{a_c{}^2}{\Gamma}$, viscous diffusion selects a specific profile very much in the same fashion as the *Lamb vortex* was selected in the axisymmetric case, among an infinite family of smooth profiles. More specifically, consider a viscous vortex embedded in the same weak uniform steady strain field (43). For convenience, the total stream function $\Psi_{tot} = \Psi^s + \Psi$ is split into the external strain part $\Psi^s = -\frac{\beta}{2}r^2 \sin 2\theta$ and the vortex contribution Ψ. Since the strain is irrotational, the vorticity Ω satisfies $\Omega = -\nabla^2\Psi$. Thereafter the dynamical equation (3) for vorticity completely defines the system. It reads in two-dimensional polar coordinates

$$\frac{1}{r}[\frac{\partial\Psi}{\partial\theta}\frac{\partial}{\partial r} - \frac{\partial\Psi}{\partial r}\frac{\partial}{\partial\theta}]\Omega = -U_r^s\frac{\partial\Omega}{\partial r} - \frac{U_\theta^s}{r}\frac{\partial\Omega}{\partial\theta} - \frac{\partial\Omega}{\partial t} + \nu\nabla^2\Omega, \tag{52}$$

where

$$U_r^s = \frac{1}{r}\frac{\partial\Psi^s}{\partial\theta} = -\beta r\cos 2\theta, \tag{53}$$

$$U_\theta^s = -\frac{\partial\Psi^s}{\partial r} = \beta r\sin 2\theta. \tag{54}$$

Equation (52) nondimensionalized with respect to the characteristic core size a_c and the velocity scale Γ/a_c thus reads

$$\frac{1}{r}\left[\frac{\partial\Psi}{\partial\theta}\frac{\partial}{\partial r} - \frac{\partial\Psi}{\partial r}\frac{\partial}{\partial\theta}\right]\Omega = \frac{\beta a_c{}^2}{\Gamma}\left[r\cos 2\theta\frac{\partial\Omega}{\partial r} - \sin 2\theta\frac{\partial\Omega}{\partial\theta}\right] - \frac{a_c{}^2}{\tau\Gamma}\frac{\partial\Omega}{\partial t} + \frac{\nu}{\Gamma}\nabla^2\Omega, \tag{55}$$

where the characteristic time scale denoted by τ is, for the moment, left unspecified. In the asymptotic regime of high circulation Reynolds number and weak uniform steady strain, two small parameters then appear: $\epsilon_1 \equiv \frac{\beta {a_c}^2}{\Gamma}$ already used in the inviscid limit and $\epsilon \equiv \frac{1}{Re_\Gamma} = \nu/\Gamma$. Assume that the inviscid equilibrium is already attained and, as in the Lamb vortex, that viscous diffusion and unsteadiness arise at the same order. Under these circumstances, τ reduces to the diffusion time scale $\tau = \frac{{a_c}^2}{\nu} = \frac{1}{\epsilon}\frac{{a_c}^2}{\Gamma}$. Finally consider the distinguished limit for which viscous effects and strain arise at the same order i.e.

$$\frac{\beta {a_c}^2}{\Gamma} = \frac{\lambda}{2}\epsilon \tag{56}$$

where λ is assumed $O(1)$. Equation (55) then takes the form

$$\frac{1}{r}\left[\frac{\partial \Psi}{\partial \theta}\frac{\partial}{\partial r} - \frac{\partial \Psi}{\partial r}\frac{\partial}{\partial \theta}\right]\Omega = \epsilon\lambda\mathcal{L}\Omega + \epsilon\mathcal{L}_0\Omega, \tag{57}$$

where $\mathcal{L}_0 \equiv -\frac{\partial}{\partial t} + \nabla^2$ and $\mathcal{L} \equiv \frac{1}{2}\left[r\cos 2\theta\frac{\partial}{\partial r} - \sin 2\theta\frac{\partial}{\partial \theta}\right]$ have been introduced. Let us now seek a perturbative solution by expanding the various fields in powers of ϵ :

$$\Psi = \sum_{i=0}^{\infty}\epsilon^i\Psi_i, \quad \Omega = \sum_{i=0}^{\infty}\epsilon^i\Omega_i. \tag{58}$$

The equation $\Omega = -\nabla^2\Psi$ is satisfied if

$$\Omega_i = -\nabla^2\Psi_i. \tag{59}$$

At zeroth order, equation (57) reduces to the inviscid equilibrium equation

$$\left[\frac{\partial \Psi_0}{\partial \theta}\frac{\partial}{\partial r} - \frac{\partial \Psi_0}{\partial r}\frac{\partial}{\partial \theta}\right]\Omega_0 = 0, \tag{60}$$

which is automatically satisfied by any axisymmetric $\Psi_0(r)$ otherwise left unspecified. At first order one obtains:

$$\frac{1}{r}\frac{\partial}{\partial \theta}\left[\Psi_1\frac{\partial \Omega_0}{\partial r} - \Omega_1\frac{\partial \Psi_0}{\partial r}\right] = \lambda\mathcal{L}\Omega_0 + \mathcal{L}_0\Omega_0. \tag{61}$$

Integrating with respect to θ from 0 to 2π yields the solvability condition

$$\mathcal{L}_0\Omega_0 = -\frac{\partial \Omega_0}{\partial t} + \nabla^2\Omega_0 = 0, \tag{62}$$

which turns out to be a simple diffusion equation for the basic axisymmetric vortex. The diffusion process is therefore dominated for large time by the Lamb vortex solution $\Omega_0 = \frac{1}{4\pi t}\exp(-\frac{r^2}{4t})$. Viscosity thus selects a particular vortex profile that slowly diffuses as in the axisymmetric case. The dynamical equation (61) now reads

$$\frac{1}{r}\frac{\partial}{\partial \theta}\left[\Psi_1\frac{\partial \Omega_0}{\partial r} - \Omega_1\frac{\partial \Psi_0}{\partial r}\right] = \frac{\lambda}{2}r\cos 2\theta\frac{\partial \Omega_0}{\partial r}. \tag{63}$$

Integrating once with respect to θ yields

$$\frac{\partial \Psi_0}{\partial r}\nabla^2\Psi_1 + \frac{\partial \Omega_0}{\partial r}\Psi_1 = \frac{\lambda}{4}r^2\sin 2\theta\frac{\partial \Omega_0}{\partial r}. \tag{64}$$

Upon taking $\Psi_1 = \lambda f(r,t)\sin 2\theta$, one finds

$$\frac{\partial^2 f}{\partial r^2} + \frac{1}{r}\frac{\partial f}{\partial r} - \frac{4}{r^2}f = q(r,t)(f - \frac{r^2}{4}), \tag{65}$$

which is identical to equation (51) in the inviscid case, provided that $f(r)$ in (51) be changed into $f(r) - \frac{r^2}{4}$ to subtract the external strain Ψ^s. The main difference with the inviscid case is that the function $q(r,t)$ is now time-dependent and of the form

$$q(r,t) = -\frac{\partial \Omega_0}{\partial r}/\frac{\partial \Psi_0}{\partial r} = -\frac{r^2}{4t^2(\exp(r^2/4t) - 1)},$$

as determined by the action of viscosity on the zeroth-order Lamb profile. Through a simple change of variable $\xi = r/\sqrt{t}$ and the rescaling $\mathcal{F} = t^{-1}f$, it is possible to reduce (65) to the steady equation

$$\frac{d^2\mathcal{F}}{d\xi^2} + \frac{1}{\xi}\frac{d\mathcal{F}}{d\xi} - \frac{4}{\xi^2}\mathcal{F} = -\frac{\xi^2}{4\left(\exp\left(\xi^2/4\right) - 1\right)}(\mathcal{F} - \frac{\xi^2}{4}). \tag{66}$$

The two boundary conditions (regularity at $r = 0$ and irrotational flow at $r = +\infty$)

$$\mathcal{F} = O(\xi^{-2}), \quad \xi \to \infty, \tag{67}$$

$$\mathcal{F} = O(\xi^{2}), \quad \xi \to 0, \tag{68}$$

uniquely determine the solution $\mathcal{F}(\xi)$ by numerical integration of equation (66). Invoking the above rescalings, the effect of a constant strain is seen to increase as the vortex diffuses and its vorticity maximum decreases. Close to the center, the flow pattern is formed of ellipses, the major axis of which is at an angle of $\pi/4$ (resp. $-\pi/4$) with respect to the stretching axis if $\Gamma \geq 0$ (resp. $\Gamma \leq 0$). The streamline ellipticity near $r = 0$ is equal to $2.5\beta/\Omega_{max}$, where Ω_{max} is the maximum vorticity precisely located at $r = 0$. Moreover, the equation (9) for an infinite vortex can be used locally if the values of the rate of strain γ and vorticity Ω are taken to be those at $r = 0$. In the weak strain limit, ellipticity in (9) approaches the ratio γ/Ω : comparing this value with the one given above indicates that the rate of strain is 2.5 times greater in the center of the deformed Lamb vortex than outside. In the present viscous case, streamlines and vortex lines are not completely identical as in a purely inviscid approximation.

Two-dimensional flows with a single axial vorticity component are obviously not sufficient to comprehend all the dynamical properties of vorticity: three-dimensional effects should be introduced. This can be done in two simple ways: either by adding an azimuthal vorticity component (section 3.3) or by keeping a unique axial vorticity and superimposing a three-dimensional potential stretching field (section 3.4). These two effects can also be simultaneously present (section 3.4). More "exotic" solutions are possible which contain multiple cell solutions (section 3.4) or display helical symmetry (section 3.5).

3.3 Three-dimensional flows: swirling jets

The *steady swirling jet* flow

$$\mathbf{U} = (0, U_\theta(r), U_z(r)) \tag{69}$$

satisfies the Euler equations for any arbitrary $U_\theta(r)$ and $U_z(r)$ profiles, a feature which generalizes the previously considered two-dimensional vortices by superimposing an axial velocity. This axial component does not influence the pressure field which is still given for inviscid flows by (36). By contrast, the associated vorticity field

$$\boldsymbol{\Omega} = (0, -\frac{dU_z}{dr}, \frac{1}{r}\frac{d(rU_\theta)}{dr}) \tag{70}$$

now possesses an *axial* Ω_z- as well as an *azimuthal* Ω_θ-component. Vortex lines are no longer straight as for two-dimensional vortices but are now coiled on cylindrical surfaces r =const. This property can be generalized for inviscid vortices of the form

$$\mathbf{U} = (U_r(r,z,t),\ U_\theta(r,z,t),\ U_z(r,z,t)) \tag{71}$$

which are still *axisymmetric* but explicitly depend on the axial variable z and time t. In such cases, *vortex lines are coiled on axisymmetric surfaces associated with a given value of the circulation.* Dynamically, *the presence of coiled vortex lines is intimately related to the existence of an axial velocity and vice versa.* Starting from the Euler equations, it is easily found [56] that, for inviscid vortices of the type (71), conservation of the local circulation $rU_\theta(r,z,t)$ reads

$$\frac{D}{Dt}[rU_\theta(r,z,t)] = 0, \tag{72}$$

where

$$\frac{D}{Dt} \equiv \frac{\partial}{\partial t} + U_r\frac{\partial}{\partial r} + U_z\frac{\partial}{\partial z} \tag{73}$$

stands for the convective derivative. Moreover, the *azimuthal* Ω_θ-component is coupled to the axial gradient of the local circulation through

$$\frac{D}{Dt}[\frac{\Omega_\theta}{r}] = \frac{1}{r^4}\frac{\partial}{\partial z}[rU_\theta(r,z,t)]^2. \tag{74}$$

As a result of the circulation conservation law (72), the presence of an axial velocity may induce a differential azimuthal rotation along the z-axis which then creates a torsion of vortex lines along this axis. Conversely the existence of an axial gradient for the quantity rU_θ generates (see equation (74)) an *azimuthal* vorticity component and consequently an axial jet velocity. Equivalently, in pressure terms, an axial velocity is generally induced by a local axial pressure gradient due to a differential azimuthal rotation along the $z-$ axis (see equation (36)).

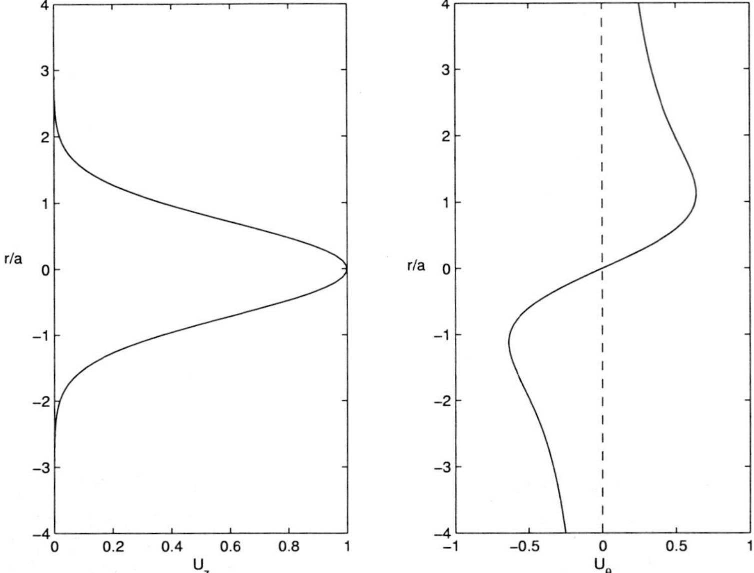

Fig. 11. (a) Axial velocity field $U_z(r)$ and (b) azimuthal velocity field $U_\theta(r)$ of the Batchelor vortex. The maximum azimuthal velocity is located at $r/a = 1.12$ and the swirl number is equal to $q_S \sim 1$.

Going back to flows of the form (69), consider as a well-known example the *Batchelor vortex* [57] obtained when a jet of core radius a and of Gaussian profile

$$U_z(r) = W \exp(-\frac{r^2}{a^2}) \ . \tag{75}$$

is added to the Lamb vortex (32) (figure 11 (a,b)). This flow is characterized by two nondimensional parameters: the Reynolds number $Re_\Gamma = \Gamma/\nu$ and the *swirl number* $q_S = \Gamma/2\pi a W$. The swirl number is seen to be the ratio between a characteristic azimuthal velocity $\Gamma/2\pi a$ and the centerline axial velocity W. As such it measures the relative magnitude of centrifugal effects and axial shear. Though the Reynolds number and especially the swirl number q_S are known to be crucial parameters, they do not completely define the flow structure. For instance, in vortex breakdown experiments [58], the downstream condition at the outlet may play an essential role in the dynamics. In vortex chamber experiments, different endwalls or openings may induce different regimes [20] such as a *rectilinear steady* vortex, a *precessing* vortex or else several *entangled* vortices !

The Batchelor solution appears in different contexts. It has often been used to fit profiles in vortex breakdown experiments [58] and it represents the structure of *trailing line vortices* far downstream of a finite wing. In the latter case, this flow comes about as a solution of the Navier-Stokes equations in the boundary layer approximation [57]. In this asymptotic framework, axial gradients are small with respect to radial ones, the radial velocity is small with respect to the axial velocity and finally the axial velocity defect is small with respect to the external velocity.

An exact Navier-Stokes solution also exists for a swirling jet confined within solid boundaries : it is an extension of the *rotating pipe flow* solution whereby a *Poiseuille profile* is added due to a superimposed constant pressure gradient :

$$U_\theta(r,t) = \frac{\Omega_z}{2} r \ , U_z(r) = W\left(1 - \frac{r^2}{a^2}\right). \tag{76}$$

In unbounded domains, one can look for viscous vortices of the form $(0, U_\theta(r,t), U_z(r,t))$ where both the axial vorticity and velocity are governed by a diffusion equation. One possible solution is the *diffusing Batchelor vortex*

$$U_\theta(r,t) = \frac{\Gamma}{2\pi r}\left[1 - \exp(-\frac{r^2}{a^2})\right], \ U_z(r,t) = W\exp(-\frac{r^2}{a^2}) \tag{77}$$

with a *time-dependent* core $a(t) = \sqrt{a_0^2 + 4\nu t}$ and a *time-dependent* centerline axial velocity $W(t) = \alpha_0/a^2(t)$, where a_0^2 and α_0/a_0^2 are respectively the initial core size and the initial axial velocity.

Non-axisymmetric vortex solutions of the Navier-Stokes equations may also be sought in the form of z-independent three-dimensional velocity fields

$$U_x(x,y,t) = \frac{\partial \Psi}{\partial y}, \ U_y(x,y,t) = -\frac{\partial \Psi}{\partial x}, \ U_z(x,y,t). \tag{78}$$

The streamfunction $\Psi(x,y,t)$ is then governed by the axial vorticity equation

$$[\frac{\partial}{\partial t} + \frac{\partial \Psi}{\partial y}\frac{\partial}{\partial x} - \frac{\partial \Psi}{\partial x}\frac{\partial}{\partial y} - \nu\Delta]\Delta\Psi = 0, \tag{79}$$

and the axial velocity component U_z is advected as a passive scalar according to

$$[\frac{\partial}{\partial t} + \frac{\partial \Psi}{\partial y}\frac{\partial}{\partial x} - \frac{\partial \Psi}{\partial x}\frac{\partial}{\partial y} - \nu\Delta]U_z = 0. \tag{80}$$

If the axial vorticity $-\Delta\Psi$ and the axial velocity U_z are further assumed to be proportional :

$$U_z = -d\Delta\Psi, \tag{81}$$

equations (79) and (80) become identical. *Any two-dimensional solution of the Navier-Stokes solutions governed by (79) may therefore be extended to three-dimensions* [59] ! The diffusing Batchelor vortex (77) falls within such a type of flows since this velocity field can be generated from the Lamb vortex using (81). Note that these three-dimensional solutions are particularly well-suited to model *rotating* flows since *Ekman pumping* near rotating boundaries does precisely impose just outside the boundary layer the constraint (81). Such solutions can be described as *Taylor-Proudman columns* with an uprising or downrising jet velocity proportional to the vorticity magnitude.

3.4 Three-dimensional flows: effect of stretching

The competition between *stretching* as a basic mechanism of vorticity enhancement, and viscosity as a basic mechanism of vorticity *damping*, has already been illustrated by the *Burgers layer* (26). In the *axisymmetric* case, the *Burgers vortex*

$$\mathbf{U} = \mathbf{U}_{vort} + \mathbf{U}_{strain} = U_\theta(r)\mathbf{e}_\theta + [-\frac{\gamma_0}{2} r\mathbf{e}_r + \gamma_0 \mathbf{e}_z] \tag{82}$$

with $\gamma_0 \geq 0$, constitutes an equivalent Navier-Stokes solution. This steady confined axisymmetric vortex which is embedded in a uniform and axisymmetric strain field, possesses a vorticity aligned with the stretching z-axis. For the vorticity field $\boldsymbol{\Omega} = \Omega(r)\mathbf{e}_z$, the vector equation (3) reduces to the simple scalar equation

$$-\frac{\gamma_0 r}{2}\frac{\partial \Omega}{\partial r} = \gamma_0 \Omega + \frac{\nu}{r}\frac{\partial}{\partial r}(r\frac{\partial \Omega}{\partial r}), \tag{83}$$

where the l.h.s. is the inward advection, the first term on the r.h.s denotes the amplification by stretching and the last term is diffusion. Equation (83) can easily be integrated and leads to a vorticity field similar to that of the *Lamb vortex* (33) with a steady core. More specifically, the azimuthal velocity becomes

$$U_\theta(r) = \frac{\Gamma}{2\pi r}[1 - \exp(-\frac{r^2}{a_0^2})] \ , a_0 = \sqrt{\frac{4\nu}{\gamma_0}}. \tag{84}$$

Similarly to the Lamb vortex, the circulation Γ which may be used to define a Reynolds number $Re = \Gamma/\nu$, is arbitrary and *independent* of stretching. This flow is now three-dimensional since it contains an explicit dependence on the axial coordinate z. Nevertheless the stretching is uniform along the vortex axis and steady, a rather unrealistic feature. However predictions based on such solutions nicely fit vortices obtained in numerical simulations of homogeneous isotropic turbulence [15].

Starting from such a flow in which the axial stretching is *axisymmetric, uniform and steady,* we extend it in various ways. A first genralization can be performed by adding a *steady nonaxisymmetric planar strain* orthogonal to the stretching axis : non-axisymmetric viscous vortices

$$\mathbf{U} = [U_r(r,\theta), U_\theta(r,\theta), \gamma z] \tag{85}$$

are then computed in the spirit of section 3.2. In a second extension, an *axial jet* is introduced while axial stretching[6] is still *uniform* but *time-dependent* [16] [17] :

$$\mathbf{U} = [U_r(r,\theta,t) - \frac{\gamma(t)r}{2}, U_\theta(r,\theta,t), \gamma(t)z + U_z(r,\theta,t)]. \tag{86}$$

[6] In equation (86), the radial part of stretching appears explicitly contrary to equations (85) and (87).

A third variation is performed by studying an even more general class of stretched vortices in which an *explicit radial* and *azimuthal* dependence for the *rate of strain* is introduced as follows:

$$\mathbf{U} = [U_r(r,\theta,t), U_\theta(r,\theta,t), \gamma(r,\theta,t)z + U_z(r,\theta,t)]. \tag{87}$$

Some of the vortices satisfying (87) may contain different radial cells where axial or azimuthal velocities are reversed [18] [19]. However all the stretched vortices considered here are quite specific and clearly far removed from realistic vortex flows since the explicit z-dependence always remains linear with infinite velocities at infinity. Nevertheless they constitute valuable local models: more realistic analytical solutions are almost non existent ! One should then resort to numerical simulations [45] to study the effect of stretching.

Let us first consider a vortex subjected to a *steady non-axisymmetric rate of strain*

$$\mathbf{U}_{strain} = [-\frac{\gamma}{2} - \beta]x\mathbf{e}_x + [-\frac{\gamma}{2} + \beta]y\mathbf{e}_y + \gamma z\mathbf{e}_z \tag{88}$$

made up of an axial axisymmetric stretching field ($\gamma > 0$) and a superimposed plane strain field. At small Reynolds numbers $Re_\Gamma = \frac{\Gamma}{\nu} << 1$ and for axial strain($\beta \leq \frac{\gamma}{2}$), a solution is easily obtained by neglecting the induction of vorticity on itself. This approximation leads to a linearized vorticity equation with a unique attracting solution [15]

$$\Omega(x,y) = \frac{\Gamma\sqrt{[(\frac{\gamma}{2})^2 - \beta^2]}}{2\pi\nu}\exp[\frac{(-\frac{\gamma}{2} - \beta)x^2 + (-\frac{\gamma}{2} + \beta)y^2}{2\nu}]. \tag{89}$$

Iso-vorticity lines are ellipses, the axes of which are aligned with the principal axes of strain. Such a branch of solutions can be extended analytically and then numerically to larger Reynolds numbers up to 100 [60]. It can be seen from such analyses that the angle of the strain axes with the ellipse axes changes with Re. More specifically, iso-vorticity lines tend to become more and more circular with increasing Reynolds number and their major axes tend to rotate anti-clockwise towards the line $x = -y$. This tendency is confirmed by asymptotic solutions at large Reynolds numbers [15] [61] through an analysis similar to that of section 3.2, as sketched below. The vorticity $\boldsymbol{\Omega} = \Omega\mathbf{e}_z$ is again assumed directed along the z-axis and still satisfies $\Omega = -\nabla^2\Psi$ where Ψ is the stream function associated to the vortex contribution, the external strain (88) being irrotational (figure 12 (a)).

Viscous diffusion is now expected to be balanced by stretching, which ensures that the time-dependent solution tends towards a steady state in contrast with the two-dimensional case where the vortex is always diffusing. There exist a characteristic length $a_c = \sqrt{\frac{\nu}{\gamma}}$ of the order of the core radius, a characteristic velocity Γ/a_c and a characteristic time scale $\tau = \frac{a_c{}^2}{\nu}$ which is associated with both diffusion and axial strain since, by definition, $\frac{a_c{}^2}{\nu} = \frac{1}{\gamma}$. When it is assumed

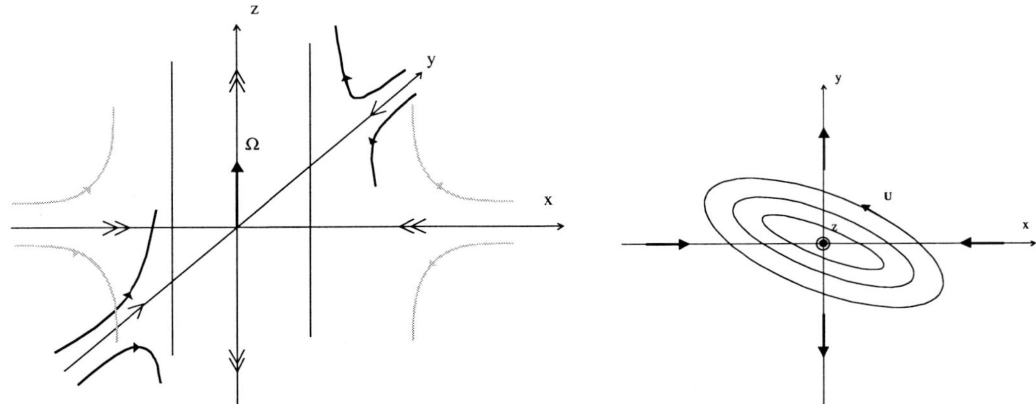

Fig. 12. Large Reynolds number vortex of axial vorticity $\boldsymbol{\Omega} = \Omega \mathbf{e}_z$ subjected to an axial external strain. (a) Axial external strain (88) with $\beta \leq \frac{\gamma}{2}$; (b) Vortex cross-section in x-y-plane.

that $\lambda \equiv \frac{2\beta}{\gamma}$ is $O(1)$, the "rate of strain" parameter $\frac{\beta {a_c}^2}{\Gamma}$ becomes proportional to $\epsilon = \frac{1}{Re_\Gamma} = \nu/\Gamma$:

$$\frac{\beta {a_c}^2}{\Gamma} = \frac{\beta}{\gamma}\epsilon. \tag{90}$$

The dynamical equations therefore display, in nondimensional form, the unique small nondimensional parameter ϵ. In polar coordinates, these equations take a form very similar to that found for a plane strain (section 3.2, equation (57)), i.e.

$$\frac{1}{r}\left[\frac{\partial \Psi}{\partial \theta}\frac{\partial}{\partial r} - \frac{\partial \Psi}{\partial r}\frac{\partial}{\partial \theta}\right]\Omega = \epsilon\lambda\mathcal{L}\Omega + \epsilon\mathcal{L}_0\Omega, \tag{91}$$

where the operator $\mathcal{L}$ has already been defined following (57), and the operator

$$\mathcal{L}_0 = 1 + \frac{1}{2}r\frac{\partial}{\partial r} - \frac{\partial}{\partial t} + \nabla^2 \tag{92}$$

is slightly modified. Note that $\mathcal{L}_0$ now contains, in addition to the viscous term and time derivative, a constant term to model vorticity enhancement by axial stretching and an advection term directed towards the center due to the inward component of the axial strain. From equations (91)-(92), it is easy to follow a similar procedure as in the plane strain case: Ψ is expanded in powers of a small parameter ϵ according to $\Psi \sim \Psi_0 + \epsilon\Psi_1 + ...$ Everything remains the same except for the solvability equation at first order which displays the viscous selection condition

$$\mathcal{L}_0\Omega_0 = \Omega_0 + \frac{1}{2}r\frac{\partial \Omega_0}{\partial r} - \frac{\partial \Omega_0}{\partial t} + \nabla^2\Omega_0 = 0. \tag{93}$$

This equation can be solved exactly and provides a general flow which asymptotically approaches the Burgers vortex. A particular solution is given by

$$\Omega_0 = \frac{1}{4\pi\delta^2}\exp[-\frac{r^2}{4\delta^2}],\ \delta^2 = 1 + (\delta_0^2 - 1)\exp(-t), \tag{94}$$

where δ_0 is a parameter given by the initial condition. When $t \to \infty$, the Burgers vortex is clearly obtained. Similarly the asymmetrical part is given by $\Psi_1 = \lambda f(r,t)\sin 2\theta$, where $f(r)$ satisfies

$$\frac{\partial^2 f}{\partial r^2} + \frac{1}{r}\frac{\partial f}{\partial r} - \frac{4}{r^2}f = q(r,t)(f - \frac{r^2}{4}), \tag{95}$$

which is identical to equation (65) in the plane strain case with the proviso that $q(r,t)$ take a form compatible with the action of viscosity and stretching. For the case of the vorticity distribution (94) one obtains

$$q(r,t) = -\frac{\partial\Omega_0}{\partial r}/\frac{\partial\Psi_0}{\partial r} = -\frac{r^2}{4\delta^4(\exp(r^2/4\delta^2) - 1)}. \tag{96}$$

By a simple change of variable $\xi = r/\delta$ and rescaling $\mathcal{F} = \delta^{-2}f$, one reduces again (95) to the steady problem (66). So the results of the *two-dimensional plane strain* problem at high Reynolds number are valid for a vortex subjected to a *non-axisymmetric three-dimensional stretching* again at high Reynolds number. In particular, the structure of the flow is comparable (figure 12 (b)), e.g. the angle of the major axis of the elliptic streamlines is still along the diagonal $x = -y$. However, the vortex is now *steady.*

We previously pointed out the apparent similarity between the two-dimensional diffusing Lamb vortex and the Burgers vortex. This is not a casual relationship but a particular case of a general relation which was first introduced by Lundgren [16] and further extended by Gibbon et al. [17]. These authors were able to reduce, through a simple change of variable, several *three-dimensional stretched* vortices (86) embedded in an axisymmetric unsteady strain field (here rewritten in Cartesian coordinates)

$$\mathbf{U} = [U_x(x,y,t) - \frac{\gamma(t)}{2}x\ ,U_y(x,y,t) - \frac{\gamma(t)}{2}y\ ,U_z(x,y,t) + \gamma(t)z\ ] \tag{97}$$

to *unstretched* and *diffusing* vortices of the type described in section 3.3. The simplest case is that of the Burgers vortex which is associated to the Lamb vortex. The stretched vortices (97) have both axial and azimuthal vorticity and *vorticity and stretching are not generically aligned* as for the simplest Burgers vortex! Since a streamfunction can be defined for the planar velocity field $(U_x(x,y,t)\ ,U_y(x,y,t))$ such that $\Omega_z(x,y,t) = -\nabla^2\psi$, the dynamics is completely characterized by their axial velocity $U_z(x,y,t)$ and axial vorticity $\Omega_z(x,y,t)$. More specifically the governing equations are:

$$\frac{DU_z}{Dt} = -\gamma U_z + \nu\Delta_{2D}U_z, \tag{98}$$

$$\frac{D\Omega_z}{Dt} = \gamma w + \nu \Delta_{2D} \Omega_z, \tag{99}$$

where Δ_{2D} is the two-dimensional Laplacian

$$\Delta_{2D} \equiv \frac{\partial^2}{\partial x^2} + \frac{\partial^2}{\partial y^2},$$

and $\frac{D}{Dt}$ stands for the convective derivative based on the basic state

$$\frac{D}{Dt} \equiv \frac{\partial}{\partial t} - \frac{\gamma(t)}{2}[x\frac{\partial}{\partial x} + y\frac{\partial}{\partial y}] + U_x\frac{\partial}{\partial x} + U_y\frac{\partial}{\partial y}. \tag{100}$$

By introducing a change of variable both in time and space defined by

$$\tau = \int_0^t s(u)du, \ \chi = \sqrt{s(t)}x, \ \eta = \sqrt{s(t)}y, \tag{101}$$

and the rescaling

$$U_z^{2D}(\chi, \eta, \tau) = s(t)U_z, \ \Omega_z^{2D}(\chi, \tau) = \frac{\Omega_z}{\sqrt{s(t)}}, \tag{102}$$

where $s(t)$ is a nondimensional quantity related to stretching

$$s(t) = \exp[\int_0^t \gamma(u)du], \tag{103}$$

equations (98)-(99) now become :

$$\frac{DU_z^{2D}}{D\tau} = \nu\tilde{\Delta}_{2D}U_z^{2D}, \tag{104}$$

$$\frac{D\Omega_z^{2D}}{D\tau} = \nu\tilde{\Delta}_{2D}\Omega_z^{2D}, \tag{105}$$

where $\tilde{\Delta}_{2D}$ and the convective derivative read :

$$\tilde{\Delta}_{2D} \equiv \frac{\partial^2}{\partial \chi^2} + \frac{\partial^2}{\partial \eta^2}, \ \frac{D}{D\tau} \equiv \frac{\partial}{\partial \tau} + U_\chi^{2D}\frac{\partial}{\partial \chi} + U_\eta^{2D}\frac{\partial}{\partial \eta}. \tag{106}$$

The quantities $U_z^{2D}(\chi)$ and $\Omega_z^{2D}(\chi)$ thus satisfy the dynamics of an *unstretched swirling jet* with the same initial conditions as the stretched vortex. In the context of *inviscid* dynamics ($\nu = 0$), an *axisymmetric* solution $U_z^{2D}(\chi)$, $\Omega_z^{2D}(\chi)$ is nothing but a steady solution (69). The corresponding inviscid stretched solution then follows by applying the transformations (101)-(102). One finally obtains the velocity field (86) or equivalently (97) for an inviscid vortex subjected to an arbitrary time-dependent stretching $\gamma(t)$ in the form

$$U_\theta(r,t) = \sqrt{s(t)}U_\theta^{2D}(\sqrt{s(t)}r) \ , \ U_z(r,t) = \frac{1}{s(t)}U_z^{2D}(\sqrt{s(t)}r)). \tag{107}$$

These solutions are attractive since (a) contrary to the Burgers vortex their *vorticity is not unidirectional* along the vortex axis and (b) there is a *dynamical exchange* between the different components of vorticity. When the strain enhances the component Ω_z along the z-axis it compresses its azimuthal counterpart Ω_θ.

For *viscous axisymmetric* vortices the above remarks are still valid. The only new effect is the equilibrium between diffusion and stretching. Consider the diffusing Batchelor vortex (77) and apply the reverse transformation associated with (101)-(102). One obtains

$$U_\theta(r,t) = \frac{\Gamma}{2\pi r}(1-\exp(-\frac{r^2}{a^2}) \ , \ U_z(r,t) = \frac{\alpha_0}{s^2 a^2}\exp(-\frac{r^2}{a^2}), \tag{108}$$

where $s(t)$ is given by (103) and the dimensional vortex core radius a, which is identical for both axial and azimuthal velocity, varies as

$$a^2(t) = \frac{a_0^2 + 4\nu\int_0^t s(u)du}{s}. \tag{109}$$

At each time t, the vortex is also characterized by the swirl number $q_S(t) = \frac{\Gamma s^2 a}{2\pi\alpha_0}$, a measure of the ratio between the characteristic azimuthal velocity $\Gamma/2\pi a$ and the centerline axial velocity α_0/s^2a^2. Two simple cases may be considered when γ is constant. If the vortex is *compressed*, then $\gamma = -\gamma_0$ with $\gamma_0 > 0$ and one obtains

$$a^2(t) = (\frac{4\nu}{\gamma_0} + a_0^2)\exp(\gamma_0 t) - \frac{4\nu}{\gamma_0}. \tag{110}$$

For large time, the characteristic radius tends to infinity and $q_S(t) = \frac{\Gamma}{2\pi\alpha_0}s^2a$ decreases to zero. In this case, the jet component increases in time with respect to the swirl component. If the vortex is *stretched*, then $\gamma = \gamma_0 > 0$ and one obtains

$$a^2(t) = \frac{4\nu}{\gamma_0} + (a_0^2 - \frac{4\nu}{\gamma_0})\exp(-\gamma_0 t). \tag{111}$$

For large time, the characteristic radius tends to a finite value $2\sqrt{\nu/\gamma_0}$ and $q_S(t) = \frac{\Gamma}{2\pi\alpha_0}s^2a$ becomes infinite. In this case, the jet component always decreases in time with respect to the swirl component and one recovers a steady stretched vortex aligned with the z-axis and subjected to a global strain field. This structure is the well-known Burgers vortex governed by the balance between stretching and viscous diffusion. A *direct connection between the Burgers vortex and the Lamb vortex* has therefore been established.

For *non-axisymmetric* vortices, the transformation (101)-(102) is still valid. One would then imagine that a similar connection might exist between the non-axisymmetric diffusing vortex studied in section 3.2 and the non-axisymmetric steady vortex of this section. Actually this is not the case : when applied to the stretched solution (94)-(95) determined in this section, its diffusive counterpart found for a plane strain is a stretched solution where the parameter of asymmetry β is decreasing in inverse proportion to time !

Let us now introduce an *explicit spatial* $r-\theta$ dependence for the rate of strain as in the velocity field (87). In this case, vorticity depends explicitly on the z variable and is again not generically aligned with a principal axis of the rate of strain. However, in that instance, an unexpected though simple decoupling is still possible [17] which makes the problem manageable : the axial velocity U_z and vorticity Ω_z still completely describe the flow since $\Omega_x = z\frac{\partial \gamma}{\partial y} + \frac{\partial U_z}{\partial y}$ and $\Omega_y = -z\frac{\partial \gamma}{\partial x} - \frac{\partial U_z}{\partial x}$. Moreover U_z and Ω_z decouple : their dynamics is still characterized by equations (98) and (99). However, the stretching term $\gamma(x, y, t)$ depends on time and on the space variables x and y, and is such that

$$\frac{D\gamma}{Dt} = -\gamma^2 - \frac{1}{\rho}\frac{\partial^2 p}{\partial z^2} + \nu \Delta_{2D}\gamma. \tag{112}$$

The main condition for this flow to be pertinent is that $\partial^2 p/\partial z^2$ be time dependent but spatially uniform: thus the axial pressure gradient must always be linear in z ! Note also that the in-plane velocity is not divergenceless :

$$\frac{\partial U_x}{\partial x} + \frac{\partial U_y}{\partial y} + \gamma = 0. \tag{113}$$

Using again the transformation (101)-(102), a *three-dimensional* problem can be recast into a *two-dimensional* problem for which standard techniques are available. In closing, note that the steady or unsteady vortices of the type

$$\mathbf{U} = (U_r(r,t), U_\theta(r,t), \gamma(r,t)z) \tag{114}$$

considered in [18] or [19] fall within the class of solutions (87). This type of configuration can display *axial flow reversal*, a feature which is interesting to model particularly in the context of vortex breakdown.

3.5 Three-dimensional inviscid flows: helical symmetry

As shown in section 3.2, a steady family of two-dimensional inviscid flows may be constructed by assuming that specific relations between vorticity and stream-function are satisfied. In this case, the problem is reduced to finding the solutions of a p.d.e. of the form $\Delta\Psi = -F(\Psi)$. An extension of this method is here introduced for three-dimensional inviscid flows which verify two additional requirements as detailed below.

As a first constraint, the flow is required to display *helical symmetry* [62]. This means that the velocity field is unaffected by a continuous one-parameter family of transformations defined by a translation of arbitrary magnitude H along a given axis followed by a rotation of angle $\theta_s = H/L$ along the same axis, where $2\pi L$ is the so-called *helix pitch* constant. As a result, the flow characteristics remain invariant along helical lines $\theta - \frac{z}{L} =$ const and are $2\pi L$-periodic along the z-axis. Note that $L > 0$ corresponds to a right-handed helix. When $L \to 0$, the case of a purely axisymmetric flow is recovered, while, for $L \to \infty$, the flow becomes two-dimensional. *Helically symmetric* vortices are of interest to describe the flow configuration in vortex chamber experiments. Indeed, in such bounded flows, helical symmetry is generally preserved away from the boundaries close to which viscous effects become predominant [20].

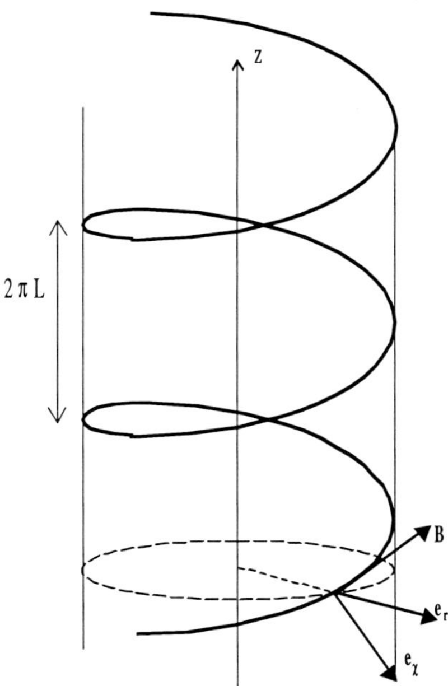

Fig. 13. Helical line located on cylindrical surface $r = r_0$ and such that $\chi \equiv \theta - z/L$ is a constant. The local orthogonal basis $(\mathbf{B}, \mathbf{e}_r, \mathbf{e}_\chi)$ is displayed.

Two types of coordinate systems are used in the following development. Cylindrical coordinates are oriented so that the swirl velocity component is in the direction of positive azimuthal angles. A second coordinate system (figure 13) is also introduced to directly implement helical symmetry. It consists of a local vector basis based on the usual radial unit vector $\mathbf{e}_r$ and the *Beltrami vector*

$$\mathbf{B} = N^2[\mathbf{e}_z + \frac{r}{L}\mathbf{e}_\theta] \tag{115}$$

with

$$N^2 = [1 + \frac{r^2}{L^2}]^{-1}. \tag{116}$$

Note that $\mathbf{B}$ is orthogonal to $\mathbf{e}_r$, directed along helical lines and, oddly enough, it is not a unit vector. From $\mathbf{B}$ and $\mathbf{e}_r$, a third vector

$$\mathbf{e}_\chi = \mathbf{B} \times \mathbf{e}_r = N^2[\mathbf{e}_\theta - \frac{r}{L}\mathbf{e}_z] \tag{117}$$

can be constructed along the direction of increasing $\chi \equiv \theta - z/L$, which completely defines the orthogonal basis $(\mathbf{B}, \mathbf{e}_r, \mathbf{e}_\chi)$. By construction, a vector field

$$\mathbf{U} = w_B \mathbf{B} + w_r \mathbf{e}_r + w_\chi \mathbf{e}_\chi \tag{118}$$

possesses helical symmetry if

$$w_B = \frac{\mathbf{U}.\mathbf{B}}{N^2} = w_z + \frac{r}{L} w_\theta, \tag{119}$$

$$w_r = \mathbf{U}.\mathbf{e}_r, \tag{120}$$

$$w_\chi = \frac{\mathbf{U}.\mathbf{e}_\chi}{N^2} = w_\theta - \frac{r}{L} w_z, \tag{121}$$

are *scalar* functions which remain constant on each helical line defined by $r = r_0$ and $\chi = \chi_0$. This condition directly translates into

$$\mathbf{B}.\nabla w_r = \mathbf{B}.\nabla w_B = \mathbf{B}.\nabla w_\chi = 0, \tag{122}$$

or, stated in a different way, solutions only depend on the two space variables r and $\chi = \theta - z/L$ instead of on the three coordinate variables r, θ and z. In terms of decomposition (118), the incompressibility condition now reads

$$\frac{\partial(r w_r)}{\partial r} + \frac{\partial w_\chi}{\partial \chi} = 0. \tag{123}$$

This equation is identical to the incompressibility condition for two-dimensional vortices provided that χ be replaced by θ. It is satisfied if there exists a stream-function $\Psi(\chi, r, t)$ such that

$$w_r = \frac{1}{r}\frac{\partial \Psi}{\partial \chi}, \quad w_\chi = -\frac{\partial \Psi}{\partial r}. \tag{124}$$

In vector form, equations (124) read

$$\mathbf{U} = w_B \mathbf{B} + \nabla \Psi \times \mathbf{B}. \tag{125}$$

In a similar fashion, the vorticity $\boldsymbol{\Omega}$ is divergenceless and can be thus decomposed into

$$\boldsymbol{\Omega} = \tilde{\omega}_1 \mathbf{B} + \nabla \tilde{\omega}_2 \times \mathbf{B}. \tag{126}$$

Note that the divergenceless character of $\mathbf{U}$ and $\boldsymbol{\Omega}$ is automatically ensured because of the relations $\nabla \times \mathbf{B} = 2\frac{N^2}{L}\mathbf{B}$ and $\nabla.\mathbf{B} = 0$. The four scalar fields $w_B(\chi, r, t)$, $\Psi(\chi, r, t)$, $\tilde{\omega}_2(\chi, r, t)$, $\tilde{\omega}_1(\chi, r, t)$ which completely describe the above velocity and vorticity fields, are not independent since the definition of vorticity $\boldsymbol{\Omega} = \nabla \times \mathbf{U}$ imposes the two additional relations

$$\tilde{\omega}_2 = w_B, \tag{127}$$

$$\mathcal{L}\Psi = \frac{2N^4}{L} w_B - N^2 \tilde{\omega}_1, \tag{128}$$

where

$$\mathcal{L} \equiv \frac{1}{r}\frac{\partial}{\partial r}(r N^2 \frac{\partial}{\partial r}) + \frac{1}{r^2}\frac{\partial^2}{\partial \chi^2} \tag{129}$$

plays the role of the two-dimensional Laplacian.

As a second constraint and following the same formulation as in the case of two-dimensional inviscid vortices, we further *postulate the two additional relations*

$$w_B = w_B(\Psi),\ \tilde{\omega}_1 = \tilde{\omega}_1(\Psi). \tag{130}$$

Equation (128) now takes the form of a single p.d.e. for the streamfunction Ψ : a complicated *three-dimensional Euler* problem has been reduced to a *two-dimensional scalar* equation with respect to r and χ very much akin to the usual two-dimensional Euler equations in θ and r ! However, these relations should be made consistent with the dynamical equations of motion. Different choices are possible. In this section, we discuss the constraint used by Alekseenko et al. [20]. In section 6.4, we will consider another constraint introduced by Dritschel [63].

For an *inviscid, incompressible* fluid, the Euler equations in the r-χ coordinate system read [20]

$$\frac{D}{Dt} w_r - \frac{N^4}{r}(w_\chi + \frac{r}{L} w_B) = -\frac{1}{\rho}\frac{\partial p}{\partial r}, \tag{131}$$

$$\frac{D}{Dt} w_\chi - N^2 w_r [2\frac{r}{L} w_B + (2 - N^{-2}) w_\chi] = -\frac{N^{-2}}{\rho r}\frac{\partial p}{\partial \chi}, \tag{132}$$

$$\frac{D}{Dt} w_B = 0, \tag{133}$$

where $\frac{D}{Dt}$ stands for the material derivative

$$\frac{D}{Dt} \equiv \frac{\partial}{\partial t} + w_r \frac{\partial}{\partial r} + \frac{w_\chi}{r}\frac{\partial}{\partial \chi}. \tag{134}$$

If the first condition in (130) takes the simple form

$$w_B = w_0 = \text{const.} \tag{135}$$

Equation (133) is then trivially satisfied. Two properties of the velocity field are thus implied :

• According to equation (119), there exists a relation between the z- and θ-components of the velocity field

$$w_z = w_0 - \frac{r}{L} w_\theta \tag{136}$$

where w_0 is effectively the uniform axial velocity on the centerline.

• According to equations (126) and (127), the vorticity field reduces to

$$\boldsymbol{\Omega} = \tilde{\omega}_1 \mathbf{B}. \tag{137}$$

It is therefore tangent to helical lines. As a consequence, helical lines and vortex lines coincide. Furthermore, according to (125) and (135), velocity and vorticity fields are always orthogonal in the Galilean reference frame moving with uniform velocity w_0 along the vortex axis ! While two-dimensional vortices can be viewed as a superposition of rectilinear vortex filaments, the *helical* vortex considered here may be regarded as a *superposition of helical vortex filaments.* As already suggested in section 3.3, the presence of helical lines and of an axial velocity are profoundly related.

We need now to specify the second function appearing in (130) i.e. a condition on $\tilde{\omega}_1$. From equations (115) and (137), the axial vorticity component reads

$$\Omega_z = N^2 \tilde{\omega}_1. \tag{138}$$

The Euler equations imply that, in unsteady flows, Ω_z remains constant along particle trajectories, i.e.

$$\frac{D}{Dt}[N^2 \tilde{\omega}_1] = 0. \tag{139}$$

This is reminiscent of the conservation of axial vorticity in two-dimensional vortex flows. By analogy with that case, equation (139) simply imposes, for steady flows, that

$$N^2 \tilde{\omega}_1 = F(\Psi). \tag{140}$$

where F is an arbitrary function. Finally (128) reduces to the two-dimensional equation

$$\mathcal{L}\Psi = \frac{2N^4}{L} w_0 - F(\Psi), \tag{141}$$

with two free parameters L and w_0 related respectively to the helix pitch and to the centerline axial velocity. It is found experimentally that such parameters are preferable to the Reynolds number and swirl number since they are much more discriminating: two experimental vortices with different flow fields may lead to almost identical Reynolds number and swirl number but distinct values of L and w_0. In addition to equation (141), boundary conditions compatible with the helical symmetry must be enforced. For instance, in the case of vortex chamber experiments, they can be applied on a cylindrical surface $r = R_0$ (this surface is clearly helically symmetric). On such boundaries, the normal velocity must vanish and according to equations (124)

$$w_r = \frac{1}{r}\frac{\partial \Psi}{\partial \chi}(R_0) = 0. \tag{142}$$

Consider some examples of such flows. Axisymmetric swirling jets of the form (69), for instance the Batchelor vortex, satisfy the constraint of *helical symmetry* since they are invariant with respect to any translation along the axis coupled with any rotation around the same axis. However all of the solutions (69) do not satisfy the additional constraints (135) and (140) : in particular their axial and azimuthal velocities are not necessarily linked by relation (136) ! In fact, it is readily shown by introducing the axisymmetry condition $\partial/\partial\chi = 0$ in the above governing equations that any flow satisfying (135) and (140) is completely defined by its axial vorticity $\Omega_z(r)$ according to the relations

$$w_\theta = \frac{1}{r}\int_0^r \Omega_z(r) r dr, \;\; w_z = w_0 - \frac{1}{L}\int_0^r \Omega_z(r) r dr, \;\; w_r = 0. \tag{143}$$

The axisymmetric helical vortex specified by (143) can be viewed as a superposition of helical lines with the *same helix pitch.* Note that the *Batchelor* and *Rankine* vortices do satisfy (143) ! Other axisymmetric vortices satisfying (143) are such that the axial vorticity Ω_z is zero except in an annular zone where it is a constant [20]. A combination of such annular vortex flows with non overlapping annular zones and opposite helix pitch can be constructed that fit steady velocity fields observed in chamber experiments. For non-axisymmetric configurations, a helical vortex with finite core can be constructed within this framework (for a discussion see [20]). In particular, a helical vortex solution is obtained which does not rotate due to the simultaneous action of an axial velocity component and a wall-induced velocity counterbalancing its self-induced motion. This flow has been experimentally observed, which emphasizes the importance of boundary conditions on the overall dynamical behaviour : unconfined vortices for which the core radius is far from boundaries and chamber vortices might be quite different.

4 Instability of unbounded linear flows

Let us now explore the properties of the *instability waves* or *neutral waves* which previously introduced basic flows may support. As in earlier sections, the flow complexity is gradually increased, starting with the stability of the primitive basic flows (7)-(8). The perturbation analysis of each of these steady flows is paradigmatic of a given phenomenon: solid body rotation $\gamma = 0$ displays *dispersive waves*, the stagnation point flow $\Omega = 0$ *hyperbolic instabilities* , the elliptic flow $0 < \gamma < \frac{\Omega}{2}$ *elliptic instability* . Finally *unbounded Couette flow* $\gamma = \frac{\Omega}{2}$ is known to be linearly stable but Kelvin [64] has shown the presence of a transient amplification which, by now, has been found in many other configurations to be connected to the so-called *non-normality* [65]. This feature is not further discussed in this paper.

Let us first consider the simplest linear flow field i.e. the solid body rotation $\mathbf{U} = (-\frac{\Omega}{2}y, \frac{\Omega}{2}x, 0)$ of uniform vorticity Ω. The dynamics of infinitesimal pressure p and velocity $\mathbf{u}$ perturbations are governed by partial differential equations

$$\frac{\partial \mathbf{u}}{\partial t} - \frac{\Omega}{2} y \frac{\partial \mathbf{u}}{\partial x} + \frac{\Omega}{2} x \frac{\partial \mathbf{u}}{\partial y} + (\mathbf{u}.\nabla)\mathbf{U} = -\frac{1}{\rho}\nabla p \tag{144}$$

resulting from a linearization of the Euler equations about this basic flow, coupled with the continuity equation $\nabla.\mathbf{u} = 0$. The above differential system is *inhomogeneous* with respect to the two space coordinates x and y, which prevents a straigthforward application of the *normal mode* method. However, in the frame of reference rotating around the z-axis with angular velocity $\Omega/2$, the basic state becomes the rest state and the entire flow field reduces to the perturbation $\mathbf{u}$. In this non-inertial frame of reference, while the continuity equation is unchanged, the linearized problem (144) becomes

$$\frac{\partial \mathbf{u}}{\partial t} + \Omega \mathbf{e}_z \times \mathbf{u} = -\frac{1}{\rho}\nabla p. \tag{145}$$

This equation is now homogeneous in both space x and y variables and contains an additional *Coriolis force.* If the pressure term were absent from (145), each fluid particle would be moving periodically along circular streamlines with period $2\pi/\Omega$. With the pressure term, these local motions become coupled and generate neutral propagating plane waves called *Kelvin waves.* Let us again stress that other authors call them inertial waves. Mathematically, plane wave solutions

$$\mathbf{u} = \mathbf{u}^0 \exp[i(\mathbf{k}.\mathbf{x} - \omega t)] + c.c. \tag{146}$$

must be searched for. Introducing the above expression into the linearized equation (145), one obtains :

$$-i\omega \mathbf{u}^0 + \Omega \mathbf{e}_z \times \mathbf{u}^0 = -\frac{ip}{\rho}\mathbf{k}. \tag{147}$$

The continuity equation, which now reads

$$\mathbf{k} \cdot \mathbf{u}^0 = 0, \tag{148}$$

indicates that these waves are transverse, motions being confined to a plane perpendicular to the wave vector $\mathbf{k}$. Moreover the two complex planar components of $\mathbf{u}^0$ are in phase quadrature as can be seen by taking the scalar product of $\mathbf{u}^0$ with equation (147). Finally a new relation is imposed on the two velocity components by taking the vector product of (147) with $\mathbf{k}$:

$$i\omega\mathbf{k} \times \mathbf{u}^0 = -(\Omega\mathbf{e}_z \cdot \mathbf{k})\mathbf{u}^0. \tag{149}$$

The two conditions (148)-(149) are compatible if a *dispersion relation*

$$\omega(\mathbf{k}) = \pm\frac{(\Omega\mathbf{e}_z \cdot \mathbf{k})}{|\mathbf{k}|} \tag{150}$$

is satisfied. Thus propagating waves only appear when the solid body rotation flow is perturbed by a *low enough forcing frequency*

$$|\omega| \leq |\Omega|\,. \tag{151}$$

This phenomenon has been demonstrated in laboratory experiments for the more realistic case of a rotating fluid column (section 6.3). Relation (150) is *scale-invariant* since the wave vector magnitude does not affect the frequency ω : hence, for a given frequency, there exists a continuous family of eigensolutions parametrized by an arbitrary spatial scale. At least before viscous effects become predominant[7], small scale structures can be directly generated from a smooth basic profile ! Because of this scale-invariance, the group velocity $\partial\omega/\partial\mathbf{k}$ remains perpendicular to the wave vector : the energy, which moves with the group velocity, propagates in a plane perpendicular to the wave vector $\mathbf{k}$, i.e. in the same plane in which the motion takes place. Going back to the laboratory frame where the fluid is in solid body rotation, these perturbations are still given by $\mathbf{u}$ written in the fixed reference frame coordinates, i.e.

$$\mathbf{u} = \mathbf{u}^0 \exp[i(\mathbf{k}(t) \cdot \mathbf{x} - \omega(\mathbf{k}))t] + c.c.\ . \tag{152}$$

The wave vector $\mathbf{k}(t)$ has a constant magnitude but it spins with the fluid around the axis of rotation according to

$$\mathbf{k}(t) = R_{oz}(\frac{\Omega}{2}t)\mathbf{k}(0) \tag{153}$$

where $R_{oz}(\chi)$ is the rotation matrix of angle χ around the z-axis. The angle θ_0 between $\mathbf{k}(t)$ and the z-axis is thus a constant, the value of which depends on the initial conditions (figure 14). A solution similar to (152) has been found for the *unbounded plane Couette flow* $\mathbf{U} = 2S(0, x, 0)$ [64]. In this case, the general viscous stability equation reads

$$\frac{\partial\mathbf{u}}{\partial t} + 2Sx\frac{\partial\mathbf{u}}{\partial y} + (\mathbf{u}.\nabla)\mathbf{U} = -\frac{1}{\rho}\nabla p + \nu\nabla^2\mathbf{u}. \tag{154}$$

[7] The introduction of viscous effects is straightforward : a damping factor $-i\nu\,|\mathbf{k}|^2$ is added to the frequency ω .

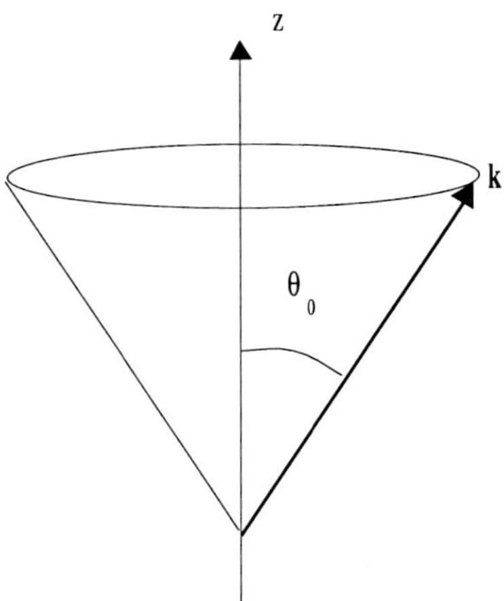

Fig. 14. Motion of the wave vector **k** for a Kelvin wave in the case of a solid body rotation around the z-axis. The initial angle θ_0 between $\mathbf{k}(t)$ and the z-axis remains constant during the time evolution.

In accordance with the usual normal mode approach, we are looking for solutions in the form of Fourier components of constant wavenumbers k_y, k_z in the y- and z-directions since these variables do not explicitly appear in (154). For the x-dependence, one chooses a Fourier component with a time varying wavenumber $k_x(t)$ so that normal mode solutions read

$$\mathbf{u}(x,t) = \mathbf{u}^0(t)\exp[i(k_x(t)x + k_y y + k_z z)] + c.c. \ . \tag{155}$$

When the wave number $k_x(t)$ is assumed to be a function of the basic shearing rate

$$k_x(t) = k_x(0) - 2Sk_y t, \tag{156}$$

the explicit x-dependence in (154) is eliminated and this equation becomes an ordinary differential equation for the amplitude $\mathbf{u}^0(t)$ which can be easily integrated. For instance, the component $u_x^0(t)$ along the x- axis can decomposed into an inviscid $u_x^E(t)$ and viscous $B_{viscous}(t)$ amplification part according to

$$u_x^0(t) \equiv B_{viscous}(t)u_x^E(t) \tag{157}$$

with

$$u_x^E(t) = \frac{|\mathbf{k}(0)|^2}{|\mathbf{k}(t)|^2}u_x^0(0), \tag{158}$$

$$B_{viscous}(t) = \exp[-\nu(k_y^2 + k_z^2)t + \nu\frac{k_x^3(t)}{6Sk_y}]. \tag{159}$$

From equation (156), it is seen that, depending on the initial condition, the wave vector magnitude $|\mathbf{k}(t)|$ may decrease during a transient period : according to equation (158), the perturbation may thus be *inviscidly amplified.* This possible *transient amplification* has been found in many other shear flows. Viscosity however kills this effect in a *super-exponential* way.

This procedure has been generalized [48] [66] to the *three-dimensional unbounded unsteady linear* flows $\mathbf{U}$ given by (10)-(11). Perturbations $\mathbf{u}$ are then of the type

$$\mathbf{u} = \mathbf{u}^0(t)F[\mathbf{k}(t).\mathbf{x} + \delta] + c.c. \tag{160}$$

where $\mathbf{u}^0(t)$ is a *complex time-dependent* vector, F an *arbitrary complex scalar field* which is not yet assumed to be a Fourier mode, and both quantities $\mathbf{k}(t)$ and δ are *real* in order for the argument of F to be real. Incompressibility imposes a transversality condition similar to (148)

$$\mathbf{k}(t).\mathbf{u}(t) = 0. \tag{161}$$

For the specific perturbations (160), condition (161) implies that the advective nonlinear term $u_j \frac{\partial}{\partial x_j}$ in the Navier-Stokes or Euler equations is identically zero : when $\mathbf{u}$ is an exact linear solution, it is automatically a *nonlinear* solution as well ! Such a property is clearly not valid for combinations of perturbations (160). Introducing the anzatz (160) into the Navier-Stokes equations and using relations (10)-(11), it is easily found (for details see [48]; the same kind of manipulations may be found in [67] in the framework of the rapid distorsion theory) that one should impose *wavenumber dynamics* reminiscent of (153) or (156), i.e.

$$\frac{dk_j}{dt} = -T_{mj}(t)k_m(t), \tag{162}$$

in order to eliminate the derivative of F. By invoking (161) and (162), the pressure field resulting from the perturbations (160) is found to be

$$p(x,t) = p^0(t)G(\mathbf{k}(t).\mathbf{x} + \delta) + c.c. \tag{163}$$

where $G(\chi) = \int F(\chi)d\chi$ and

$$p^0(t) \equiv 2\rho\frac{T_{ml}(t)u_l^0(t)k_m(t)}{k_ik_i} \tag{164}$$

is a function of $\mathbf{k}(t)$ and $\mathbf{u}^0(t)$. Using the latter expression, the equation for the amplitude $\mathbf{u}^0(t)$

$$\frac{du_i^0}{dt} + T_{ij}(t)u_j^0 = \nu k_m k_m \frac{F''}{F}u_i^0 + 2\frac{T_{lj}k_l u_j}{k_m k_m}k_i \tag{165}$$

is obtained. In the *inviscid* case ($\nu = 0$), equation (165) does not depend on space variables. With (162), it thus completely defines how the total velocity field $\mathbf{U}+\mathbf{u}$ evolves for any *arbitrary* function F. The nonlinear non-homogenous partial differential problem has thus been reduced to *two ordinary differential* equations (162)-(165) ! In the *viscous* case, the function F is not arbitrary but must be chosen so that $\frac{F''}{F}$ remains constant. This constant can be rescaled to ± 1 by a simple redefinition of the wave vector $\mathbf{k}$. Considering that it should remain bounded at infinity, F must be a sum of two Fourier modes

$$F[\mathbf{k}(t).\mathbf{x}+\delta] = F_1 \exp[i(\mathbf{k}(t).\mathbf{x}+\delta)] + F_2 \exp[-i(\mathbf{k}(t).\mathbf{x}+\delta)]. \tag{166}$$

The viscous term in (165) now simplifies to $-\nu\,|\mathbf{k}|^2\, u_i^0$ and the perturbation $\mathbf{u}^0(t) \equiv B_{viscous}(t)\mathbf{u}^E(t)$ can be split into a viscous damping factor

$$B_{viscous}(t) = \exp[-\nu \int_0^t |\mathbf{k}|^2\, dt]. \tag{167}$$

and an inviscid contribution $\mathbf{u}^E$ which satisfies

$$\frac{du_i^E}{dt} = [-T_{ij}(t) + 2\frac{T_{kj}k_k k_i}{k_m k_m}]u_j^E. \tag{168}$$

The time evolution of $\mathbf{u}^E(t)$ *a priori* departs from the usual *normal mode* exponential behaviour ! These *nonseparable* flows in space and time (160) are extremely valuable since they are exact Navier-Stokes solutions. However several undesirable features are noteworthy : they are only valid for unbounded domains in the three directions; the superposition of such modes is clearly a linear solution around the basic state $\mathbf{U}(\mathbf{x},t)$ but it is no longer an exact nonlinear solution ! Nonetheless it will be seen that basic instability features of more realistic flows are captured. Depending on the velocity fields (10)-(11), the inviscid contribution $\mathbf{u}^E$ may induce *no growth* as in Kelvin waves, a *transient growth* as in unbounded Couette flow or an *asymptotic growth* as in the unstable elliptic flow considered in section 7. In this paper, we exploit the above formulation to successively examine the instability of a hyperbolic point (section 5.1), the elliptic instability (section 7) and the influence of stretching (section 9).

5 Instability of layers

5.1 Instability of a pure hyperbolic point

The stability of a *steady two-dimensional stagnation point*

$$U_1 = \alpha x_1 \ , \ U_2 = -\alpha x_2 \ , \ U_3 = 0 \ ; \ \alpha > 0, \tag{169}$$

is now considered [21] [48] [68] [69]. This flow constitutes a specific example of velocity field (14) which is locally encountered in the Taylor four-roll mill experiment [70]. In less controlled situations, it may be associated with *secondary instabilities* in shear layers as outlined below. According to the primary inflexion point instability, a vortex layer first rearranges its initial vorticity into an array of *co-rotating spanwise Kelvin-Helmholtz billows*. When these primary vortices have reached a sufficient amplitude, most of the spanwise vorticity is contained inside their core. The so-called *braid region* located between two such co-rotating rolls, is almost deprived of *spanwise* vorticity : a *hyperbolic* point region thus appears between two consecutive spanwise rolls, precisely in a region where *streamwise counter-rotating* vortices are experimentally observed. The latter structures plausibly originate from an instability process of this hyperbolic point flow. Hyperbolic points are also pervasive in turbulent flows and their instability constitutes a possible mechanism of intense vortex creation in addition to the classical Kelvin-Helmholtz roll-up associated with vortex layers. Note, however, that the Kelvin-Helmholtz process leads to an array of co-rotating instead of counter-rotating vortices. In section 5.2, we examine the mixed case of vortex formation due to the instability of a *stretched vortex layer*.

As demonstrated in section 4, the linear stability of the basic flow (169) can be recast into two ordinary differential equations. The wave vector evolution which is governed by equation (162), is easily obtained in the form

$$k_1(t) = k_1(0)\exp(-\alpha t), \tag{170}$$

$$k_2(t) = k_2(0)\exp(\alpha t), \tag{171}$$

$$k_3(t) = k_3(0). \tag{172}$$

Since α is assumed positive, the wavenumber $k_2(t)$ along the compression x_2-axis increases to infinity, i.e. the associated wavelength decreases towards small scales : such a time dependence could have been intuitively guessed by following the motion of two particles along the compression axis. Moreover the viscous factor $B_{viscous}$ in (167) now reads

$$B_{viscous} = \exp(-\nu k_3^2(0)t - \nu k_1^2(0)[\frac{1-\exp(-2\alpha t)}{2\alpha}] - \nu k_2^2(0)[\frac{\exp(2\alpha t)-1}{2\alpha}]). \tag{173}$$

Any perturbation along the compression axis such that $k_2(0) \neq 0$ is thus damped by viscosity in a *super-exponential* way. It can be checked that the inviscid dynamics governed by (168) cannot counterbalance this damping factor. Hence, in the subsequent linear analysis, we only consider the case $k_2(0) = 0$, i.e. perturbations homogeneous along the compression x_2-axis for all time. Since the wavenumber $k_1(t)$ along the stretching x_1-axis decreases to zero by virtue of (170), these disturbances become, for large time, homogeneous along the stretching axis as well, while keeping the same initial wavenumber $k_3(0)$ along the neutral axis (see equation (172)). According to equation (173), viscosity diffuses $k_3(0) \neq 0$ perturbations in a standard fashion. As seen by a direct integration of (168), the inviscid components $u_1^E(t)$, $u_3^E(t)$ for $k_2(0) = 0$ tend towards zero while $u_2^E(t) \sim u_2^E(0)\exp(\alpha t)$ for large time. The disturbance $\mathbf{u}(t)$ thus behaves asymptotically, up to a multiplicative constant, as

$$\mathbf{u}(t) \sim \exp[(\alpha - \nu k_3^2(0))t]\mathbf{e}_2, \tag{174}$$

and its vorticity becomes gradually aligned with the stretching x_1-axis. Equation (174) expresses the competition between vorticity amplification by stretching and vorticity diffusion by viscosity[8]. When the initial wavenumber $k_3(0)$ along the neutral x_3-axis is large enough, vorticity diffusion predominates upon vorticity enhancement by stretching and perturbations are attenuated; when the wavenumber $k_3(0)$ is small enough, i.e.

$$|k_3(0)| < \sqrt{\frac{\alpha}{\nu}}, \tag{175}$$

the opposite situation prevails and the $k_2(0) = 0$ modes are amplified. A *hyperbolic stagnation point flow is thus always unstable* with respect to *three-dimensional disturbances* in the bandwith $|k_3(0)|\sqrt{\frac{\nu}{\alpha}} < 1$.

Unstable perturbations tend to become *homogeneous along the stretching axis* with amplified vorticity along the same axis and decaying vorticity in a plane perpendicular to the same axis. Both these linear trends can be exploited in the *nonlinear* regime as exemplified in the following development. Furthermore nonlinear perturbations initially independent of the stretching axis variable x_1 and with no perturbation velocity component along this axis – the stretching component αx_1 is not included here since it belongs to the basic flow–, will keep these properties during the subsequent time evolution. From these considerations, it is therefore tempting to use a procedure analogous to the one applied in section 3.4 : the reduction of the three-dimensional case to the analysis of a two-dimensional problem for nonlinear perturbations homogeneous along the stretching axis and periodic along the neutral axis of period $\frac{2\pi}{k_3}$ (figure 15). No specific assumption is made regarding the dependence along the compression x_2-direction. Furthermore the nonlinear saturation of the hyperbolic instability is surmised to lead to a *steady* flow field.

We may hence search for a steady perturbation velocity field characterized by a streamfunction $\Psi(x_2, x_3)$ such that

$$\mathbf{u} = (0, \frac{\partial \Psi}{\partial x_3}, -\frac{\partial \Psi}{\partial x_2}). \tag{176}$$

When written in nondimensional form with respect to the length scale $1/k_3$ and the velocity scale νk_3, the vorticity

$$\boldsymbol{\Omega} = \Omega \mathbf{e}_1 \ , \ \Omega(x_2, x_3) = -(\frac{\partial^2}{\partial x_2^2} + \frac{\partial^2}{\partial x_3^2})\Psi \tag{177}$$

which is aligned with the stretching axis, satisfies the governing equation

$$[\frac{\partial \Psi}{\partial x_3}\frac{\partial}{\partial x_2} - \frac{\partial \Psi}{\partial x_2}\frac{\partial}{\partial x_3}]\Omega = R[x_2\frac{\partial \Omega}{\partial x_2} + \Omega] + [\frac{\partial^2}{\partial x_2^2} + \frac{\partial^2}{\partial x_3^2}]\Omega, \tag{178}$$

[8] It is hence no surprise to find the length scale $\sqrt{\frac{\nu}{\alpha}}$ appear in the specification of the Burgers layer or the Burgers vortex which are precisely based on such a balance.

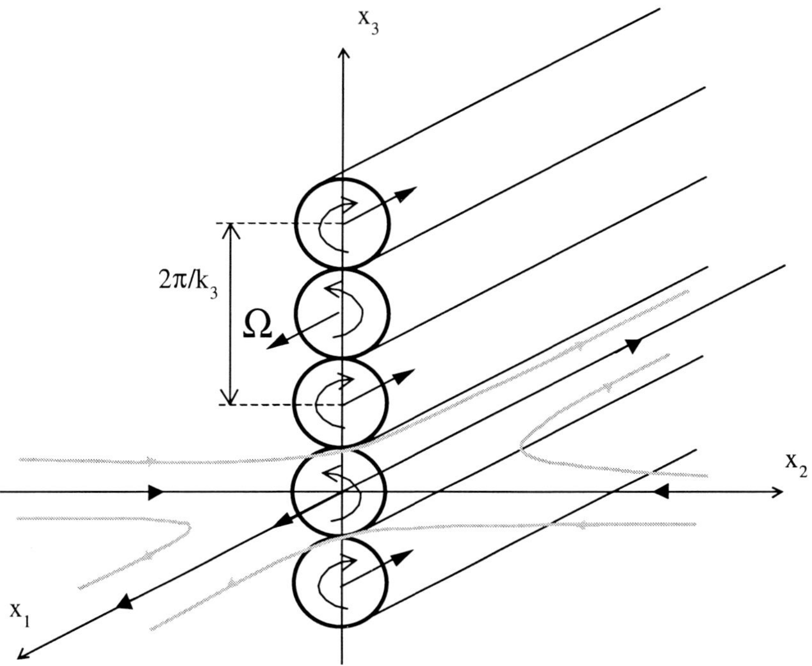

Fig. 15. Flow produced by the hyperbolic instability of a two-dimensional stagnation point (169). Streamwise counter-rotating vortices are aligned with the stretching axis.

where the single nondimensional parameter $R \equiv \alpha/(\nu k_3^2)$ has already been encountered in the linear context (see equation (175)). This number quantifies the ratio between the periodicity $\frac{2\pi}{k_3}$ of the steady nonlinear structures and the thickness $\sqrt{\frac{\nu}{\alpha}}$ of the stretched viscous layer (section 3.1). The following Fourier expansions are assumed for the periodic functions Ω and Ψ :

$$\Omega(x_2, x_3) = \sum_{n=1}^{\infty} [a_n(x_2)\cos(2n-1)x_3 + b_n(x_2)\sin 2nx_3], \tag{179}$$

$$\Psi(x_2, x_3) = \sum_{n=1}^{\infty} [c_n(x_2)\cos(2n-1)x_3 + d_n(x_2)\sin 2nx_3]. \tag{180}$$

The choice of harmonics originates from the form of the nonlinear terms in equation (178) and relation (177). Moreover one may impose on these steady solutions symmetry conditions

$$\Psi(-x_2, -x_3) = \Psi(x_2, x_3)\ ;\ \Psi(-x_2, x_3 - \pi) = -\Psi(x_2, x_3) \tag{181}$$

which are compatible with the governing equations. These constraints can be restated in terms of Fourier coefficients : $a_n(x_2)$ and $c_n(x_2)$ (resp. $b_n(x_2)$ and $d_n(x_2)$) are even (resp. odd) functions of their argument x_2. This automatically implies the following conditions at $x_2 = 0$:

$$\frac{da_n}{dx_2}(0) = b_n(0) = \frac{dc_n}{dx_2}(0) = d_n(0) = 0. \tag{182}$$

For localized structures in x_2, the quantities Ψ and Ω decay as $x_2 \to \pm\infty$, which implies that $a_n(x_2)$, $b_n(x_2)$, $c_n(x_2)$, $d_n(x_2)$ go to zero at infinity.

Introducing expansions (179)-(180) in (177)-(178) provides an infinite system of ordinary differential equations for $a_n(x_2)$, $b_n(x_2)$, $c_n(x_2)$, $d_n(x_2)$. This system, coupled with the above conditions at infinity and the boundary conditions (182) at $x_2 = 0$, is reminiscent of the *nonlinear eigenvalue problem* derived to compute the primary nonlinear branches arising from the linear instability of plane Poiseuille flow [37]. The *linearized* version of this problem takes the form of a system of decoupled equations for $a_n(x_2)$ and $b_n(x_2)$:

$$\frac{d^2a_n}{dx_2^2} + Rx_2\frac{da_n}{dx_2} + [R-(2n-1)^2]a_n = 0, \tag{183}$$

$$\frac{d^2b_n}{dx_2^2} + Rx_2\frac{db_n}{dx_2} + [R-(2n)^2]b_n = 0. \tag{184}$$

If the coefficient in front of a_n (resp. b_n) in equation (183) (resp. (184)) is negative, a_n (resp. b_n) does not possess a positive maximum, such that $a_n > 0$, $\frac{da_n}{dx_2} = 0$ and $\frac{d^2a_n}{dx_2^2} < 0$, or a negative minimum,i.e. such that $b_n < 0$, $\frac{da_n}{dx_2} = 0$ and $\frac{d^2a_n}{dx_2^2} > 0$. As a consequence, when $R < 1$ infinitesimal modes cannot decay at infinity, which prevents the existence of a nontrivial bounded localized nonlinear solution. Thus $R \geq 1$ is a *necessary* condition[9] for this system to possess a *localized nonlinear solution* [21]. When $R = 1$, the linear system admits the solution $\Omega = \Psi = \cos x_3$ which is not localized but nonetheless bounded. This linear flow also satisfies the *nonlinear* system since it has already been found in section 4 with $k_3(0) = \sqrt{\alpha/\nu}$. For $R \geq 1$, numerical computations indicate that nonlinear solutions exist : the parameter R plays the role of a *bifurcation parameter for the branch of nonlinear solutions*, in the same way as the Reynolds number for finite amplitude states in plane Poiseuille flow,. These nonlinear solutions, even with strong amplitudes, preserve the structure of linear modes : a small number of modes n, e.g. n = 3-6, is sufficient to capture the velocity field. Otherwise stated, the structure of the nonlinear field does not evolve much as the perturbation amplitude is increased from small to large values.

The nonlinear solutions (179)-(181) describe *a periodic array of counter-rotating vortices aligned with the stretching axis.* As in the case of the Burgers vortex, each counter-rotating vortex is characterized by an *arbitrary* circulation.

---

[9] It can be shown that it is a *sufficient* condition as well.

This flow is thus dependent on two parameters : the vortex strength, i.e. the circulation, and the parameter R which quantifies the magnitude of stretching. Two asymptotic limits have been considered in [21] : large stretching intensity $R >> 1$ and large vortex strength. In the first instance, one is almost in the linear regime and the strong compression confines the vorticity in a layer of thickness $\sqrt{\frac{\nu}{\alpha}}$ reminiscent of the Burgers layer. In the large vortex strength case, the leading order problem is purely inviscid and it gives rise to the classical functional relationship $\Omega = F(\psi)$ between streamfunction and vorticity as in equation (42). It is found numerically that this function is nonetheless quite close to that of the Burgers vortex [21]. When a *unique* vortex instead of an array of vortices is embedded in a hyperbolic point (169), i.e. a bi-axial strain, an exact steady solution cannot exist because of vorticity leakage in the direction of the neutral axis [15]. For the periodic array, the presence of near-by vortices of opposite sign prevents such a leakage. Moreover, for a given stretching intensity α, there is a critical separation distance below which vortices with too small a periodicity, are damped by diffusion.

The experimental geometry which is not accounted for in this infinite model, does play a role in selecting the vortex scale. If such a scale is small compared to the critical separation, the undisturbed hyperbolic flow is observed. In real stagnation flow experiments, a critical strain is thus measured for the appearance of this hyperbolic point instability [69]. Furthermore, the vortex strength is no longer arbitrary but depends on the boundary conditions, e.g. the angular velocity of the rotating cylinders in Taylor's four-roll mill experiment. Finally note that experimental flows become time-dependent for sufficiently large strain rates.

5.2 Instability of a stretched layer

In section 5.1, the instability of a *pure hyperbolic* point was considered and a mechanism for vortex creation was identified. It is known that the *Kelvin-Helmholtz* instability of a pure parallel shear layer is another such mechanism that engenders vorticity tubes. In this section, we study an intermediate case : *a time dependent modulated stretched vortex layer.* This example has been repeatedly analyzed [23] [24] [25] [71] to account for the appearance of streamwise vortices in the braid region between two Kelvin-Helmholtz billows. The mechanism concomitantly proposed by Corcos and Lin [23] and Neu [24], slightly differs from the hyperbolic instability presented in section 5.1 in which counter-rotating vortices were *directly* produced by the *hyperbolic stagnation point* instability. Here this vortex formation appears following a slightly different sequence of phenomena. First, as in the previous mechanism, the initial Kelvin-Helmholtz instability concentrates the vorticity of the initially parallel shear layer into spanwise rolls while the braid regions in-between are depleted of this same component. *Streamwise vorticity* however is assumed still present in these braid regions with a periodicity along the span *a priori* of the same order as the *primary rolls spacing.* Such patterns have their vorticity aligned along the stretching direction and are extremely *flat* since their spanwise size is related to the primary roll spacing while their thickness, associated to the compression part of the strain, scales as in the Burgers layer. Their dynamics can be consequently analyzed in the more general terms of a stretched vortex layer with a modulated strength [10]. As a final step, such flat patterns collapse, thereby forming *counter-rotating secondary streamwise vortices.*

Two features can be encountered in the general case of a *stretched modulated vortex layer* : a *self-focusing* mechanism which collapses these patterns into concentrated quasi-circular *counter-rotating* vortices as mentioned above; a *modified Kelvin-Helmholtz* instability which splits a stretched layer into various *co-rotating* vortices that can then be subjected to pairing. In this section, we first present the asymptotic analysis of the evolution of a time-dependent shear layer [23] [24] [25]. This method is then applied to various cases to introduce the self-focusing mechanism and the modified Kelvin-Helmholtz instability. Finally the modified Kelvin-Helmholtz case is re-examined in the context of the linear instability of the steady Burgers layer solution [71].

A *time-dependent stretched vortex layer* is a *two-dimensional plane* shear layer characterized by a vorticity field of the form

$$\boldsymbol{\Omega} = \Omega(x_2, x_3, t)\mathbf{e}_1 \tag{185}$$

on which is superimposed a hyperbolic stagnation point flow $\mathbf{U} = (\alpha x_1, -\alpha x_2, 0)$. For these layers, the typical scale L of variations in the x_3-direction is assumed to be large compared to the typical scale $\delta(x_3, t)$ of variations in the x_2-direction. Steady or unsteady Burgers layers(section 3.1) constitute the extreme case of solutions (185) : they are homogenous in the x_3-direction. For large but not infinite scale variations, the stretched modulated layer can be regarded as a *modulated Burgers layer* [25]

$$\Omega(x_2, x_3, t) = \frac{\sigma(x_3, t)}{\delta(x_3, t)} G[\frac{x_2 - \eta(x_3, t)}{\delta(x_3, t)}], \tag{186}$$

where $G(\chi) = \frac{1}{\sqrt{2\pi}} \exp(-\frac{\chi^2}{2})$ is taken to be the Gaussian profile of the Burgers layer. The quantity $\sigma(x_3, t)$ stands for the local strength of the stretched layer of thickness $\delta(x_3, t)$ located near the surface $x_2 = \eta(x_3, t)$. For the uniform Burgers layer (26), it is readily seen that the *strength* is equal to the shear $\sigma_0 = U_2 - U_1$, the *thickness*[11] is $\delta(x_3, t) = \sqrt{\nu/\alpha}$ and the *position* $\eta(x_3, t) = 0$. The form (186) effectively reduces the determination of the layer dynamics to the evolution of these three quantities which are related to the first three vorticity momenta

$$I_0 \equiv \int_{-\infty}^{\infty} \Omega(x_2, x_3, t) dx_2, \; I_1 \equiv \int_{-\infty}^{\infty} x_2 \Omega(x_2, x_3, t) dx_2,$$
$$I_2 \equiv \int_{-\infty}^{\infty} x_2^2 \Omega(x_2, x_3, t) dx_2. \tag{187}$$

[10] This layer has nothing to do with the initial shear layer.
[11] This definition differs from a factor $\sqrt{2}$ from the one used in section 3.1.

according to the expressions

$$\sigma(x_3,t) \equiv I_0,\ \eta(x_3,t) \equiv \frac{I_1}{I_0},\ \delta(x_3,t) \equiv \sqrt{\frac{I_2}{I_0} - (\frac{I_1}{I_0})^2}. \tag{188}$$

An asymptotic analysis based on a small parameter defined as the ratio between the scale variations in x_2 and x_3 leads when the second order approximation is included, to a well-posed system of three nonlinear differential equations which couples the local strength $\sigma(x_3,t)$, the thickness $\delta(x_3,t)$ and the surface deformation $\eta(x_3,t)$. At second order [25], the influence of a finite thickness on the layer evolution can be thus taken into account. This ingredient is necessary in order to stabilize the small scales as in the usual Kelvin-Helmholtz instability. We refer the reader to [24] [25] for details and simply mention the main results of the leading order approximation in which the dynamical equation for the layer strength $\sigma(x_3,t)$ and position $\eta(x_3,t)$ does not depend on the thickness of the layer $\delta(x_3,t)$ [24], as in the case of an unstretched vortex sheet. In addition, for the sake of simplicity, the deformation $\eta(x_3,t)$ is assumed to be slaved to the layer strength $\sigma(x_3,t)$ [24]. The problem then reduces to the unique *nonlinear conservation equation*

$$\frac{\partial \sigma}{\partial t} + \frac{\partial F(\sigma)}{\partial x_3} = 0 \tag{189}$$

for the local strength σ, where the flux term F reads

$$F(\sigma) = \frac{\sigma^2}{4\alpha}\frac{\partial \sigma}{\partial x_3} - \nu\frac{\partial \sigma}{\partial x_3}. \tag{190}$$

Equation (189) with (190) is ill-posed since the nonlinear term induces a *negative diffusion*. However basic results may be obtained for the linear and the initial non linear phases. It may be shown that the ill-posedness is cured at the next level of approximation as mentioned above.

Equations (189)-(190) demonstrate the tendency of increasing transport of σ towards regions of higher absolute amplitude. This *self-focusing* mechanism therefore results in vorticity concentration for a modulated layer

$$\sigma(x_3,0) = \sigma_0 \cos kx_3, \tag{191}$$

initially composed of flat counter-rotating vortices ($k\delta << 1$). Two-dimensional direct numerical simulations confirm this result : a self-induced rotation of the initially flat vortices and a subsequent vorticity collapse lead to quasi-circular counter-rotating vortices. Together with the hyperbolic instability, this scenario constitutes the second possible mechanism for the generation of *streamwise counter-rotating vortices.*

Equations (189)-(190) also describe how the *large scale Kelvin-Helmholtz linear instability is modified in the presence of stretching.* This case corresponds to the evolution of disturbances

$$\sigma_{tot}(x_3,0) = \sigma_0 + \epsilon\sigma(x_3) \tag{192}$$

of small amplitude ϵ developing on a uniform Burgers layer of constant strength σ_0. The linearization of (189) around $\sigma = \sigma_0$ leads to the *heat equation*

$$\frac{\partial \sigma}{\partial t} = [\nu - \frac{\sigma_0^2}{4\alpha}]\frac{\partial^2 \sigma}{\partial x_3^2}. \tag{193}$$

When $[\nu - \frac{\sigma_0^2}{4\alpha}] > 0$, all wavelengths are stable : there may thus exist a *critical Reynolds number* $Re_c \equiv \frac{\sigma_0}{2\sqrt{\alpha\nu}} = 1$ for the modified Kelvin-Helmholtz instability. It is recalled that the classical Kelvin-Helmholtz instability is unstable at all Reynolds numbers [14]. When $[\nu - \frac{\sigma_0^2}{4\alpha}] \leq 0$, the stretched layer is unstable. The growth rate however scales as k^2 which should be compared to the scaling in k for the usual Kelvin-Helmholtz instability. In the present model, *viscosity* by itself is incapable of smoothing the small scales. Fortunately, the study of the second order approximation confirms the existence of a critical Reynolds number $Re_c = 1$ and predicts as an additional feature a *short wave length cut-off*. Moreover, for the complete system at second order, the ensuing nonlinear regime does not lead to blow up as in (189) : an asymptotic state is reached where the *vorticity is concentrated into strained vortex tubes.*

In order to consider wavelengths of any scale e.g. of scale comparable to the layer thickness, a different approach is required, which is free from the large scale modulation hypothesis. A linear stability calculation [71] has been performed on the steady Burgers layer. The problem is viewed from a different standpoint where only *two-dimensional infinitesimal periodic* perturbations with a stream-function of the form

$$\Psi(x_2, x_3, t) = \Phi(x_2)\exp[i(kx_3 - \omega t)] + c.c. \; . \tag{194}$$

are considered. The classical *normal mode* analysis that ensues leads to a *modified Orr-Sommerfeld* equation with an additional term due to the flow compression in the direction orthogonal to the main stream. Upon enforcing exponential decay at $x_2 = \pm\infty$, one is led to a *generalized eigenvalue problem* i.e. a dispersion relation $\omega = \omega(k)$. Numerical computations as well as asymptotic calculations [71] confirm the existence of a critical Reynolds number $Re_c = 1$ and a short wave-length cut-off. A good fit for the neutral curve in the *Re*-k plane is given by

$$k(Re) = 0.733 - \frac{0.863}{Re} + O(\frac{1}{Re^2}) \tag{195}$$

when $Re \geq Re_c$.

6 Instability of bounded axisymmetric vortices

The finite extent of the vorticity field brings about several new instability features. In this section, we focus, in the context of *inviscid dynamics*, on three distinct phenomena : *centrifugal instability*, *shear instability* and, for centrifugally and shear stable vortices, the existence of neutrally stable *dispersive inertial waves* (again some authors use the term Kelvin waves). In section 6.1, the centrifugal instability and the associated *Rayleigh criterion* are introduced, through an original approach [73]. Section 6.2 is devoted to the analysis of a shear instability analogous to its counterpart in parallel inflectional shear flows. Neutral waves propagating along elongated stable vortex tubes are discussed in section 6.3. The dispersion relation for the Rankine vortex (30) and the rotating vortex column in a pipe (29) are obtained. Corresponding nonlinear analyses of these dispersive waves are briefly alluded to. Finally, exact Euler solutions are given in section 6.4.

Let us first consider the dynamics of infinitesimal pressure p and velocity $\mathbf{u} = (u, v, w)$ perturbations about a basic flow field $\mathbf{U} = (0, U_\theta(r), 0)$ which is only characterized by an azimuthal velocity field $U_\theta(r)$ i.e. a purely axial vorticity $\Omega = \frac{1}{r}\frac{d}{dr}(rU_\theta)$. Here the analysis is restricted to be inviscid since it can be checked that viscosity is not of prime significance. The *linear stability* equations governing infinitesimal disturbances read, in cylindrical coordinates nondimensionalized with respect to a characteristic vortex core radius and characteristic azimuthal velocity :

$$\frac{Du}{Dt} - 2\frac{U_\theta}{r}v = -\frac{1}{\rho}\frac{\partial p}{\partial r}, \tag{196}$$

$$\frac{Dv}{Dt} + (\frac{\partial U_\theta}{\partial r} + \frac{U_\theta}{r})u = -\frac{1}{\rho r}\frac{\partial p}{\partial \theta}, \tag{197}$$

$$\frac{Dw}{Dt} = -\frac{1}{\rho}\frac{\partial p}{\partial z}, \tag{198}$$

$$\frac{\partial u}{\partial r} + \frac{u}{r} + \frac{1}{r}\frac{\partial v}{\partial \theta} + \frac{\partial w}{\partial z} = 0, \tag{199}$$

where D/Dt stands for the convective derivative associated with the basic state $\mathbf{U}$. The above system is homogeneous with respect to time and the two space coordinates θ, z but inhomogeneous with respect to the radial coordinate r. A normal mode analysis is thus applied to this partial differential system, which amounts to searching for solutions of the form

$$\mathbf{u}(r, \theta, z, t) = \tilde{\mathbf{u}}(r) \exp[i(kz + n\theta - \omega t)] + c.c., \tag{200}$$

$$p(r, \theta, z, t) = \rho\tilde{p}(r) \exp[i(kz + n\theta - \omega t)] + c.c., \tag{201}$$

where the wavenumber k is real. These Fourier modes which are all decoupled, are, by construction, periodic along the z-axis and also 2π-periodic in θ, which imposes that the azimuthal wavenumber n be an integer. The case $n = 0$ corresponds to axisymmetric perturbations. For non axisymmetric $n \neq 0$ disturbances, the phase remains constant along the lines $\theta = \theta_0 - kz/n$: perturbations for $n > 0$, $k > 0$ (resp. $n < 0$, $k > 0$) define a left-hand (resp. right-hand) screw. The reverse holds if $k < 0$. Note that (a) to the complex frequency $\omega = \omega_r + i\omega_i$, are associated the growth rate ω_i and the phase velocity ω_r/k; (b) the wave rotates according to the sign of ω_r/n. Upon making the substitutions $\frac{\partial}{\partial t} \to -i\omega$, $\frac{\partial}{\partial z} \to ik$, $\frac{\partial}{\partial \theta} \to in$, the partial differential equations (196)-(199) now reduce to the system of ordinary differential equations

$$-i(\omega - n\frac{U_\theta}{r})\tilde{u} - 2\frac{U_\theta}{r}\tilde{v} = -\frac{d\tilde{p}}{dr}, \tag{202}$$

$$-i(\omega - n\frac{U_\theta}{r})\tilde{v} + (\frac{dU_\theta}{dr} + \frac{U_\theta}{r})\tilde{u} = -\frac{in\tilde{p}}{r}, \tag{203}$$

$$-i(\omega - n\frac{U_\theta}{r})\tilde{w} = -ik\tilde{p}, \tag{204}$$

$$\frac{d\tilde{u}}{dr} + \frac{\tilde{u}}{r} + \frac{1}{r}in\tilde{v} + ik\tilde{w} = 0, \tag{205}$$

which generally cannot be tackled by purely analytical means. Because of the singular nature of cylindrical coordinates, one must impose that the velocity $\mathbf{u}(r, \theta, z, t)$ and pressure $p(r, \theta, z, t)$ be smooth on the centerline, in other words

$$\frac{\partial \mathbf{u}}{\partial \theta}(0, \theta, z, t) = 0 \ ; \ \frac{\partial p}{\partial \theta}(0, \theta, z, t) = 0. \tag{206}$$

In terms of the eigenfunctions $\tilde{\mathbf{u}}(r)$ and $\tilde{p}(r)$, these conditions can be expressed as follows:

• For *axisymmetric (also referred to as varicose) modes* $n = 0$, quantities $\tilde{p}(0)$ and $\tilde{w}(0)$ are finite while the radial $\tilde{u}(0)$ and azimuthal $\tilde{v}(0)$ velocities vanish. These perturbations are characterized by a fixed vortex axis and an *axisymmetric bulging of the vortex core.*

• For *non-axisymmetric modes* $n \neq 0$ with $n \neq \pm 1$, velocities and pressure perturbations vanish on the axis. The vortex axis does not move but its inner core structure is modified in a non-axisymmetric fashion.

• For *non-axisymmetric (also referred to as helical) modes* $n = \pm 1$, quantities $\tilde{p}(0)$ and $\tilde{w}(0)$ are zero but radial $\tilde{u}(0)$ and azimuthal $\tilde{v}(0)$ velocities do not necessarily vanish. However, one must satisfy

$$\tilde{u}(0) + n\tilde{v}(0) = 0. \tag{207}$$

In this case, the *vortex axis does move.* Some of the *helical modes* are even solely associated to vortex axis deformations, the inner core structure remaining unchanged. Furthermore, specific helical modes such as the *rotating helix* or *the sinuous vortex* contained in a plane, admit nonlinear counterparts within the cut-off theory [17], as outlined in section 6.3.

Other conditions should be applied at the outer boundaries. For instance, in the case of an impermeable cylindrical wall of radius $r = R_{out}$, one must enforce

$$\tilde{u}(R_{out}) = 0. \tag{208}$$

Upon using equations (202) and (203), this constraint reduces to the equivalent form for pressure

$$-i(\omega - n\frac{U_\theta}{r})\frac{d\tilde{p}}{dr} + 2in\frac{U_\theta}{r^2}\tilde{p} = 0, \ r = R_{out}. \tag{209}$$

In the case of an *unbounded domain*, perturbations vanish at infinity:

$$\tilde{u}(r), \tilde{v}(r), \tilde{w}(r) \to 0, \ r \to \infty. \tag{210}$$

Equivalently, pressure satisfies

$$\tilde{p}(r) \to 0, \ r \to \infty. \tag{211}$$

When the above boundary conditions at $r = 0$ and $r \to \infty$ are imposed, the system (202)-(205) defines an *eigenvalue problem.* For a given azimuthal wavenumber n, nontrivial solutions exist only if the complex frequency satisfies a *dispersion relation* $\omega = \omega^l(k, n)$, where the integer superscript l refers to a specific branch. Such a relation constitutes the essential ingredient that determines whether a given disturbance will grow or decay.

6.1 Centrifugal instability

When the basic axial vorticity $\Omega(r)$ varies along the radial coordinate, a differential angular momentum $\rho r U_\theta(r)/2$ is present. *Axisymmetric* perturbations can then be viewed as an exchange of angular momentum between various radial positions. By invoking the angular momentum conservation implied by the Euler equations, Rayleigh [14] devised in 1916 a *sufficient* condition for stability, which was extended by Synge in 1933 to a *necessary* condition. For an *axisymmetric vortex* characterized by a unique azimuthal velocity $U_\theta(r)$, this condition can be stated as follows: there is stability with respect to *axisymmetric infinitesimal* perturbations if and only if the square of the circulation $\Gamma(r) = rU_\theta(r)$ monotonically increases with radius, i.e.

$$\frac{d\Gamma^2}{dr} > 0. \tag{212}$$

Note that a profile satisfying the so-called *Rayleigh criterion* (212), can nonetheless be unstable to non-axisymmetric perturbations. Such a case is precisely analyzed in section 6.2. The criterion (212) is violated, for instance, by a vorticity profile in which a ring of negative vorticity near the radial position r_0 is embedded in a positive vorticity field. In that case, the variation of circulation $\delta\Gamma$ created near r_0 is such that

$$\delta\Gamma = 2\omega(r_0)\pi r_0 \delta r \leq 0. \tag{213}$$

This argument works as well for a ring of positive vorticity embedded in a negative vorticity field. Conversely if the vorticity is always of the same sign, the circulation always satisfies (212).

The general procedure followed by Synge is a standard *normal mode* approach which reduces the eigenvalue problem (202)-(205) for $n = 0$ to a Sturm-Liouville equation where general results are available regarding the eigenvalues. This mathematically rigorous method can be found in several authoritative texts [14] [72]. Here we present a less mathematically rigorous method [73] which only applies to the *necessary* condition for stability. This local procedure is based on the construction of a single eigenvalue and eigenvector. The first step, similar to the usual normal mode approach, restricts the eigenvalue problem to axisymmetric perturbations $n = 0$ governed by

$$-i\omega\tilde{u} = -[-2\Omega_\theta]\tilde{v} - \frac{d\tilde{p}}{dr}, \tag{214}$$

$$-i\omega\tilde{v} = -[\frac{1}{r}\frac{d(r^2\Omega_\theta)}{dr}]\tilde{u}, \tag{215}$$

$$-i\omega\tilde{w} = -ik\tilde{p}, \tag{216}$$

$$\frac{1}{r}\frac{d(r\tilde{u})}{dr} + ik\tilde{w} = 0, \tag{217}$$

where Ω_θ exceptionally denotes the angular velocity U_θ/r instead of the azimuthal vorticity. Terms within brackets arise from the linearization of the convective derivative and they effectively define a 3×3 matrix operator $\mathbf{L}$ applied to the vector $\tilde{\mathbf{u}} = (\tilde{u}, \tilde{v}, \tilde{w})$ with only non zero elements $L_{\tilde{u}\tilde{v}} = -2\Omega_\theta$ and $L_{\tilde{v}\tilde{u}} = \frac{1}{r}\frac{d(r^2\Omega_\theta)}{dr}$. Contrary to Rayleigh's approach, this *local* method does not need to take into account the boundary conditions and builds an eigenvalue and eigenvector by hand which makes it rather attractive.

In order to prove that the Rayleigh criterion (212) is a *necessary condition for stability* or, equivalently, a sufficient condition for instability, we assume that it is violated near a given radius located away from the boundary and demonstrate that the flow is then unstable. Under such circumstances, there exists a position r_{max} for which the square of the circulation reaches a maximum $\frac{d\Gamma^2}{dr} = 0$. It is then possible to build a localized unstable mode around r_{max}. We make first a bold move and assume that pressure and consequently the continuity equation – which is so intimately associated to pressure–, can be discarded in system (214)-(217)! The new system

$$i\omega \tilde{\mathbf{u}} = \mathbf{L}\tilde{\mathbf{u}} \tag{218}$$

may quite easily be diagonalized since the operator $\mathbf{L}$ appearing on the r.h.s. is purely local. The eigenvalue spectrum $\sigma \equiv i\omega$ is easily computed to read :

$$\sigma_1(r_0) = \sigma(r_0),\ \sigma_2(r_0) = -\sigma(r_0),\ \sigma_3(r_0) = 0, \tag{219}$$

where

$$\sigma(r_0) \equiv \sqrt{[-\Phi(r_0)]},\ \Phi(r) = \frac{1}{r^3}\frac{d(r^2\Omega_\theta)^2}{dr} = \frac{1}{r^3}\frac{d\Gamma^2}{dr}. \tag{220}$$

The function $\Phi(r)$ is the so-called Rayleigh discriminant. Note that the eigenvalues depend on the discrete index $i = 1, 2, 3$ and on the continuous parameter r_0 specifying the radial location of the corresponding eigenvector

$$\mathbf{a}_1(r_0) = (1, \frac{\sigma(r_0)}{2\Omega_\theta}, 0),\ \mathbf{a}_2(r_0) = (1, -\frac{\sigma(r_0)}{2\Omega_\theta}, 0),\ \mathbf{a}_3(r_0) = (0, 0, 1). \tag{221}$$

For a given wavenumber k, the pressureless equations possess an infinite number of eigenvectors

$$\delta(r - r_0)\mathbf{a}_i(r_0), \tag{222}$$

where $\delta(r - r_0)$ stands for the Dirac delta function and $i = 1, 2, 3$.

Let us now find a "true" eigenvector solution localized around a point r_0 which satisfies the equations (214)-(217) with pressure and the continuity equation reintroduced. The velocity field is decomposed on the local eigenvector basis according to

$$\tilde{\mathbf{u}}(r) = \tilde{u}_1(r)\mathbf{a}_1(r) + \tilde{u}_2(r)\mathbf{a}_2(r) + \tilde{u}_3(r)\mathbf{a}_3(r), \tag{223}$$

where the components are connected to their cylindrical coordinate counterparts *via* the relations $\tilde{u}_3(r) \equiv \tilde{w}(r)$ and

$$\tilde{u}_1(r) \equiv \frac{\tilde{u}(r)}{2} + \frac{\Omega_\theta(r)}{\sigma(r)}\tilde{v}(r), \tag{224}$$

$$\tilde{u}_2(r) \equiv \frac{\tilde{u}(r)}{2} - \frac{\Omega_\theta(r)}{\sigma(r)}\tilde{v}(r). \tag{225}$$

In the new basis, the eigenvalue problem (214)-(217) now reads

$$[-i\omega - \sigma(r)]\tilde{u}_1 = -\frac{1}{2}\frac{d\tilde{p}}{dr}, \tag{226}$$

$$[-i\omega + \sigma(r)]\tilde{u}_2 = -\frac{1}{2}\frac{d\tilde{p}}{dr}, \tag{227}$$

$$-i\omega\tilde{u}_3 = -ik\tilde{p}, \tag{228}$$

$$\frac{1}{r}\frac{d(r\tilde{u}_1 + r\tilde{u}_2)}{dr} + ik\tilde{u}_3 = 0. \tag{229}$$

The localized eigenvector near an arbitrary radial station r_0 away from the boundaries, is sought by resorting to a large wavenumber approximation scheme whereby

$$\omega \sim i\sigma(r_0) - \frac{1}{k^{n_s}}\omega_1(r_0) + ..., \tag{230}$$

$$\tilde{u}_1(r) \sim U_1(\eta) + ..., \tag{231}$$

$$\tilde{u}_2(r) \sim k^{n_2}U_2(\eta) + ..., \tag{232}$$

$$\tilde{u}_3(r) \sim k^{n_3}U_3(\eta) + ..., \tag{233}$$

$$\tilde{p}(r) \sim k^{n_p}P(\eta) + ..., \tag{234}$$

where η is the rescaled radial variable

$$\eta = k^n(r - r_0). \tag{235}$$

Note that the leading order term is surmised to approach the eigenvalue $i\sigma(r_0)$ and eigenvector (222): the integers n and n_s are taken to be *positive* to ensure the local character of this solution. Aside from these constraints, the various exponents are as yet undetermined and the functions $U_1(\eta)$, $U_2(\eta)$, $U_3(\eta)$, $P(\eta)$ are supposed $O(1)$. Without loss of generality, $\tilde{u}_1$ is $O(1)$ in equation (231) : the problem being linear, the set of eigenfunctions is defined up to an arbitrary multiplicative constant. From equation (228), one obtains the first relation $n_3 = n_p + 1$ and equation (227) provides the additional constraint $n_2 = n_p + n$. Equations (226) and (227) impose $n_2 < 0$. Finally relation (229) provides the relationship $n = n_3 + 1$. Since we have defined 3 relations for 5 unknowns, two additional conditions remain to be found.

Let us now specify r_0 to coincide with the location r_{max} for which the square of the circulation reaches a maximum. This implies, in turn, that $d\sigma/dr(r_0) = 0$ and $\sigma_2 = d^2\sigma/dr^2(r_0) < 0$. A local expansion of $\sigma(r)$ around $r_0 = r_{max}$ now reads

$$\sigma(r) = \sigma_0 + \frac{1}{2}\sigma_2(r - r_0)^2 + .. \tag{236}$$

with $\sigma_0 \equiv \sigma(r_0)$. Upon resorting to expansion (236) and scalings (230) to (235), the l.h.s. of (226) can be rewritten as

$$[-i\omega - \sigma(r)]\tilde{u}_1 \sim [(\sigma_0 + \frac{1}{k^{n_s}} i\omega_1 + ...) - (\sigma_0 + \frac{1}{2}\sigma_2 \frac{\eta^2}{k^{2n}})]U_1, \tag{237}$$

while the r.h.s. reads

$$-\frac{1}{2}\frac{d\tilde{p}}{dr} = -\frac{k^{n+n_p}}{2}\frac{dP}{d\eta}. \tag{238}$$

The principle of dominant balance imposes that all these terms be of the same order, thereby providing the two missing relations $n_s = 2n$ and $-n_s = n + n_p$. The correct scalings are therefore determined and read:

$$n_s = 1,\ n = \frac{1}{2},\ n_2 = -1\ n_3 = -\frac{1}{2},\ n_p = -\frac{3}{2}. \tag{239}$$

Upon introducing these scalings into the expansions (231)-(234) and substituting into the governing equations (226)-(229), one obtains at leading order:

$$-(-i\omega_1 + \frac{1}{2}\sigma_2\eta^2)U_1 = -\frac{1}{2}\frac{dP}{d\eta}, \tag{240}$$

$$\sigma_0 U_3 = -iP, \tag{241}$$

$$\frac{dU_1}{d\eta} + iU_3 = 0. \tag{242}$$

This system can further be reduced to a single equation for U_1 :

$$\frac{\sigma_0}{2}\frac{\partial^2 U_1}{\partial \eta^2} + (-i\omega_1 + \frac{1}{2}\sigma_2\eta^2)U_1 = 0. \tag{243}$$

Since $\sigma_2 < 0$, a localized solution of (243) is possible only if $-i\omega_1$ is real and positive. For instance a Gaussian function

$$U_1(\eta) = \exp[-\frac{\lambda\eta^2}{2}] \tag{244}$$

with $\lambda = \sqrt{-\sigma_2/\sigma_0}$ is such a solution. It corresponds to the leading order eigenvector associated with the eigenvalue

$$\omega \sim i\sigma_0 - \frac{i}{k}\frac{\lambda\sigma_0}{2} + ... \tag{245}$$

of positive growth rate $\sigma_0 = \sqrt{[-\Phi(r_0)]} > 0$. This procedure has therefore led to the identification of an unstable mode localized around r_0. We leave it to the reader to check that any other choice of r_0 does not yield such a localized eigenfunction.

A similar procedure can be applied to two-dimensional nonaxisymmetric flows with closed streamlines [73]. In that case, the convective operator $\mathbf{L}$ is diagonalized within the framework of Floquet theory to account for explicit periodicity with respect to azimuthal direction. In the vicinity of the streamline where the absolute circulation reaches a maximum, the problem reduces again to an equation of the form (243). The following general statement holds : a *sufficient condition for centrifugal instability of two-dimensional non-axisymmetric vortices is that circulation along the closed streamlines be locally decreasing outwards.*

6.2 Shear instability

The centrifugal instability described in section 6.1, prevents any axisymmetric profile satisfying the Rayleigh criterion from being observed. Since only axisymmetric waves $n = 0$ periodic along the vortex axis $k \neq 0$ are examined, a profile with increasing circulation outwards can nonetheless be unstable to *non-axisymmetric* waves $n \neq 0$. In this section, a standard *normal mode* approach is applied to demonstrate the existence of another instability mechanism for non-axisymmetric ($n \neq 0$) but homogeneous along the z-axis ($k = 0$) waves. For such perturbations $\mathbf{u} = (u, v, w)$, the radial u- and azimuthal v-velocities are decoupled from the axial component w which merely evolves like a passive scalar advected by the two-dimensional field $(0, U_\theta, 0)$. As a consequence, w does not play a role in the linear instability and it is thereafter not involved. A two-dimensional disturbance $\mathbf{u} = (u, v, 0)$ of the form (200)-(201) admits a streamfunction $\psi(r, \theta, t)$ such that

$$u = \frac{1}{r}\frac{\partial \psi}{\partial \theta} = \frac{in}{r}\psi \ , \ v = -\frac{\partial \psi}{\partial r}, \tag{246}$$

where the streamfunction is such that

$$\psi = \tilde{\psi}(r) \exp[i(kz + n\theta - \omega t)] + c.c. \ . \tag{247}$$

From (202)-(203), the ordinary differential equation

$$(\omega - \frac{n}{r}U_\theta)[(\frac{d}{dr} + \frac{1}{r})\frac{d}{dr} - \frac{n^2}{r^2}]\tilde{\psi}(r) + \frac{n}{r}\frac{d\Omega}{dr}\tilde{\psi}(r) = 0 \tag{248}$$

is obtained together with boundary conditions for $\tilde{\psi}(r)$ as detailed at the beginning of section 6. For $n > 1$, the conditions read $\tilde{\psi}(0) = \tilde{\psi}(\infty) = 0$. Note the formal analogy with the celebrated Rayleigh equation

$$(\omega - kU)[\frac{d^2}{dy^2} - k^2]\tilde{\psi}(y) - k\frac{d\Omega_z}{dy}\tilde{\psi}(y) = 0 \tag{249}$$

pertaining to the inviscid instability of parallel shear flows of the type $\mathbf{U} = U(y)\mathbf{e}_x$. For a vortex of core radius a and for small scale perturbations $\frac{n}{r}a >> 1$, the basic vortex flow is seen locally as a parallel shear flow where r plays the role of y and $\frac{n}{r}$ that of the streamwise wavenumber k in (249). In both cases, the basic vorticity is directed along the z-axis and its local vorticity gradient $d\Omega/dr$ is found in (249) as well as in (248), the minus sign being due to the axis orientation. The inflexion point theorem is thus likely to be applicable to this rotating flow. This seemingly rough argument can be put on firm grounds by following the same procedure as in the plane case. *For an axisymmetric vortex, a necessary condition for instability to non-axisymmetric disturbances is that the gradient of the basic vorticity change sign at least once in the flow domain.*

6.3 Inertial waves

When a two-dimensional axisymmetric vortex is stable to centrifugal and shear instabilities, it generally acts as a *waveguide* for dispersive neutral waves [8]. Two complementary analyses help in defining their main features: *linear instability theory* and the *nonlinear cut-off filament approximation. Linear instability theory* follows the path presented at the beginning of section 6 : it is restricted to *infinitesimal perturbations* and seeks to determine the dispersion relation as a function of the azimuthal n and axial k wavenumbers. The *cut-off approach* focuses on *nonlinear long waves* for which the vortex axis is deformed, leaving the inner structure unchanged, at least to leading order. More specifically, the velocity field $\mathbf{U}(\mathbf{x},t)$ induced by the vortex tube at a position $\mathbf{x}$, is given by the Biot and Savart law

$$\mathbf{U}(\mathbf{x},t) = \frac{1}{4\pi}\int \frac{\boldsymbol{\Omega}(\mathbf{y}) \times (\mathbf{x}-\mathbf{y})}{|\mathbf{x}-\mathbf{y}|^3} d^3\mathbf{y}. \tag{250}$$

When the ratio between the distance $|\mathbf{x}-\mathbf{y}|$ and the vortex core radius a, supposed uniform along the filament, is large, expression (250) may be expanded with respect to this large parameter. In such a case, the expansion coefficients depend on the vorticity momenta. For instance, the first term takes the form of the line integral

$$\mathbf{U}(\mathbf{x},t) = \frac{\Gamma}{4\pi}\int_L \frac{\mathbf{t} \times (\mathbf{x}-\mathbf{y}(s))}{|\mathbf{x}-\mathbf{y}(s)|^3} ds \tag{251}$$

along the vortex axis parametrized by the curvilinear coordinate s, $\mathbf{t}$ denoting a tangent unit vector along the filament axis. The vortex circulation Γ appears as a constant parameter. This procedure is similar to the one used for stretched or unstretched shear layers as outlined in section 5.2. In contrast to the case of layers, the far-field expression (251) cannot be directly used to compute the vorticity transport in the vortex core : without modifications, it leads to a singular behaviour when $\mathbf{x}$ is located on the filament. However, such a singularity can be removed if the integration is interpreted by introducing a cut-off distance δ around the point $\mathbf{x}$. In the ensuing discussion, the core is assumed to be characterized by an azimuthal velocity $U_\theta(r)$ and an axial velocity $U_z(r)$. The cut-off distance generally scales as the core radius and depends on the core velocity structure [17] according to

$$\ln(\frac{2\delta}{a}) = \frac{1}{2} + \frac{8\pi^2}{\Gamma^2}\int_0^a (U_z^2 - \frac{U_\theta^2}{2})rdr. \tag{252}$$

At leading order, equation (252) constitutes the only relation between the internal core structure and the filament self-induced motion. Note, in particular, that the dynamics inside the core are not taken into account. This theory which has been put on firm grounds [17] [75] [76] [77], leads to simplified evolution equations for the interaction between several vortex tubes. For a single filament, this model is much improved when compared to the local induction approximation [78]. The latter induction scheme is capable of describing families of vortex filaments such as the kinked solitary waves or helical filaments. However the cut-off theory appears to be more appropriate in order to obtain quantitative predictions. Consider for instance the dynamics of *helical waves*. More specifically, let

$$\mathbf{x} = D\mathbf{e}_r + \frac{\omega t - \theta}{k}\mathbf{e}_z \tag{253}$$

denote the nonlinear left-handed vortex filament of wavelength $2\pi/k > 0$, circulation $\Gamma > 0$ and finite amplitude D. It satisfies the *nonlinear dispersion relation* [17] [75]

$$\omega(k) = -\frac{\Gamma k^2}{4\pi}[\ln(\frac{2}{ka}) - C + \frac{8\pi^2}{\Gamma^2}\int_0^a (\frac{U_\theta^2}{2} - U_z^2)rdr], \tag{254}$$

where $C = .57$ is the Euler constant. The left-handed helical filament thus propagates along the negative z- direction. The equivalent nonlinear right-handed vortex filament simply propagates in the opposite direction with the same frequency $|\omega|$. The basic flows (30)-(31) or (32) are often used in the context of the cut-off approach : for the Rankine vortex, equation (254) reads

$$\omega(k) = -\frac{\Gamma k^2}{4\pi}[\ln(\frac{2}{ka}) - C + \frac{1}{4}]. \tag{255}$$

The above result for left- or right-handed nonlinear filaments, is also obtained in the context of the *linear instability analysis* of the Rankine vortex [17]. In that instance, the dispersion relation (255) corresponds to specific modes $n = \pm 1$ for infinitesimal helical waves (see below). Note that the phase or group velocity based on (254) is small compared to the velocity inside the vortex : it is of order Γk i.e. much smaller than Γ/a since, by assumption, this formulation is only valid for long waves $ka << 1$. The cut-off approach is attractive for the following reasons : (a) true nonlinear solutions can be considered and (b) the results do not rely on a specific inner core structure. Consequently, more general statements can be made than in the context of instability theory where each dispersion relation is numerically computed anew. Nevertheless important phenomena such as *fast varicose* waves are completely out of reach of the cut-off theory. We now present the instability analysis for vortices with no axial flow, which predicts the behaviour of infinitely many other waves with very different phase speeds and for which the vortex core structure is altered.

The *linear instability theory* of a *two-dimensional vortex* has already been presented in the beginning of section 6. Though a rigorous proof of completeness is not yet available (see however [79] where this problem is considered for the Rankine vortex), *infinitesimal* perturbations may generally be written as a superposition of neutral Fourier modes (200)-(201) propagating along the vortex axis. The discussion is restricted here to such neutral *inertial waves*. An eigenvalue problem is effectively defined by (202) to (205) together with the boundary conditions specified at the beginning of section 6 for various azimuthal wavenumbers n. The velocity components $\tilde{u}$, $\tilde{v}$, $\tilde{w}$ are related to $\tilde{p}$ without any differentiation so that the system (202)-(205) can be reduced to the single ordinary differential equation for pressure

$$\frac{d^2\tilde{p}}{dr^2}+[\frac{1}{r}-\frac{1}{G}\frac{dG}{dr}]\frac{d\tilde{p}}{dr}+[\frac{2n}{rF}(\frac{\Omega_\theta}{G}\frac{dG}{dr}-\frac{d\Omega_\theta}{dr})+\frac{k^2G}{F^2}-\frac{n^2}{r^2}]\tilde{p}=0, \tag{256}$$

where

$$F(r)\equiv\omega-n\Omega_\theta(r),\ G(r)\equiv 4\Omega_\theta^2+2r\Omega_\theta\frac{d\Omega_\theta}{dr}-F^2, \tag{257}$$

and Ω_θ exceptionally denotes the angular velocity U_θ/r in (256)-(257). Corresponding boundary conditions for $\tilde{p}$ have been given for $r=0$ at the beginning of section 6 and at $r=R_{out}$ or $r\to\infty$ in (209), (211). As in previous instances, this eigenvalue problem generally gives, for a given wavenumber pair (k,n), an infinite but denumerable set of modes of complex frequency $\omega^l(k,n)$ where l is an integer that discriminates between various modes with the same frequency. It is easily seen that (256) is invariant with respect to the transformation

$$n\to -n,\ \omega\to-\omega. \tag{258}$$

As a consequence, the dispersion relation $\omega^l(k,n)$ for $n\neq 0$ satisfies

$$\omega^l(k,-n)=-\omega^l(k,n), \tag{259}$$

which simply means that mode $n>0$ and its counterpart $n<0$ propagate in opposite directions, as for the left- or right-handed nonlinear helical filaments in the cut-off approach. Moreover, equation (256) is also invariant with respect to the transformation $k\to-k$, $\omega\to\omega$, hence

$$\omega^l(k,n)=\omega^l(-k,n). \tag{260}$$

In particular *varicose* modes may propagate in both directions along the vortex axis. These general invariances (259)-(260) ensure the existence of *rotating plane waves* formed by two opposite helical waves $n=1$ and $n=-1$. For instance, in the case of the disturbed Rankine vortex, the flow is characterized by a disturbed core boundary shown in figure 16. In section 7, the existence of such plane waves is particularly important to account for a *global resonance mechanism.* We thereafter present the stability analysis of two specific basic vortex flows : a *confined* vortex column in a *rotating pipe* (29) and a vortex in an *unbounded domain*, i.e. the *Rankine vortex* (30)-(31).

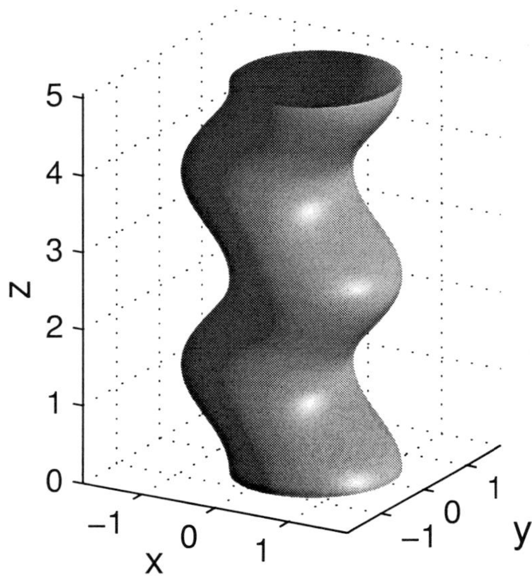

Fig. 16. In-plane wave motion produced, in the case of the disturbed Rankine vortex, by the combination of two opposite helical waves $n = 1$ and $n = -1$. The vortex core boundary is displayed. Courtesy of C.Eloy.

In *rotating pipe flow*, the basic vorticity Ω is uniform. The equation for pressure (256) very much simplifies and reads

$$\frac{d^2\tilde{p}}{dr^2} + \frac{1}{r}\frac{d\tilde{p}}{dr} + [\beta^2 - \frac{n^2}{r^2}]\tilde{p} = 0, \tag{261}$$

where

$$\beta^2 \equiv (\frac{\Omega^2}{F^2} - 1)k^2 \tag{262}$$

can be interpreted as a kind of radial wavenumber and

$$F \equiv \omega - n\frac{\Omega}{2} \tag{263}$$

is constant and stands for the frequency of the mode in a frame rotating with the fluid (see (200)-(201)). The general solution of equation (261) is given by

$$\tilde{p}(r) = P_1 J_n(\beta r) + P_2 Y_n(\beta r), \tag{264}$$

where $J_n(\chi)$and $Y_n(\chi)$ are Bessel functions of first and second kind of order $|n|$. The constant P_2 must vanish in order for the pressure to be finite on the axis and P_1 is left unspecified since the eigenfunction is determined up to an arbitrary multiplicative constant. Finally the constraint (209) on pressure at the outer boundary

$$F\frac{d\tilde{p}}{dr} - n\frac{\Omega}{a}\tilde{p} = 0, \; r = a \tag{265}$$

provides an implicit dispersion relation between k, n and ω

$$\beta F\frac{dJ_n}{d\chi}(\beta a) - n\frac{\Omega}{a}J_n(\beta a) = 0. \tag{266}$$

The roots of this equation yield all the possible frequencies. It may be shown that, for given k and n, the infinite denumerable set of real frequencies $\omega^l(k,n)$ is located in the frequency interval

$$(n-2)\frac{\Omega}{2} < \omega < (n+2)\frac{\Omega}{2}. \tag{267}$$

The index l is ordered so that the structure of the pressure field becomes more complex as l is increased. Note that, when $ka \to 0$, all the branches accumulate towards the frequency $n\frac{\Omega}{2}$. For *axisymmetric perturbations*, i.e. $n = 0$, equation (266) simplifies: βa must be one of the denumerable zeroes b_l of the function $\frac{dJ_0}{d\chi}(\chi) \equiv J_1(\chi)$. Note that these zeroes are ordered so that they are increasing towards ∞. From equation (262), one finally obtains:

$$\omega^l(k,0) = \pm\frac{\Omega}{\sqrt{1+(\frac{b_l}{ka})^2}}. \tag{268}$$

These axisymmetric inertial modes may propagate along the vortex axis in both directions as expected. However they exist only when the rotating fluid column is perturbed by a low enough forcing frequency, since $|\omega^l(k,0)| < \Omega$ for any k, l. This property is reminiscent of the Kelvin waves in an unbounded domain (see (151)). For helical waves $n = 1$ (resp. $n = -1$), solid (resp. dashed) lines in figure 17 correspond to the first twelve branches $\omega^l(k,1)$ (resp. $\omega^l(k,-1)$). Two features are noteworthy: the existence of stationary disturbances $\omega = 0$ for short waves $ka = O(1)$ and crossing between modes $n = 1$ and $n = -1$ which may result in resonance phenomena (section 7). Inertial waves can be generated experimentally in a rotating cylinder by moving a small disk up and down along the vortex centerline [80]. It is also possible to generate more complex flows corresponding to helical modes $n = \pm 1$ by rotating a top-end cap [81] or by forcing the cylinder to precess [82] [83].

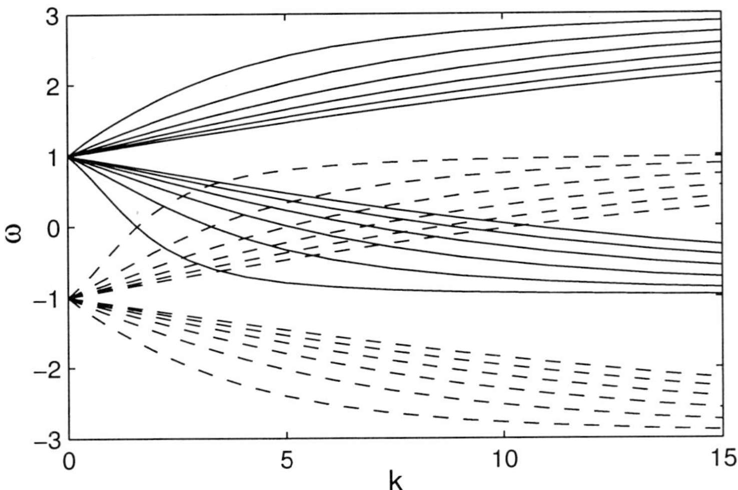

Fig. 17. First twelve branches $\omega^l(k,1)$ (resp. $\omega^l(k,-1)$) of helical waves $n=1$ (resp. $n=-1$) in rotating pipe flow. Solid (resp. dashed) lines refer to $\omega^l(k,1)$ (resp. $\omega^l(k,-1)$). Courtesy of C.Eloy.

The inertial waves of a *Rankine vortex* (30) have a different dispersion relation. Whether vorticity is confined by solid boundaries as in a rigidly rotating pipe, or simply surrounded by a potential field as in the Rankine vortex, perturbations exhibit different phase velocities. In the core region $r < a$, the basic vorticity Ω is still uniform: equations (261) to (262) are hence valid as well as $\tilde{p}(r) = P_1 J_n(\beta r)$. In the unbounded potential region, the flow is perturbed by an irrotational field which derives from a potential $\tilde{\phi}(r)\exp[i(kz+n\theta-\omega t)]$. Incompressibility imposes that this potential satisfy the Laplace equation which, for normal modes, reads

$$\frac{d^2\tilde{\phi}}{dr^2} + \frac{1}{r}\frac{d\tilde{\phi}}{dr} - [k^2 + \frac{n^2}{r^2}]\tilde{\phi} = 0. \tag{269}$$

According to equation (204) and the relation $\tilde{w} = ik\tilde{\phi}$, the pressure is given by

$$\tilde{p}(r) = -(-i\omega + in\frac{\Gamma}{2\pi r^2})\tilde{\phi}(r) \tag{270}$$

which is nothing but a linearized form of the Bernoulli equation. The general solution of (269) is a combination of Bessel functions

$$\tilde{p}(r) = Q_1 I_n(|k|r) + Q_2 K_n(|k|r), \tag{271}$$

where $I_n(\chi)$and $K_n(\chi)$ are the modified Bessel functions of first and second kind of order $|n|$. The constant Q_1 must vanish in order for the pressure to satisfy the condition at infinity (211). We need now to impose a matching condition across the vorticity jump that separates the rotational core and the outer potential zone of the Rankine vortex. This is done by imposing the continuity of pressure and particle displacement on the vortex boundary. In the linear approximation, these conditions are imposed at leading order on the undisturbed surface $r = a$. A system of two linear homogeneous equations for P_1 and Q_1 is thereby obtained. Non trivial solutions exist only when the determinant of this system is zero, which provides the dispersion relation. For given k and n, an infinite denumerable set of frequencies is found which are real and located, as in the case of rotating pipe flow, in the frequency interval defined by (267). Similarly to the confined vortex case, most branches accumulate towards the frequency $n\frac{\Omega}{2}$ when $ka \to 0$. However, for $n \neq 0$, there also exists a unique branch that starts at $[n - \mathrm{sgn}(n)]\Omega/2$ which has no counterpart in the rotating pipe case. At given k and n, the perturbations inside the core are still given by $\tilde{p}(r) = P_1 J_n(\beta r)$ but β as well as the frequency $\omega^l(k,n)$ are different than in the rotating pipe flow case. For varicose modes $n = 0$, one finds, for each dimensionless wavenumber ka, an infinite denumerable set of radial wave numbers $\beta = d_l(ka)/a$, where the function $d_l(ka)$ remains bounded when $ka \to \infty$. In such a case, the dispersion relation takes the form

$$\omega^l(k,0) = \pm \frac{\Omega}{\sqrt{1 + (\frac{d_l(ka)}{ka})^2}}. \tag{272}$$

These solutions, which may be viewed as periodic core expansions, have a behaviour qualitatively close to that of the confined case (268) : for short waves $k \to \infty$ they all tend toward Ω; for long waves $k \to 0$, they tend towards zero. The group velocity $d\omega/dk$ increases with wavelength : its maximum is attained for $ka \to 0$ and $l = 1$:

$$\frac{d\omega^1(k,0)}{dk} \to 0.83\Omega a. \tag{273}$$

It is seen that the group velocity scales with the characteristic core rotation velocity. The vortex resists any local bulging by generating such fast disturbances which are then damped by viscous diffusion. For *helical waves* $n = 1$ (resp. $n = -1$), solid (resp. dashed) lines in figure 18 correspond to the first ten branches $\omega^l(k,1)$ (resp. $\omega^l(k,-1)$). Several features are shared with the confined case : the existence of stationary disturbances $\omega = 0$ for short waves and crossing between modes $n = 1$ and $n = -1$ which may again result in resonance phenomena. For helical modes $n = \pm 1$, there exists a unique branch that starts at $\omega = 0$ for $ka = 0$. This slow mode which is suppressed in rotating pipe flow, by the presence of finite distance walls, can be expanded for long axial helical waves $ka \to 0$. One thus recovers the dispersion relation (255) already obtained in the cut-off approximation theory for helical filaments. As a consequence, the time scale in which a vortex filament deforms due to external flow (helical modes), is much longer than the one in which its core radius evolves (varicose modes). This is why, in the cut-off theory, the vortex core may be assumed uniform along its axis but not constant in time. Finally, note that combinations of various inertial waves may lead to complex structures (figures 19 and 20).

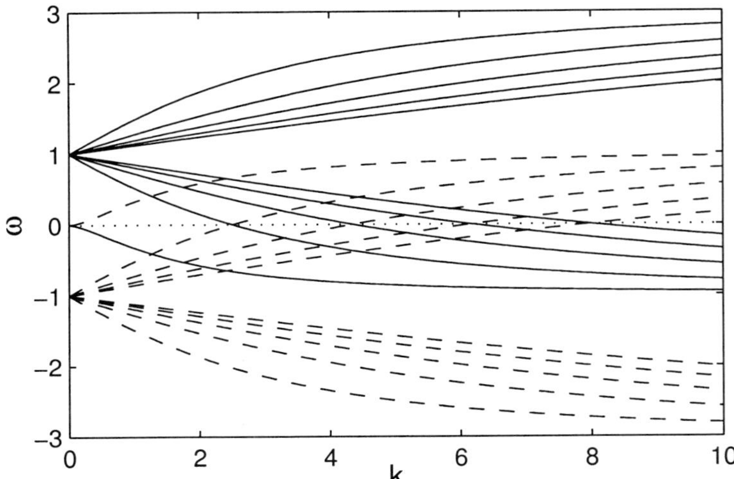

Fig. 18. First ten branches $\omega^l(k,1)$ (resp. $\omega^l(k,-1)$) of helical waves $n = 1$ (resp. $n = -1$) in Rankine vortex flow. Solid (resp. dashed) lines refer to $\omega^l(k,1)$ (resp. $\omega^l(k,-1)$). Courtesy of C.Eloy.

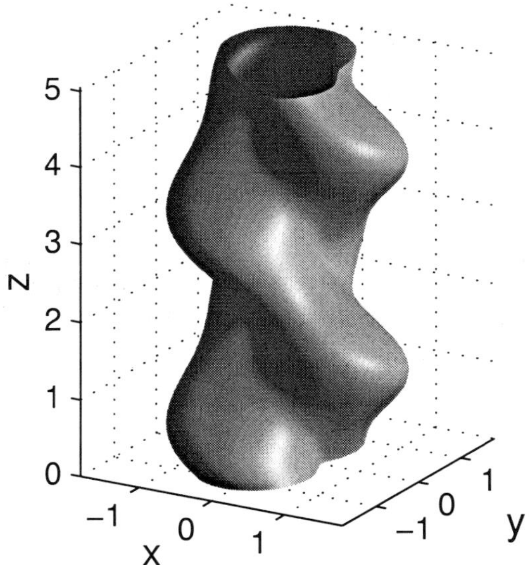

Fig. 19. Combination of inertial modes $n = 2$ and $n = -1$ riding on a Rankine vortex. The vortex core boundary is displayed. Courtesy of C.Eloy.

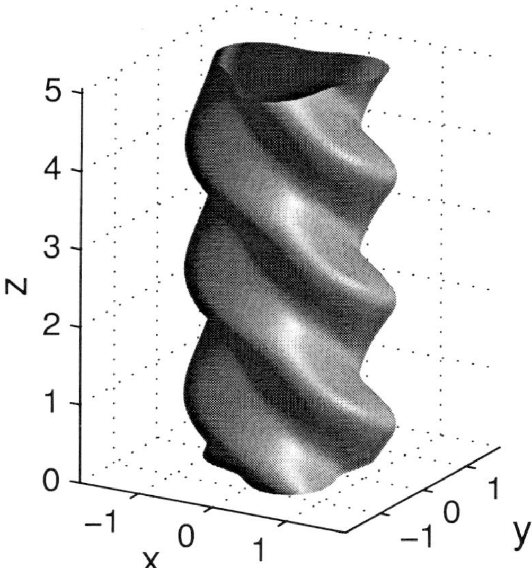

Fig. 20. Combination of inertial modes $n = 3$ and $n = 0$ riding on a Rankine vortex. The vortex core boundary is displayed. Courtesy of C.Eloy.

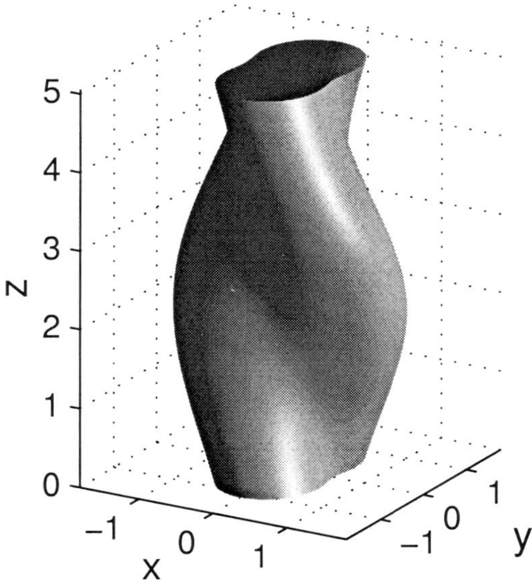

Fig. 21. Combination of inertial modes $n = 2$ and $n = 0$ riding on a Rankine vortex. The vortex core boundary is displayed. Courtesy of C.Eloy.

These classical linear concepts have found a new life in more recent nonlinear studies. For instance perturbations $|n| > 1$ of high enough amplitude may lead to the appearance of multiple vortices (figure 21) [1]. The flow structure can also be modified by finite-amplitude varicose inertial waves at the origin of vortex core bulging. Core dynamics analyses [56] use direct numerical simulations of the axisymmetric Navier-Stokes equations to recover basic features of inertial waves and to show the effect of viscous damping on vortex core oscillations. Let us mention also another original numerical approach [85] which models a vortex tube as a finite number of vortex filaments. The motion of each filament is determined in the framework of cut-off theory: this approach hence preserves the simplicity of this theory while allowing bulging of vortex tubes ! This trick was used in [85] to understand the destruction of a *finite-length vortex tube*: at $t = 0$, the vortex tube is taken to be finite since its core abruptly expands at both ends. As time evolves, two fronts propagate from each end along the vortex, until they finally meet thereby destroying the vortex tube. The two fronts delineate two regions in which the vortex core abruptly expands and filaments are twisted. This behaviour is a clear sign of the presence of an axial flow (cf twisted filaments) inside the vortex core. As already emphasized in section 3.3, the axial flow arises from the coupling between differential rotation along the vortex and axial velocity. More specifically, a jet velocity is generated by the pressure gradient in the transition zone where the core radius abruptly increases and this pressure gradient is due to the differential rotation along the vortex axis generated by core expansion. The front velocities have been shown to equal the group velocity (273). The same basic coupling between differential rotation and axial flow comes into play, together with viscous diffusion, for initial vorticity fields made up of short vorticity packets [45]. As time evolves, these short vortex elements tend to combine into longer tube-like structures.

Inertials waves are observed experimentally either by forcing [81] [83], or as a result of an elliptic instability (section 7). In both situations, it is observed, in specific Reynolds number ranges, that finite-amplitude inertial waves rapidly result in the complete destruction of vortex tubes into disordered small scales. Though the origin of this *collapse* has not yet been elucidated, the existence of a *secondary instability* of inertial waves constitutes a possible explanation [86] [9] [14].

6.4 Generalized Beltrami Flows

Kelvin waves initially obtained in a linear framework for an unbounded solid body rotation, turn out, because of transversality conditions, to be also solutions of the nonlinear problem (see section 4). Surprisingly enough, the same feature holds for their inertial wave counterpart in rotating pipe flow : the linear *inertial waves* found by Kelvin are also nonlinear solutions of the Euler equations with the same frequency. As shown below, this result arises because inertial waves verify both a *helical symmetry* and a *generalized Beltrami flow* condition.

Let us first briefly recall the notion of *Beltrami flows*. By using incompressibility, the Euler equations can be rewritten as follows:

$$\frac{\partial \mathbf{U}}{\partial t} + \boldsymbol{\Omega} \times \mathbf{U} = -\nabla h \ , \tag{274}$$

with

$$h = \frac{p}{\rho} + \frac{\mathbf{U}^2}{2} \ . \tag{275}$$

Three-dimensional steady flows thus exist when $\mathbf{U} \times \boldsymbol{\Omega}$ is equal to the gradient of the function -h. This condition is automatically satisfied when the vorticity $\boldsymbol{\Omega}$ and velocity $\mathbf{U}$ are parallel: $\boldsymbol{\Omega} = -\alpha(\mathbf{x})\mathbf{U}$. *Beltrami flows* constitute even more specific cases in which the function $\alpha(\mathbf{x})$ is constant in space and time :

$$\boldsymbol{\Omega} = -\alpha \mathbf{U}. \tag{276}$$

In such an instance, the Euler equations are replaced by a scalar Helmholtz equation for the three components of the velocity U_j:

$$\Delta U_j = -\alpha^2 U_j \ . \tag{277}$$

One can solve (277) for two components and use the incompressibility condition for the last one. However the main difficulty lies in also verifying the boundary conditions !

This procedure can be slightly generalized [63] to provide *steadily rotating Euler flows* instead of steady solutions ! This is simply obtained by imposing the same kind of constraints to a *helically symmetric* flow but in a *non-inertial* reference frame. Consider the velocity field $\mathbf{U}_{tot}$ which is the combination of an overall solid body rotation $\mathbf{U}_0 = (-\frac{\Omega_0}{2}y, \frac{\Omega_0}{2}x, 0)$ and a time-dependent field $\mathbf{u}(t)$. In the fixed reference frame, the total vorticity thus reads $\boldsymbol{\Omega}_{tot} = \boldsymbol{\Omega}_0 + \boldsymbol{\Omega} = \Omega_0 \mathbf{e}_z + \boldsymbol{\Omega}$. In the non-inertial reference frame rotating with angular velocity Ω_0, the total velocity and total vorticity are respectively given by $\mathbf{u}$ and $\boldsymbol{\Omega}$. As demonstrated in section 3.5 (equations (125)-(127)), the helical symmetry leads to the following relations (here written in the non-inertial frame):

$$\mathbf{u} = w_B(\chi, r, t)\mathbf{B} + \nabla \Psi \times \mathbf{B}, \tag{278}$$

$$\boldsymbol{\Omega} = \tilde{\omega}_1 \mathbf{B} + \nabla w_B(\chi, r, t) \times \mathbf{B}, \tag{279}$$

where it is recalled that $\chi \equiv \theta - z/L$. The scalar fields $w_B(\chi, r, t)$, $\Psi(\chi, r, t)$, $\tilde{\omega}_1$ are not independent but satisfy (128), i.e.

$$\mathcal{L}\Psi = \frac{2N^4}{L} w_B - N^2 \tilde{\omega}_1, \tag{280}$$

where $\mathcal{L}$ is given by (129). In order to close equation (280), two additional requirements of the type (130) are necessary to isolate a precise class of flows.

These conditions are here chosen to be the *generalized Beltrami condition* $\boldsymbol{\Omega} = -\alpha\mathbf{u}$, i.e. in component form

$$w_B(\chi,r,t) = -\alpha\Psi(\chi,r,t),\ \tilde{\omega}_1 = -\alpha w_B(\chi,r,t). \tag{281}$$

The Beltrami condition (276) has here been imposed in the non-inertial frame i.e. on the *perturbations*. The relations (280)-(281) imply that

$$\tilde{\omega}_1 = \alpha^2\Psi, \tag{282}$$

$$\frac{1}{N^2}\mathcal{L}\Psi + [\alpha^2 + \frac{2N^2}{L}\alpha]\Psi = 0, \tag{283}$$

The above constraints have to be made consistent with the non-inertial Euler equations

$$\frac{\partial\mathbf{u}}{\partial t} + \boldsymbol{\Omega}\times\mathbf{u} + \boldsymbol{\Omega}_0\times\mathbf{u} = -\nabla[\frac{p}{\rho} + \frac{|\mathbf{u}|^2}{2} - \frac{\Omega_0^2 r^2}{8}], \tag{284}$$

or in terms of vorticity

$$\frac{\partial\boldsymbol{\Omega}}{\partial t} + \nabla\times(\boldsymbol{\Omega}\times\mathbf{u}) = \boldsymbol{\Omega}_0.\nabla\mathbf{u}. \tag{285}$$

Because of the *generalized Beltrami condition* $\boldsymbol{\Omega} = -\alpha\mathbf{u}$, equivalently (281), the above vector equation reads

$$\mathcal{J}[\mathbf{u}] = 0, \tag{286}$$

where the operator $\mathcal{J}$ stands for

$$\mathcal{J} \equiv [\alpha\frac{\partial}{\partial t} + \Omega_0\frac{\partial}{\partial z}]. \tag{287}$$

Invoking equations (278) and (281), and noting that, according to definition (115), $\frac{\partial}{\partial t}\mathbf{B} = \frac{\partial}{\partial z}\mathbf{B} = 0$, equation (286) can be rewritten as

$$-\alpha\mathcal{J}[\Psi]\mathbf{B} + \nabla(\mathcal{J}[\Psi])\times\mathbf{B} = 0. \tag{288}$$

This relation is automatically satisfied when $\mathcal{J}[\Psi] = 0$ or using the dependence of Ψ on $\chi \equiv \theta - \frac{z}{L}$

$$[\frac{\partial}{\partial t} - \frac{\Omega_0}{\alpha L}\frac{\partial}{\partial\chi}]\Psi = 0. \tag{289}$$

The compatibility between the generalized Beltrami condition and the Euler equations in the non-inertial frame therefore reduces to the simple time-shift

$$\Psi(r,\chi,t) = \Psi(r,\phi), \tag{290}$$

where the phase ϕ is defined by

$$\phi = \theta - \frac{z}{L} + \frac{\Omega_0}{\alpha L}t. \tag{291}$$

By substitution of (290) into (283), an eigenvalue problem for $\Psi(r,\phi)$ with eigenvalue α is obtained. Since there is no explicit dependence on χ in the coefficients of the operator $\mathcal{L}$ (see equation (129)), one may look for solutions

$$\Psi(r,\phi) = A(r)\exp(in\phi) + c.c. = A(r)\exp[i(n\theta - \frac{n}{L}z + \frac{n\Omega_0}{\alpha L}t)] + c.c.\,. \tag{292}$$

In the *specific case of rotating pipe flow*, such a family of solutions must also satisfy the impermeability boundary conditions

$$\frac{\partial}{\partial \chi}\Psi = \frac{\partial}{\partial \phi}\Psi = 0 \tag{293}$$

on the pipe wall. Note that these conditions are compatible with helical symmetry. The combination of (292)-(293) constitutes a simplified eigenvalue problem where the frequency eigenvalue $\omega \equiv -(n\Omega_0)/(\alpha L)$ has to be determined for each axial wavenumber $k \equiv -n/L$ and azimuthal wavenumber n. It should be emphasized that this eigenvalue problem is identical to the one arising in the linear instability of *rotating pipe flow* (see section 6.3). Remarkably, the linear inertial waves living in *rotating pipe flow* are also true nonlinear waves !

7 Elliptic instability of two-dimensional vortices

Let us now consider the instability [31] [17] [2] [91] [92] occurring in *elliptic flows* that are assumed stable with respect to the extended centrifugal instability described at the end of section 6.1. Typically this analysis is concerned with *non-axisymmetric two-dimensional vortices* experimentally observed in shear flows [37] or in counter-rotating vortices [35].

To avoid centrifugal instability, the circulation of such elliptic flows should be increasing outwards [73]. For instance, this is the case for the two-dimensional unbounded linear models (7)-(8) with $0 < \gamma < \frac{\Omega}{2}$ (figure 2 (a)) considered in section 2.2. Results of section 4 can be directly applied to the *local instability* of such unbounded elliptic flows : Fourier modes (160)-(166) of wave vector $\mathbf{k}(t)$ and amplitude $\mathbf{u}^0(t) \equiv B_{viscous}(t)\mathbf{u}^E(t)$ are solutions of the linear instability equations provided that (a) $\mathbf{k}(t)$, $\mathbf{u}^E(t)$ satisfy the two evolution equations (162), (168) and (b) the viscous factor $B_{viscous}(t)$ be given by (167). In this instance, the wave vector satisfies

$$\frac{dk_1}{dt} = -(\frac{\Omega}{2} - \gamma)k_2, \ \frac{dk_2}{dt} = (\frac{\Omega}{2} + \gamma)k_1, \ \frac{dk_3}{dt} = 0, \tag{294}$$

the solution of which is the time periodic vector

$$\mathbf{k}(t) = K_0(\sin\theta_0 \cos Q(t-t_0),\ E\sin\theta_0 \sin Q(t-t_0),\ cos\theta_0). \tag{295}$$

The quantities t_0, K_0, θ_0 are free parameters set by the initial conditions. They respectively stand for a time shift or phase shift, a wave vector magnitude and the angle between the wave vector and the vortex x_3-axis at $t = t_0$. The aspect ratio E is given in (9) and

$$Q = \sqrt{(\frac{\Omega}{2})^2 - \gamma^2} \tag{296}$$

scales as an angular velocity. Similarly to the usual *Kelvin waves* in *unbounded solid body rotation* examined in section 4, the wavenumber component k_3 along the vortex axis remains constant and the wavevector $\mathbf{k}$ spins around the x_3-axis with an angular frequency Q. However, during its time evolution, $\mathbf{k}$ now follows an ellipse the major axis of which is perpendicular to the major axis of the elliptically shaped basic streamlines. When the wavevector (295) is substituted into evolution equation (168), a system of ordinary differential equations is obtained for the inviscid amplitudes in the form

$$\frac{du_i^E}{dt} = QF_{ij}(Qt)u_j^E. \tag{297}$$

The matrix elements $F_{ij}(\chi)$ are 2π-periodic functions and depend on two parameters: the aspect ratio E which is directly linked to the basic state through equation (9), and the angle θ_0 fixed by the initial conditions. The spatial scale of perturbations specified by K_0 does not affect the inviscid evolution : as for Kelvin waves, *scale invariance* is thus obtained. A self-similar behaviour is then expected : there exists a continuous family of eigenmodes which differ by their shape and dynamics only through their respective scale. It is known from *Floquet theory* that the solutions of the third-order system (297) can be written as a combination of three ($m = 1, 2, 3$) independent solutions

$$\mathbf{u}_m^E(t) = \exp(\sigma_m t)\mathbf{R}_m(t - t_0), \tag{298}$$

where the vector $\mathbf{R}_m(t)$ is periodic of period $T = 2\pi/Q$ and the complex coefficients σ_m are the so-called *Floquet exponents.* All we need as far as instability is concerned, is contained in the three Floquet exponents σ_m : If one σ_m possesses a positive real part, its associated velocity field $\mathbf{u}_m^E$ is amplified and the basic flow (7)-(8) with $0 < \gamma < \frac{\Omega}{2}$ is thus unstable. For a given pair (E, θ_0), these exponents are computed by numerically integrating system (297) during one period $T = 2\pi/Q$. The integration provides the Poincaré map that relates the solution at time t with the solution at time $t + T$. The Poincaré map is here a 3×3 matrix, the eigenvalues of which should be equal to $\exp(\sigma_m T)$ (see equation (298)). One of these eigenvalues is always zero while the other two are nontrivial and, according to (297), of the form

$$\sigma_m = \frac{Q}{2\pi}\tilde{\sigma}_m(E, \theta_0). \tag{299}$$

In the plane (E, θ_0), unstable regions correspond to points where one of the nonzero σ_m possesses a positive real part. When $E = 1$ we recover the usual

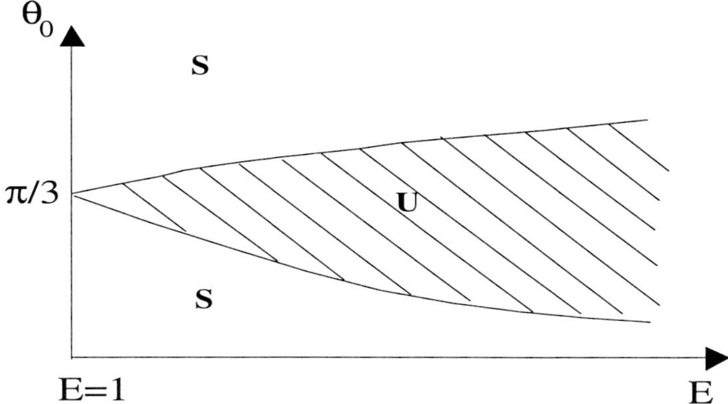

Fig. 22. Parametric instability of Kelvin waves : domains of instability (hatched area U) and stability (clear area S) in (E, θ_0) plane.

neutral Kelvin waves and the flow is neutrally stable. However, as soon as $E \neq 1$, there always exists a domain of instability for some range of angles (figure 22) ! For aspect ratios close to unity $E \sim 1$ or weak strain $\gamma/\Omega << 1$, the Floquet exponents can be computed analytically [31] and the unstable range is located near $\theta = \pi/3$. In this limit, the maximum growth rate

$$\sigma_{max} = \frac{9}{8}\frac{\gamma}{\Omega}, \tag{300}$$

is proportional to the rate of strain imposed on the unbounded elliptic flow. A generic *inviscid parametric instability* called the *elliptic instability* thus arises as soon as vortex streamlines become elliptic. Three-dimensional unstable waves exist that may deform the original vortex tube. In order to see how this study is relevant to finite extent two-dimensional elliptic vortices, let us build a localized solution made up of the previous plane waves [2] [92]. More specifically, in the weak rate of strain limit, Fourier modes of the form (160)-(166) and (295)-(298) are superposed with different t_0 but equal axial wavenumber $k_3 = K_0 \cos\theta_0$ and identical angle $\theta_0 = \pi/3$ corresponding to the maximum growth rate. The latter condition determines the spatial scale K_0. When these modes are all taken to be of equal amplitude A_0, the total velocity field thus obtained reads

$$\mathbf{u}^E(t) = A_0 \exp(\sigma_{max} t) \exp(ik_3 z) \int_0^T \exp(ik_1 x_1 + ik_2 x_2) \mathbf{R}_{max}(t - t_0) dt_0. \tag{301}$$

By using (295), this integral may be easily rewritten in polar coordinates in terms of an infinite series of Bessel functions which decrease to zero at infinity. Furthermore the component of vorticity perpendicular to the vortex axis is predominantly aligned with the stretching axis. According to (3), this feature is responsible for the exponential growth of the unstable mode : vorticity is always directed so as to be enhanced by stretching (for details see [92]). Furthermore the velocity field (301) is an analytical counterpart of the solution found in inviscid numerical computations of two-dimensional eddies [17]. Growth rates obtained in [17] are close to those of the *parametric instability* analysis. As far as *viscosity* is concerned, it only appears through the viscous factor

$$B_{viscous}(t) = \exp[-\nu \int_0^t |\mathbf{k}|^2 \, dt] \tag{302}$$

which always dampens the perturbations [91]. From (295), it is easily seen that the inviscid growth rate should be decreased by a factor proportional to $-\nu K_0^2$. This *viscous diffusion* provides a *small scale cut-off*, breaks the scale invariance with respect to K_0 but it is *incapable, by itself, of dampening the elliptic instability* at large wavelengths.

Inertial waves can destabilize an inviscid two-dimensional vortex when it is subjected to a *large scale strain.* This *global resonance* mechanism [17] leads to a velocity field quite similar to the flow (301) : in the rotating pipe [31] or Rankine vortex flow [7] (figure 16), it consists of a planar mode formed by two helical inertial waves $n = 1$ and $n = -1$. This observation indicates a close connection between the previous local *elliptic instability* and this *global resonance.* To be more specific, consider the instability analysis of the elliptic inviscid vortex (46)-(51) in the weak external strain limit (see section 3.2 for notations). By invoking the expansion of the basic state (46) in terms of the small parameter $\epsilon_1 = \frac{\beta {a_c}^2}{\Gamma}$, the associated linear instability operator [17] [15] may be expanded in terms of the same parameter ϵ_1 :

$$\mathbf{L} \sim \mathbf{L}_0 + \epsilon_1 \mathbf{L}_1 + ... \tag{303}$$

The operator $\mathbf{L}_0$ is associated to the axisymmetric part of the vortex (46) while the operator $\mathbf{L}_1$ contains the non-axisymmetric part which depends on the second harmonic $\exp(2i\theta)$. If linear perturbations are sought in terms of the expansion

$$\tilde{\mathbf{u}} \sim \tilde{\mathbf{u}}_0 + \epsilon_1 \tilde{\mathbf{u}}_1 + ..., \tag{304}$$

the first order term $\tilde{\mathbf{u}}_0$ satisfies $\mathbf{L}_0 \tilde{\mathbf{u}}_0 = 0$ which is similar to equation (256) studied in section 6.3 : it is hence a *combination of neutral inertial waves* propagating along the undisturbed axisymmetric vortex. As demonstrated in the case of the *rotating pipe* and the *Rankine vortex* (section 6.3), there exists a family of such modes that have the *same axial wavenumber and frequency* ω and for which $n_1 - n_2 = 2$. This is in particular true for the sum of two steady helical waves ($n_1 = 1$, $n_2 = -1$, $\omega(k, \pm 1) = 0$)[12]. At second order, the problem reads

$$\mathbf{L}_0 \tilde{\mathbf{u}}_1 = -\mathbf{L}_1 \tilde{\mathbf{u}}_0. \tag{305}$$

[12] It actually works for combinations of other modes such as ($n_1 = 2$, $n_2 = 0$).

Since $\mathbf{L}_1$ contains the second harmonic, the r.h.s of equation (305) can generate, when combined with the helical waves at first order, an inertial helical wave since the resonance relation $n_1 - n_2 = 2$ is satisfied. The effective computation (for details see [17] or [15]) indicates that this mechanism leads to a linear instability. In a way similar to the case of the local elliptic instability, the vorticity component perpendicular to the vortex axis is mainly directed along the stretching axis (figure 23) thereby explaining the origin of the global resonance instability.

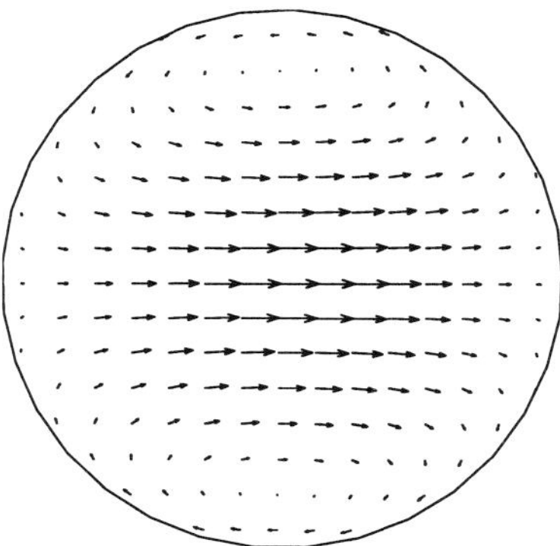

Fig. 23. Instability mode formed by the combination of two opposite helical waves $n = 1$ and $n = -1$ generated by a global resonance mechanism. Only the vorticity component perpendicular to the vortex z-axis is shown. It is predominantly aligned with the stretching y-axis. Courtesy of C.Eloy.

By invoking equation (300) in which the rate of strain γ and vorticity Ω are computed at the center of the vortex, a growth rate is estimated and it is found to be identical to the one computed for the global resonance of two helical modes subjected to the same large scale strain [30]. For instance, this result has been checked for the *deformed Lamb vortex* in which the rate of strain is 2.5 times greater at the center of the vortex than at infinity (section 3.2) [7]. Note that other unstable resonances exist which cannot be found in the context of the parametric instability, such as $n_1 = 2$, $n_2 = 0$. It is worth mentioning that a generalization of this linear instability to *multipolar strain fields* $\Psi_{strain} \sim \frac{r^p}{p} \sin(p\theta)$ is possible for $p = 3$ and $p = 4$ [94]. Finite strain effects do not fundamentally alter the weak strain results, as shown in the study of the linear instability of the elliptic patch (40) [93]. The size of resonance regions and the maximum growth rate simply increase in magnitude. The weak strain approximation still provides good estimates for growth rates.

The *nonlinear saturation* of the elliptic instability has been studied in the framework of amplitude equations : saturation occurs *via* a phase shift effect [31] which implies that vorticity and stretching are no more aligned. In this saturated state, the vortex core axis is periodically distorted. However it is experimentally observed that this flow generally undergoes a *violent collapse* that occurs as in the case of inertial wave forcing (see discussion at the end of section 6.3).

The *elliptic instability* is *generic* : it originates from various effects e.g. interactions with other vortices or boundaries. In shear flows it also appears as a *secondary instability* of the array of vortices arising from the primary instability. Direct numerical simulations [95] performed with an elliptic vortex confined by walls or periodic boundaries compare fairly well with the above local analysis.

8 Swirling jet instability

When the azimuthal velocity $U_\theta(r)$ and axial velocity $U_z(r)$ are both present, new instability features appear that differ from that of a jet or of a two dimensional vortex. For instance, the addition to a basic *unstable jet*, of a weak *centrifugally stable* azimuthal velocity component may *considerably increase the growth rate* of the instability waves. Furthermore it introduces a distinction between positive and negative azimuthal wavenumbers n.

The *inviscid* instability analysis of the basic steady flow (69) is similar to that performed for a two-dimensional axisymmetric vortex. The velocity field is still invariant with respect to any translation or rotation along the z-axis and Fourier modes in z and θ are hence decoupled. A solution of the form (200)-(201) is thus introduced in the linearized Euler equations. When derivatives with respect to time (resp. axial z and angular θ) variables are replaced by $-i\omega$ (resp. ik and in), the linear partial differential equations become a simpler system of ordinary differential equations

$$-i(\omega - n\frac{U_\theta}{r} - kU_z)\tilde{u} - 2\frac{U_\theta}{r}\tilde{v} = -\frac{d\tilde{p}}{dr}, \tag{306}$$

$$-i(\omega - n\frac{U_\theta}{r} - kU_z)\tilde{v} + (\frac{dU_\theta}{dr} + \frac{U_\theta}{r})\tilde{u} = -\frac{in\tilde{p}}{r}, \tag{307}$$

$$-i(\omega - n\frac{U_\theta}{r} - kU_z)\tilde{w} + \frac{dU_z}{dr}\tilde{u} = -ik\tilde{p}, \tag{308}$$

$$\frac{d\tilde{u}}{dr} + \frac{\tilde{u}}{r} + \frac{1}{r}in\tilde{v} + ik\tilde{w} = 0. \tag{309}$$

If the azimuthal velocity U_θ happens to be negligible, one recognizes the *generalized Kelvin-Helmholtz instability* problem for jets, due to the shear component dU_z/dr. On the contrary, if this same shear term (second term on the l.h.s. of equation (308)) is omitted, the above system become quite close to (202)-(204) pertaining to the instability of two-dimensional axisymmetric vortices, provided that the term $(\omega - n\frac{U_\theta}{r})$ be replaced by $(\omega - n\frac{U_\theta}{r} - kU_z)$. This suggests that system (306)-(309) may be regarded as describing a *generalized form of centrifugal instability*. Indeed, in the small wavelength approximation, it can be shown [32], using a local system of coordinates, that this analogy is pertinent.

By eliminating the variables $\tilde{v}$, $\tilde{w}$, $\tilde{p}$ from system (306)-(309), one obtains the so-called *Howard-Gupta* equation [96] for the radial perturbation field $\tilde{u}(r)$

$$F_1^2 \frac{d}{dr}[F_2 \frac{d\tilde{u}}{dr} + F_2 \frac{\tilde{u}}{r}] - F_3 \tilde{u} = 0, \tag{310}$$

with

$$F_1(r) \equiv (\omega - n\frac{U_\theta}{r} - kU_z), \ F_2(r) \equiv \frac{r^2}{n^2 + k^2 r^2}, \tag{311}$$

$$F_3(r) \equiv F_1^2 + rF_1 \frac{d}{dr}[\frac{F_2}{r}(\frac{dF_1}{dr} - 2n\frac{U_\theta}{r^2})] - 2k\frac{F_2 U_\theta}{r^2}[kr(\frac{dU_\theta}{dr} + \frac{U_\theta}{r}) - n\frac{dU_z}{dr}]. \tag{312}$$

Boundary conditions at $r = 0$ and $r = \infty$ for the velocity and pressure fields remain given by (206), (210), (211), as in the two-dimensional vortex case. These relations completely define the *eigenvalue problem* leading to the determination of the dispersion relation for infinitesimal waves. Only the main results of the *temporal* linear analysis are summarized here (for additional details, the reader is referred to [97]).

- A *necessary* condition for *instability* with respect to *axisymmetric* modes $n = 0$ reads

$$\Phi(r) < \frac{1}{4}(\frac{dU_z}{dr})^2, \tag{313}$$

where $\Phi(r) = \frac{1}{r^3}\frac{d(rU_\theta)^2}{dr}$ is the *Rayleigh discriminant*. This inequality extends the Rayleigh criterion (212) pertaining to the centrifugal instability of two-dimensional vortices.

- A *sufficient* condition for *instability* with respect to *three-dimensional perturbations* reads [32] [98]

$$U_\theta \frac{d\Omega_\theta}{dr}[\frac{d\Omega_\theta}{dr}\frac{d(r^2\Omega_\theta)}{dr} + (\frac{dU_z}{dr})^2] < 0, \tag{314}$$

where Ω_θ exceptionally denotes the angular velocity U_θ/r instead of the azimuthal vorticity.

Aside from these general criteria, results are mostly obtained by numerical analyses of the *Batchelor vortex* (section 3.3) . The usual jet instability for $q_S = 0$ is modified by a small addition of swirl. Positive azimuthal wavenumbers $n > 0$ are stabilized. For instance, the mode $n = 1$ mode which is unstable for $q_S = 0$, becomes stable for a very small amount of swirl $q \sim 0.0739$ [99] [100]. *A contrario*, the instability of negative azimuthal wavenumbers $n < 0$ is enhanced by swirl. When $q > q_c \sim 1.5$, the swirling jet is completely stabilized [99]. Within the unstable regime $0 < q < q_c$, the maximum growth rate increases with azimuthal wavenumber n. No inviscidly axisymmetric $n = 0$ unstable modes have been reported.

Paradoxically, *viscosity* generates instability modes [100] [101]. These viscous modes however possess smaller growth rates compared to their inviscid counterparts. Finally note that an ***absolute–convective*** transition has been identified both in the inviscid [16] and viscous case [6] [104]. This transition is thought to be linked to the *vortex breakdown* phenomenon [58]. The connection between this brutal event and vortex instability is neither ascertained nor completely dismissed : the only clear point is that the *helical waves* observed downstream of the breakdown point are due to the instability of the mean flow field.

9 Instability and Stretching

The presence of *stretching* may affect the instability processes occurring in a vortex. However such a phenomenon is difficult to consider even in the simplest case of the *axisymmetric* Burgers vortex (84) and only partial results are presently available. For the *Burgers vortex*, a global energy analysis [105] indicates that no finite critical viscosity is found, to ensure *global monotonic stability* : for any given viscosity, an admissible perturbation may be found the energy of which is not monotonically decreasing. For small Reynolds numbers, the stability of two-dimensional perturbations, i.e. with no dependence along the vortex axis, has been obtained [60]. This result has thereafter been extended numerically, again for two-dimensional stability, to higher Reynolds numbers [106]. Finally it has been proved [107] that the discrete part of the temporal spectrum is only associated with two-dimensional perturbations : *no unstable three-dimensional* modes $k \neq 0$ can then arise from the discrete part of the spectrum.

The *elliptic instability* is modified by the presence of stretching. This result has been shown for a generic model, i.e. an *unbounded time-dependent linear flow* [33], and later on for the *nonaxisymetric* Burgers vortex [7].

Let us briefly mention the effect of stretching on unbounded linear flows. The analysis follows a method based on the *parametric instability* of Kelvin waves as in section 7. The amplification or attenuation of vorticity by axial stretching is illustrated by examining the superposition of a time-dependent uniform axial vorticity field (15) and a three-dimensional uniform strain (14). In such a case, vorticity is time-dependent and evolves according to (16). An extension of the method introduced in section 4 indicates that the axial wavenumber k_3 in the stretching direction is no longer constant but exponentially decreases towards zero : perturbations are thus more and more homogeneous along the x_3-axis. Moreover, the plane wavevector (k_1, k_2) increases in magnitude and rotates more and more rapidly around the x_3-axis : the angle θ_0 of $\mathbf{k}(t)$ with the x_3-axis thus increases towards $\pi/2$ (figure 24). Since the elliptic instability is only active in the shaded area of the figure, perturbations only lie within the unstable region during a finite time interval. The width of this unstable domain is proportional to γ/Ω and the rate at which the angle increases is $O(1/\gamma)$. This time interval

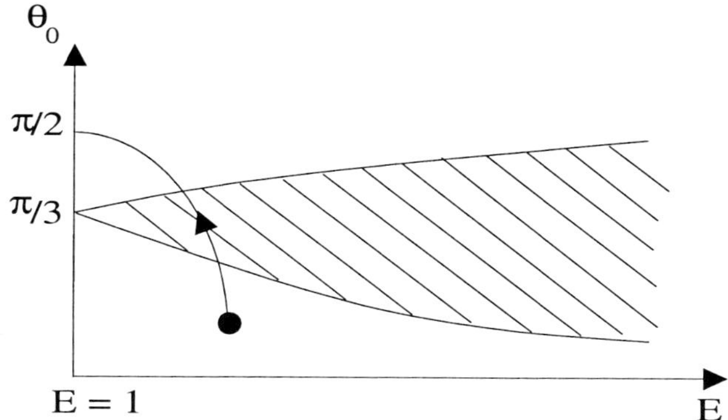

Fig. 24. Parametric instability of Kelvin waves in the presence of stretching : trajectory of Kelvin wave in (E, θ_0) plane. Hatched area determines the domain of pure elliptic instability.

consequently scales as $O(1/\Omega)$. The amplitude is inviscidly amplified with a growth rate $O(\gamma/\Omega)$. Therefore an exponential factor of the form $\exp(B\frac{\gamma}{\Omega^2})$, where B is a constant, is expected for the total gain during the transient phase within the unstable region : heuristically, the basic flow is *stable* if $\frac{\gamma}{\Omega^2} << 1$ and *unstable* if $\frac{\gamma}{\Omega^2} >> 1$. *Axial stretching* may hence suppress the *elliptic instability.* The case of the Burgers vortex [7] is quite similar : it is based on *the global resonance* of inertial waves described in section 7.

Finally the *Batchelor vortex* subjected to *axial stretching* has also been recently studied *via* numerical investigation [108] of equations (98) and (99). A *singularity* has been found that leads to blow-up in finite time. It is also possible to show how stretching and the three-dimensional swirling jet instability can cooperate to further destabilize the Batchelor vortex [109].

10 Conclusion

Vortex dynamics is still a very active field of research as evidenced by the collection of papers in the present volume [41]. The birth and death of vortices clearly involve processes that are very much connected to the global problem of turbulence. In this review, we have explored various possibilities : birth can arise through shear layer instabilities whether of the stretched or unstretched Kelvin-Helmholtz variety or directly through a hyperbolic instability. Death of vortices may also occur in very different ways : centrifugal instability, elliptic instability followed by a secondary inertial wave collapse, swirling jet instability or an interplay of all these effects coupled with stretching.

In order to build good turbulence models, other vortex structures and their stability should deservedly be studied : spiral vortices [16], conical flows [110] [111] or other non-columnar configurations [112]. Additional effects may also be significant in the physics of natural phenomena : propagation of nonlinear waves [113], Coriolis forces [114] [115] [116], ground interactions as in tornadoes [110] [111] and finally the two-way coupling between the rate of strain tensor and the vorticity. Furthermore, single vortex dynamics can be influenced by effects that cannot be described by the incompressible Euler or Navier-Stokes equations : for instance, compressibility [117], thermal or stratification effects. Going back to pure hydrodynamics, other topics of interest, not explicitly considered in this paper, might include the interaction of multiple vortices, secondary instabilities (see for instance [118]) and transient effects. Catastrophic events observed in vortex flows such as vortex breakdown or the collapse of resonant inertial waves are still lacking a proper or complete explanation. Finally vortex control is still in its infancy as shown by the desperate attempts of aircraft manufacturers to destroy trailing vortices ! The above list is far from complete but it suggests that we shall still be swirling around the problem of turbulence for some time !

11 Acknowledgments

Gratuitousness is yet one of the few valuable items which cannot be merchandized. It is thus a pleasure to express my gratitude to Patrick Huerre for his free, accurate and diligent reading of the first, second .. drafts of this manuscript. His efforts greatly improved this review. I would like to thank Christophe Eloy for providing some of the figures of his PhD dissertation. Thanks to Agnès Maurel for helping me so joyfully in drawing many of the figures ! Finally thanks to both Agnès Maurel and Philippe Petitjeans for giving me the opportunity to write this paper.

References

1. Poincaré H., *Théorie des tourbillons.* Editions J.Gabay (1990).
2. Batchelor G.K., *An Introduction to Fluid Dynamics.* Cambridge University Press (1967).
3. Acheson D.J., *Elementary Fluid Dynamics.* Clarendon Press Oxford (1990).
4. Saffman P.G., *Vortex Dynamics.* Cambridge University Press (1992).
5. Tanaka M. and Kida S., "Characterization of vortex tubes and sheets" Phys. Fluids A **5**, 2079-2082 (1993).
6. Cadot O., Douady S. and Couder Y., "Characterization of the low pressure filaments in three-dimensional turbulent shear flow." Phys. Fluids **7**, 630-646 (1995).
7. Villermaux E., Sioux B. and Gagne Y. "Intense vortical structures in grid-generated turbulence." Phys. Fluids **7**, 2008-2013 (1995).
8. Vincent A. and Meneguzzi M., "The spatial structure and statistical properties of homogeneous turbulence." J. Fluid Mech. **225**, 1-25 (1991).
9. Jiménez J., Wray A. A., Saffman P. G. and Rogallo R. S., "The structure of intense vorticity in homogeneous isotropic turbulence." J. Fluid Mech. **255**, 65-90 (1993).

10. Jiménez J. and Wray A. A., "On the characteristics of vortex filaments in isotropic turbulence."J. Fluid Mech. **373**, 255-285 (1998).
11. Pullin D.I. and Saffman P.G., "Vortex dynamics in turbulence."Ann.Rev.Fluid Mech. **30**, 31-51 (1998).
12. Hatakeyama N. and Kambe T., "Statistical laws of random strained vortices in turbulence."Phys.Rev.Lett. **79**, 1257-1260 (1997).
13. Misra A. and Pullin D.I., "A vortex based subgrid stress model for large eddy simulation."Phys. Fluids **9**, 2443-2454 (1997).
14. Drazin P.G. and Reid W.H., *Hydrodynamic Stability*. Cambridge University Press (1981).
15. Huerre P. and Rossi M., *Hydrodynamic instabilities in open flows.* in Hydrodynamics and Nonlinear Instabilities, pp. 81-294, Editors Godrèche C., Manneville P.; Cambridge University Press (1998).
16. Lundgren T.S., "Strained spiral vortex model for turbulent fine structure. " Phys. Fluids **25**, 2193-2203 (1982).
17. Gibbon J.D., Fokas A.S. and Doering C.R., "Dynamically stretched vortices as solution of the 3D Navier-Stokes equations." Physica D **132**, 497-510 (1999).
18. Donaldson C.D. and Sullivan R.D., "Behaviour of solutions of the Navier-Stokes equations for a complete class of three-dimensional viscous vortices."In Proc.Heat Transfer and Fluid Mech. Inst. Stanford University (1960).
19. Bellamy-Knights P.G., "An unsteady vortex solution of the Navier-Stokes equations. " J. Fluid Mech. **41**, 673-687 (1970).
20. Alekseenko S.V., Kuibin P.A., Okulov V.L. and Shtork S.I. , "Helical vortices in swirl flow." J.Fluid Mech. **382**, 195-243 (1999).
21. Kerr O.S. and Dold J.W., "Periodic steady vortices in a stagnation-point flow." J. Fluid Mech. **276**, 307-325 (1994).
22. Corcos G.M. and Lin S.J., "The mixing layer: deterministic models of a turbulent flow. Part 2 . The origin of the three-dimensional motion." J. Fluid Mech.**139**, 67-95 (1984).
23. Lin S.J. and Corcos G.M. , "The mixing layer: deterministic models of a turbulent flow. Part 3 . The effect of plane strain on the dynamics of streamwise vortices." J. Fluid Mech. **141**, 139-178 (1984).
24. Neu J.C., "The dynamics of stretched vortices." J. Fluid Mech. **143**, 253-276 (1984).
25. Passot T., Politano H., Sulem P.L. Angilella J.R. and Meneguzzi M., "Instability of strained vortex layers and vortex tube formation in homogeneous turbulence." J. Fluid Mech. **282**, 313-338 (1995).
26. Maxworthy T., Hopfinger E.J. and Redekopp L.G., "Wave motions on vortex cores." J. Fluid Mech. **151**, 141-165 (1985).
27. Malkus W.V.R., "An experimental study of global instabilities due to tidal (elliptical) distorsion of a rotating elastic cylinder." Geophys. Astrophys. Fluid Dyn. **48**, 123-134 (1989).
28. Gledzer E.B., Dolzhansky F.V., Obukhov A.M. and Ponomarev V.M., "An Experimental and theoretical study of the stability of motion of a liquid in an elliptical cylinder." Izv.Atmos.Ocean.Phys. *11*, 617-622 (1975).
29. Gledzer E.B. and Ponomarev V.M., "Instability of bounded flows with elliptical streamlines." J.Fluid Mech. **240**, 1-30 (1992).
30. Eloy C., "Instabilité multipolaire de tourbillons." Thèse de l'université Aix-Marseille II (2000).
31. Waleffe F., "The three-dimensional instability of a strained vortex and its relation to turbulence." MIT thesis (1989).

32. Emanuel K.A., "A note on the stability of columnar vortices." J.Fluid Mech. **145**, 235-238 (1984).
33. Le Dizès S., Rossi M. and Moffatt H.K., "On the three-dimensional instability of an elliptical vortex subjected to stretching." Phys. Fluids **8**, 2084-2090 (1996).
34. Eloy C. and Le Dizès S., "Three-dimensional instability of Burgers and Lamb-Oseen vortices in a strain field." J.Fluid Mech. **378**, 145-166 (1999).
35. Leweke T. and Williamson C.H.K., "Cooperative elliptic instability of a vortex pair." J. Fluid Mech. **360**, 85-119 (1998).
36. Sipp D., "Instabilités dans les écoulements tourbillonnaires." Thèse de l'Ecole Polytechnique (1999).
37. Bayly B.J., Orszag S.A. and Herbert T., "Instability mechanisms in shear-flow transition.", Ann. Rev. Fluid Mech. **20**, 359-39 (1988).
38. Leweke T. and Williamson C.H.K., "Three-dimensional instabilities in wake transition." Eur. J. Mech. B/ Fluids **17**, 571-586 (1998).
39. Green S.I, *Fluid Vortices.* Ed. Green S.I, Kluwer Academic Publishers (1995).
40. Le Dizès S. (Ed), *Dynamics and Statistics of Concentrated Vortices in Turbulent Flow.*, Euromech Colloquium 364, Eur. J. Mech. B / Fluids **17**, 4 (1998).
41. Maurel A. and Petitjeans P. (Eds), *Structure and Dynamics of Vortices.*, Lecture notes in Physics. Springer Verlag (2000).
42. Ohkitani K. and Kishiba S., "Nonlocal nature of vortex stretching in an inviscid fluid." Phys. Fluids **7**, 411-421 (1995).
43. Tsinober A., "Is concentrated vorticity that important ?" Eur. J. Mech. B / Fluids **17**, 4, 421-449 (1998).
44. Kida S. and Miura H., "Analysis of vortical structures." Eur. J. Mech. B / Fluids **17**, 4, 471-487 (1998).
45. Verzicco R., Jimenez J. and Orlandi P., "On steady columnar vortices under local compression." J.Fluid Mech. **299**, 367-388 (1995).
46. Moffatt H. K., Kida S. and Ohkitani K., "Stretched vortices- the sinews of turbulence; large-Reynolds-number asymptotics." J. Fluid Mech. **259**, 241-264 (1994).
47. Ting L. and Tung C., "Motion and decay of a vortex in a nonuniform stream." Phys. Fluids **8**, 1039-1051 (1965).
48. Craik A.D.D. and Criminale W.O., "Evolution of wavelike disturbances in shear flows: a class of exact solutions of the Navier-Stokes equations." Proc.R. Soc. Lond. A. **406**, 13-26 (1986).
49. Craik A.D.D. and Allen H.R., "The stability of three-dimensional time-periodic flows with spatially uniform strain rates." J.Fluid Mech. **23**, 613-627 (1992).
50. Foster G.K. and Craik A.D.D., "The stability of three-dimensional time-periodic flows with ellipsoidal stream surfaces." J.Fluid Mech. **324**, 379-391 (1996).
51. Rossi L.F. and Graham-Eagle J., "On the existence of two-dimensional, localized, rotating self-similar vortical structures." Preprint (2000).
52. Dritschel D., "On the persistence of non-axisymmetric vortices in inviscid two-dimensional flows." J.Fluid Mech. **371**, 141-155 (1998).
53. Kida S., "Motion of an elliptic vortex in a uniform shear flow." J. Phys.Soc. Japan **50**, 3517-3520 (1981).
54. Moore D.W. and Saffman P.G., "The instability of a straight vortex filament in a strain field." Proc. R. Soc. Lond. A. **346**, 413-425 (1975).
55. Jiménez J., Moffatt H.K. and Vasco C., "The structure of the vortices in freely decaying two-dimensional turbulence." J. Fluid Mech. **313**, 209-222 (1996).
56. Melander M.V. and Hussain F., "Core Dynamics on a vortex column." Fluid Dynamics Research **13**, 1-37 (1994).

57. Batchelor G.K., "Axial flow in trailing line vortices ." J.Fluid Mech. **20**, 645-658 (1964).
58. Leibovitch S., "Vortex stability and breakdown : survey and extension." AIAA J. **22**, 1192-1206 (1984).
59. Khomenko G. and Babiano A., "Quasi-three-dimensional flow above the Ekman Layer." Phys.Rev.Lett. **83**, 1, 84-87 (1999).
60. Robinson A.C. and Saffman P.G., "Stability and structure of stretched vortices." Stud. Appl. Math. **70**, 163-81(1984).
61. Moffatt H.K., "Vortices subjected to non-axisymmetric strain -unsteady asymptotic evolution." in Asymptotic Modelling in Fluid Mechanics, 29-35, Springer Verlag (1994).
62. Landman M.J., "On the generation of helical waves in circular pipe flow." Phys. Fluids **2**, 738-747 (1990).
63. Dritschel D., "Generalized helical Beltrami flows in hydrodynamics and magnetohydrodynamics." J.Fluid Mech. **222**, 525-541 (1991).
64. Kelvin (Lord), "Stability of fluid motion: rectilinear motion of viscous fluid between two parallel plates." Phil. Mag. **24**, 188-196 (1887).
65. Farrell B.F. and Ioannou P.J., "Generalized stability theory. Part I: autonomous operator." J. Atm. Sci. **53**, 2025-2040 (1996).
66. Craik A.D.D., "The stability of unbounded two- and three-dimensional flows subject to body forces : some exact solutions." J.Fluid Mech. **198** 275-292 (1989).
67. Cambon C., Teissedre C. and Jeandel D., "Étude d'effets couplés de déformation et de rotation sur la turbulence homogène." J.Méc.Th.Appl. **4**, 629-657 (1985).
68. Lagnado R.R., Phan-Thien N. and Leal L.G., "The stability of two-dimensional linear flows." Phys. Fluids **27**, 1094-1101 (1984).
69. Andreotti B., "Action et réaction entre étirement et rotation : du laminaire au turbulent." Thèse Paris VII (1999).
70. Taylor G.I. "The formation of emulsions in definable fields of flow." Proc. R.Soc.Lond.A **146**, 501-523 (1934).
71. Beronov K.N and Kida S., "Linear two-dimensional stability of a Burgers vortex layer." Phys. Fluids **8**, 1024-1035 (1996).
72. Swinney H.L and Gollub J.P. (Eds) *Hydrodynamic instabilities and the transition to turbulence.* Springer Verlag (1981).
73. Bayly B.J., "Three-dimensional centrifugal-type instability in an inviscid two-dimensional flow." Phys. Fluids **31**, 56-64 (1988).
74. Kelvin (Lord), "Vibrations of a columnar vortex." Phil. Mag. **10**, 155-168 (1880).
75. Moore D.W. and Saffman P.G., "The motion of a vortex filament with axial flow." Phil. Trans. R. Soc. Lond. A. **272**, 403-429 (1972).
76. Callegari A.J.and Ting L., "Motion of a curved vortex filament with decaying vortical core and axial velocity." SIAM. J. Appl. Math **35**, 148-175 (1978).
77. Fukumoto, Y. and Miyazaki, T., "Three-dimensional distorsions of a vortex filament with axial velocity." J.Fluid Mech. **222** , 369-416 (1991).
78. Kida S., "A vortex filament moving without change of form." J.Fluid Mech. **112**, 397-409 (1981).
79. Arendt S., Fritts D.C. and Andreassen O., "The initial value problem for Kelvin vortex waves." J.Fluid Mech. **344**, 181-212 (1997).
80. Fultz, D., "A note on the overstability of the elastoid-inertia oscillations of Kelvin, Solberg and Bjerknes." J. Met. **16** , 199-208 (1959).
81. McEwan A.D., "Inertial oscillations in a rotating fluid cylinder." J.Fluid Mech. **40**, 603-640 (1970).

82. Manasseh J.J., "Breakdown regimes of inertia waves in a precessing cylinder." J. Fluid Mech. **243**, 261-296 (1992).
83. Kobine J.J., "Inertial wave dynamics in a rotating and precessing cylinder." J. Fluid Mech. **303**, 233-252 (1995).
84. Arendt S., Fritts D.C. and Andreassen O., "Kelvin twist waves in the transition to turbulence." Eur. J. Mech. B / Fluids **17**, 595-604 (1998).
85. Samuels D.C., "A finite length instability of vortex tubes."Eur. J. Mech. B / Fluids **17**, 4, 587-594 (1998).
86. Lifschitz A. and Fabijonas B., "A new class of instabilities of rotating fluids.", Phys. Fluids **8**, 2239-2241 (1996).
87. Kerswell R.R., "Secondary instabilities in rapidly rotating fluids : inertial wave breakdown." J.Fluid Mech. **382**, 283-306 (1999).
88. Mason D.M. and Kerswell R.R., "Nonlinear evolution of the elliptical instability : an example of inertial breakdown." J.Fluid Mech. **396**, 73-108 (1999).
89. Pierrehumbert R.T., "Universal short-wave instability of two-dimensional eddies in an inviscid fluid." Phys.Rev.Lett. **57**, 2157-2160 (1986).
90. Bayly B.J., "Three-dimensional instability of elliptical flow.", Phys. Fluids **57**, 2160-2163 (1986).
91. Landman M.J. and Saffman P.G., "The three-dimensional instability of strained vortices in a viscous fluid." Phys. Fluids **30**, 2339-2342 (1987).
92. Waleffe F., "On the three-dimensional instability of strained vortices." Phys. Fluids A **2**, 76-80 (1990).
93. Robinson A.C. and Saffman P.G., "Three-dimensional stability of an elliptical vortex in a straining field. " J.Fluid Mech. **142**, 451-466 (1984).
94. Le Dizès S. and Eloy C., "Short-wavelength instability of a vortex in a multipolar strain field." Phys. Fluids **11**, 500-502 (1999).
95. Lundgren T.S. and Mansour M.N., "Transition to turbulence in an elliptic vortex." J. Fluid Mech. **307**, 43-62 (1996).
96. Howard L.N. and Gupta A.S., "On the hydrodynamic and hydromagnetic stability of swirling flow." J.Fluid Mech. **14**, 463-476 (1962).
97. Ash R.L. and Khorrami M.R., "Vortex stability." In Fluid Vortices, ed. Green S.I, Chap. VIII, 317-372 Kluwer (1995).
98. Leibovitch S. and Stewarston K., "A sufficient condition for the instability of columnar vortices." J. Fluid Mech. **126**, 335-356 (1983).
99. Lessen M., Singh P.J. and Paillet P., "The stability of a trailing line vortex. Part 1 Inviscid theory" J. Fluid Mech. **63**, 753-763 (1974).
100. Mayer E.W. and Powell K.G., "Viscous and inviscid instabilities of a trailing line vortex." J. Fluid Mech. **245**, 91-114 (1992).
101. Khorrami M.R., "On the viscous modes of instability of a trailing line vortex." J.Fluid Mech. **225**, 197-212 (1991).
102. Olendraru C., Sellier A., Rossi M. and Huerre P., "Inviscid instability of the Batchelor vortex: absolute-convective transition and spatial branches." Phys. Fluids **11**, 1805-1820 (1999).
103. Delbende I., Chomaz J.-M. and Huerre P., "Absolute/ convective instabilities in the Batchelor vortex: a numerical study of the linear impulse response." J. Fluid Mech. **355**, 229-254 (1998).
104. Olendraru C., "Etude spatio-temporelle de jets et sillages tournants." Thèse de l'Ecole Polytechnique (1999).
105. Leibovitch S. and Holmes P., "Global stability of a vortex subjected to stretching." Phys. Fluids **24**, 548-549 (1981).

106. Prochazka A. and Pullin D. I., "On the two-dimensional stability of the axisymmetric Burgers vortex." Phys. Fluids **7**, 1788-1790 (1995).
107. Rossi M. and Le Dizès S., "Three-dimensional temporal spectrum of stretched vortices." Phys.Rev.Lett. **78**, 2567-69 (1997).
108. Ohkitani K. and Gibbon J. D., "Numerical study of singularity formation in a class of Euler and Navier-Stokes flows." preprint (2000).
109. Delbende I., Le Dizès S. and Rossi M., "Three-dimensional linear stability of stretched vortices." In preparation.
110. Fernandez-Feria R., Fernandez de La Mora J., Perez-Saborid and Barrero A. "Conically similar swirling flows at high Reynolds number." Q.J.Mech.Appl. Math **52**, 1-53 (1999).
111. Shtern V. and Hussain F., "Collapse, symmetry breaking, and hysteresis in swirling flows." Ann. Rev. Fluid Mech. **31**, 537-566 (1999).
112. Wang S. and Rusak Z., "On the stability of non-columnar swirling flows." Phys. Fluids **8**, 1017-1023 (1996).
113. Leibovitch S. and Kribus A., "Large amplitude wavetrains and solitary waves in vortices." J. Fluid Mech. **216**, 459-504 (1990).
114. Hopfinger E.J. and van Hejist G.J.F., "Vortices in rotating fluids." Ann.Rev. Fluid Mech., **25** 241-289 (1993).
115. Carnevale G.F. et al, "Three-dimensional perturbed vortex tube in a rotating flow."J. Fluid Mech. **341**, 127-163 (1997).
116. Leblanc S., "Instabilités tridimensionnelles dans un fluide en rotation." Thèse de l'Ecole Centrale de Lyon (1997).
117. Bershader D., "Compressible Vortices." in Fluid Vortices. Ed. Green S.I, Kluwer Academic Publishers (1995).
118. Loiseleux Th., "Instabilités dans les jets tournants." Thèse de l'Ecole Polytechnique (1999).

Visualizations of Stretched Vortices

Frédéric Bottausci and Philippe Petitjeans

Laboratoire Physique et Mécanique des Milieux Hétérogènes, UMR CNRS 7636, ESPCI, 10 rue Vauquelin, 75005 Paris, France

Abstract. We study the structure, dynamics, and the instabilities of a stretched vortex in two different experiments: The first experiment generates a vortex by stretching the vorticity of a laminar boundary layer on a plane wall in a water channel. The second experiment generates a vortex between two co-rotating discs with a suction through a hole in the center of each disc. These vortices model the filaments of vorticity of turbulent flow. Visualizations are presented which show the structure and the instabilities of these stretched vortices.

1 Introduction

The analysis of vortex structures and dynamics is very important in fluid mechanic for many reasons. Structures concentrating the vorticity are observed in lots of systems in laminar as well as in turbulent flows [1], [2] It is now understood that these structures play an important role in the intermittence observed in turbulence. In such flows, the filaments of vorticity are generated by thestretching of local vorticity sheets (or a shear). Two experiments devoted to this study are presented in this paper. In each experiment, a vortex is generated by the stretching of initial vorticity. In these experimental set-ups, a structure with all the ingredients of vorticity filaments is isolated in the absence of turbulence. The objective is to extract the fundamental mechanisms at the origin of the dynamics and the instabilities of a stretched vortex. The parameters such as the stretching or the initial vorticity are controlled in order to measure their influence on the vortex dynamics. In these experiments, velocity measurements are performed and the behaviour of the velocities is still under analysis. This work should be published soon. Pressure measurements are also conducted, and are presented in another paper in this book. In this paper, we present some visualizations performed in the two experiments in order to illustrate the subject of this book. The structure as well as some instabilities are pointed out in particular.

1.1 The channel

This experimental set-up is a low-velocity water channel constructed of Plexiglas with well controlled flow. The flow is generated by a constant head pressure tank. A diffuser keeps the flow laminar with a minimum of perturbation. The key elements of the channel are presented in figure 1. In this straight part (60 cm x 12 cm x 7 cm) the flow develops laminar boundary layers with a velocity on the order of a few cm/s. The Reynolds number based on the boundary layer thickness is on the order of a few hundred. At these values of Re, no shear-type instability can be observed. The boundary layers on the lower and upper walls form the initial vorticity. A flow slot, made of 5 superimposed 6 mm diameter holes on each lateral wall is used to stretch the vorticity sheet parallel to the vorticity vector. The total flow rate through the channel is controlled as well as the flow rate through the slots; consequently, the initial vorticity can be controlled (via the thickness of the boundary layer) as well as the stretching. The formation of strong periodic vortices by roll-up of fluid sheets is obtained when the stretching and the main flow are large enough (ref exp in fluids et applied...) [3], [4]. In the experiments described in this paper, the stretching is produced only through the hole near the bottom, so that only one vortex is generated on the bottom (and none on the top). To stabilize and amplify the initial shear, a step of 1.1 cm high is placed before the holes. When the main flow rate is below a threshold, the vortex remains attached to the slots with a slow precessing motion. Above this threshold, a periodic emission of vortices is observed [5].

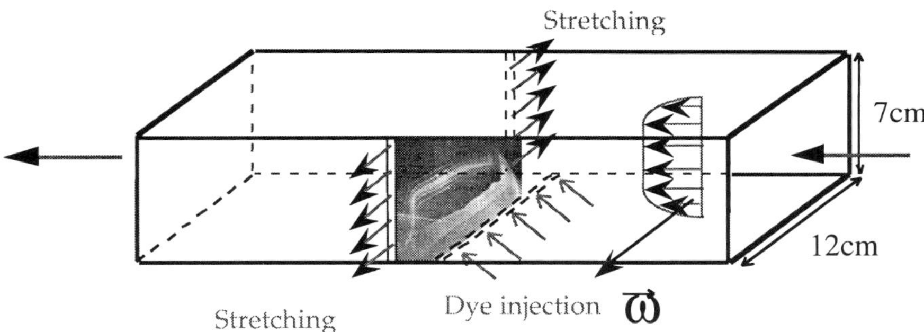

Fig. 1. Experimental set-up

The first visualisation (Fig. 2) shows a cross section of a vortex. Fluorescent dye (fluoresceine) is injected as a sheet just before the vortex (see Fig. 1). A light sheet is produced with an argon laser beam passing through a small cylindrical lens. The vortex being stretched by suction, mass conservation implies a radial velocity towards the axis of the vortex. Combined with the rotation, this gives a spiral picture. The path line describes a logarithmic spiral, leading to a Kolmogorov dimension DK = 1 (i. e. a self-similar structure) [6]. When the flow is not stationary, the path lines are not identical to the streamlines. In that case, a possible alternative is to extract the position of the leading-edge of the dye sheet as a function of time, in order to extract the streamlines. This work has been presented in [7] [5]. In Figure 2, the dye sheet remains extremely thin, although strong velocity gradients are presented. Indeed, the flow is extremely laminar; the maximum velocity Vq is of order 20 cm/s and the diameter of the vortex is of order 5 mm. The core of the vortex can be seen in this picture where the light is much lower due to a smaller concentration of dye because of the stretching in the axial direction (perpendicular to the picture). This technique is well-adapted to follow the position of the axis of the vortex to study its precession. The precession has been measured to be of very low frequency (of order 0.1 Hz) and low amplitude (of order 5 mm). These characteristics, which depend on the stretching and on the other parameters of the experiments, will be described in the future.

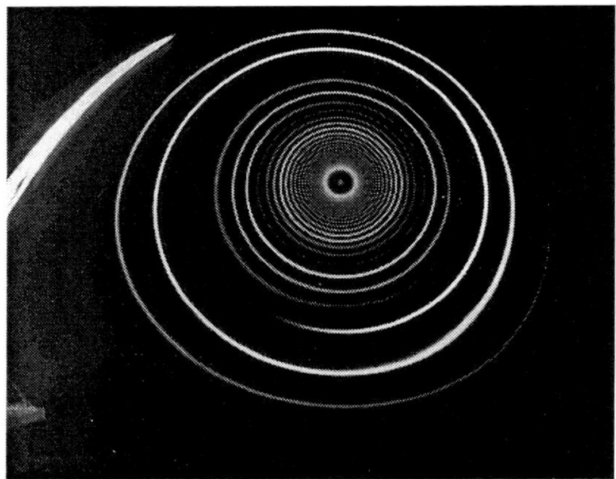

Fig. 2. View of a cross section of a stretched vortex. Fluorescent dye (fluoresceine) is injected upstream, and an argon laser light sheet allows the visualisation of the pathlines

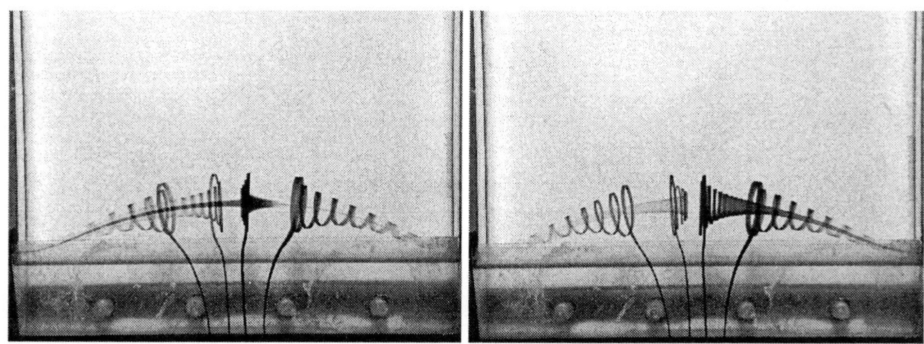

Fig. 3. Visualisation of a stretched vortex by injection of very small jets of dye (with the same initial velocity than the velocity of the flow). The flowrate downstream is zero, and the stretching is 5.7 s^{-1} (corresponding to a flow-rate of 5 l/min at each suction).

Another technique consists of injecting very small dye jets upstream enough to avoid perturbations of the flow, with very small injectors (0.5 mm diameter tubes). The dyes are water coloured by methyl blue or amaranth (red). The density is very close to 1, so that no buoyancy effects are visible in this experiment. The Figure 3 is an example of such a visualisation. In that case, there is no flow downstream, and the total flow-rate goes symmetrically through both suction holes. The vortex has a banana shape and is attached to the suction holes. An important point is to not confound the pathlines visualised in this picture with the streamlines. Indeed, the flow is not stationary because of the precession and because of the instabilities which will be described later. In addition, the pathlines do not show the shape of the vortex: it does not have the long diamond shape observed in the picture with a fat part in the center and a thinner on the sides. Its diameter is almost constant, except at its extremities where it is smaller. The dye jet, that had a circular cross section at the beginning, takes a ribbon shape later when it enters into the velocity field of the vortex, particularly the effect of the stretching is visible. The third dye jet (blue) of Fig. 3-a goes sometimes right, and sometimes left. This means that the stagnation point has an unstable motion along the axis of the vortex, and does not remain fixed. This is connected to the instability that is observed in this experiment, and what is usually known as vortex breakdown . In this text, this pattern is referred to as "bulb" because of its shape. A sequence showing the formation and the evolution of this instability is given Figure 4 in a quite stable case: the two "bulbs" remain more or less at the same position.

Figure 5 is a sketch of this flow. There are three stagnation points along the axis as explained in this Figure. The stable case, without bulbs, would have only one central stagnation point because of the symmetry of this flow.

In the unstable case, presented in Figure 4, two extra stagnation points are located on the axis as shown Figure 5. This case can be itself unstable if one bulb is stronger than the other one. This second instability can be observed in the sequence of Figure 6.

Figure 7 shows a sketch of this instability where the bulb at the left part becomes stronger than the bulb at the right. The central stagnation point is moved to the right until it disappears together with the stagnation point number 3. Then the whole pattern (bulbs) is sucked through the right suction hole, or is destroyed without breaking the vortex. It is the reason why the dye injected at the left (resp. right) part of the vortex can leave through the right (resp. left) suction hole.

Figure 8 is a sequence where the vortex breaks as a turbulent burst. In the experiment shown here, the flow downstream is not zero. The advection of the vortex by the main flow elongates the vortex, decreases the stretching, increases its curvature, and finally breaks it. Work is in progress in order to qualitatively describe this dynamics. The final step before the disintegration is the detachment of the vortex from one of its ends, where the stretching is cut. The flow enters then in the vortex (because of the smaller pressure in its core), and promotes the explosion.

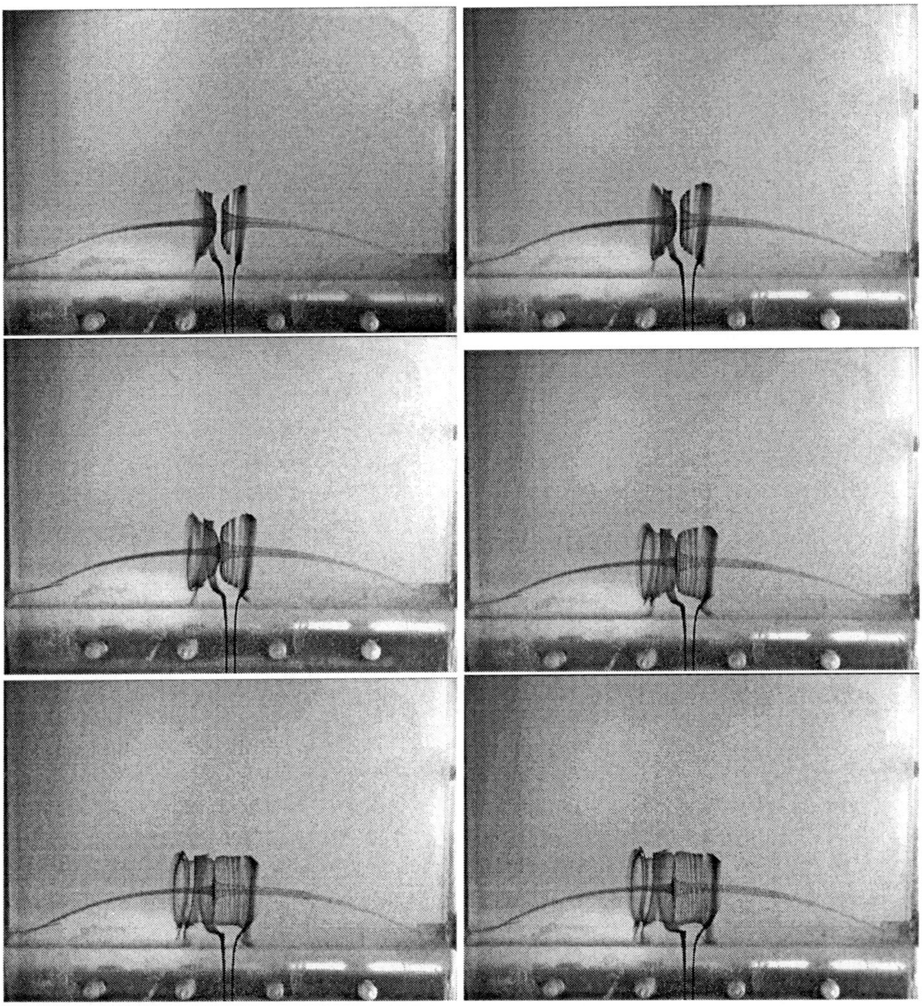

Fig. 4. Sequence of the development of two symmetric bulbs. The time interval between two consecutive images is 0.08 s.

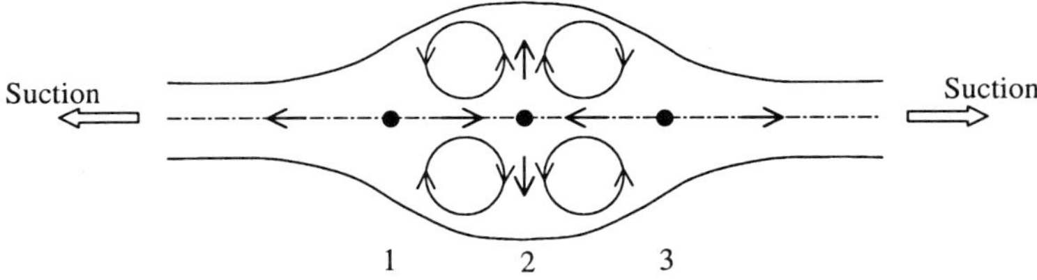

Fig. 5. Sketch of symmetric bulbs.

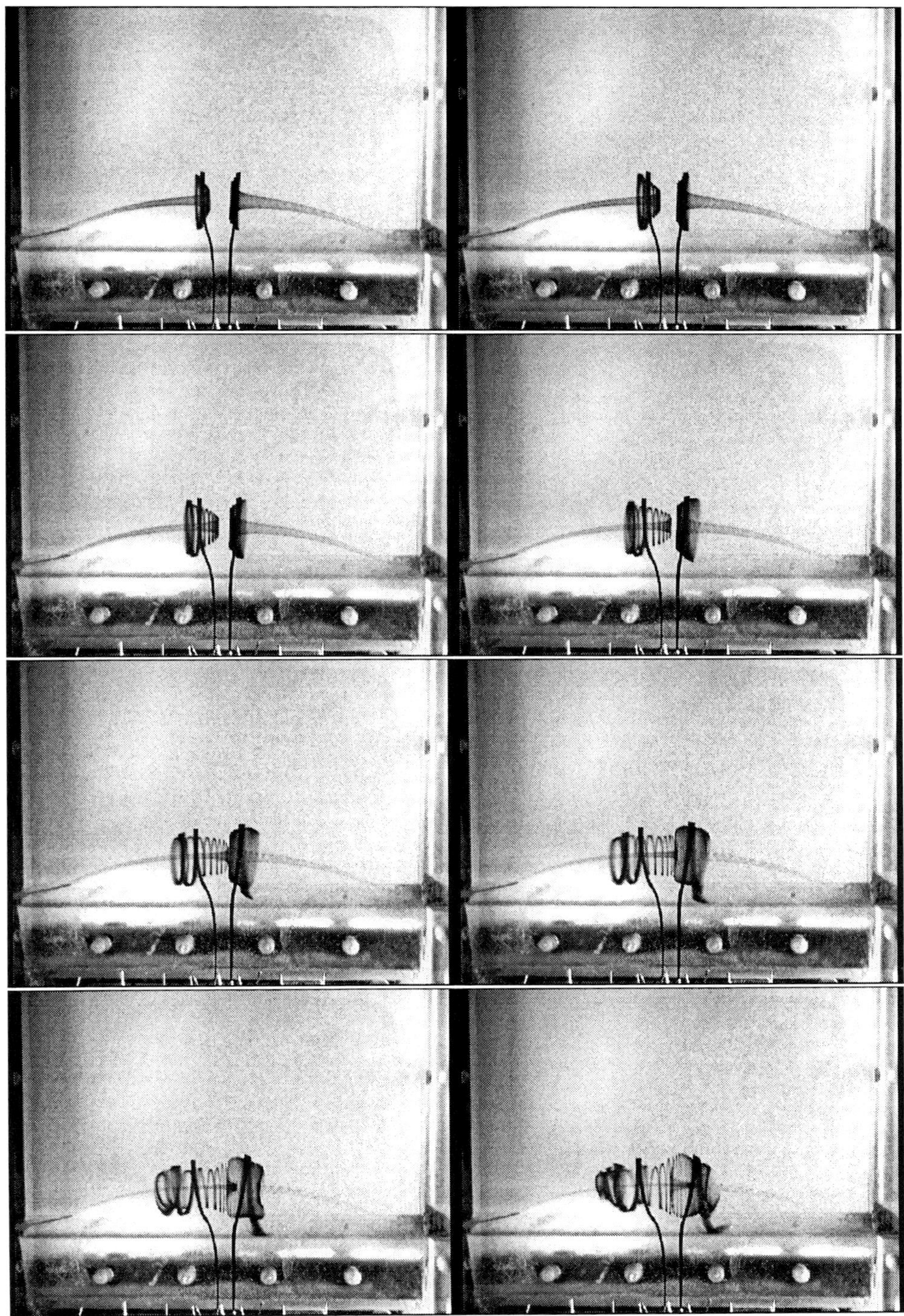

Fig. 6. Sequence of the development of two unstable bulbs showing the displacement of the central stagnation point to the right. The time stop between two consecutive images is 0.08 s.

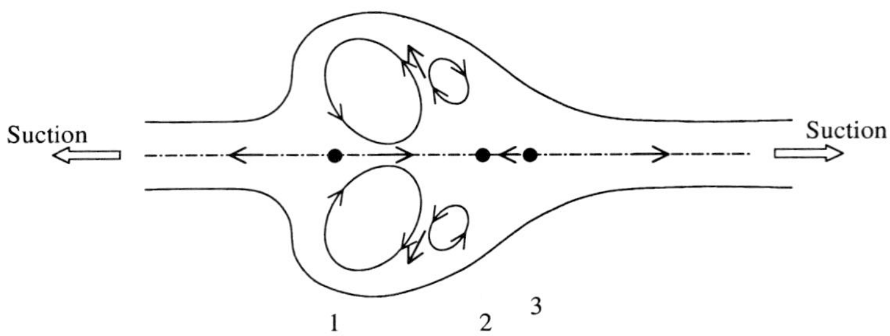

Fig. 7. Sketch of unstable bulbs.

1.2 The discs

The second experiment produces a vortex between 2 discs next to each other, co-rotating, and drilled in the center by a 5 mm diameter hole through which a symmetric suction is applied.

A sketch of the experimental set-up is given Figure 9. The initial vorticity is given by the rotation of the discs. The Ekmann recirculation brings the rotation of the discs in the flow between the discs. The stretching is produced by a symmetric suction, and the flow rate of this suction is measured. The distance h between the water level of the big tank and the water level of the small tank gives the initial pressure drop which produces the stretching. Although the initial flow rate is fixed by the distance h, it can freely vary, particularly with the pressure drop inside the vortex (see Moisy-Petitjeans). The diameters of the discs are 5, 10, or 15 cm, the distance between them is controlled by a motor and can vary from 0 to 20 cm. The rotation of the discs is fixed by another motor between 0 and 1000 rotations per minute. Similar visualisations can be obtained with dye injected with two small injectors of 0.7 mm diameter (Figure 10), although it should be noticed that their presence perturbs the vortex [Pressure measurements in a stretched vortex, F. Moisy and P. Petitjeans, in this book]. The same instabilities as those observed in the first experiment (precession, bulbs, stagnation point instability,...) are found here.

An interesting detection of the vortex by air bubbles is presented in Figure 11. When the vortex is strong enough, the pressure drop in its core is large enough to capture the micro-bubbles of the water. A long bubble can be observed which remains in the vortex core without been absorbed by the suction when the velocity is about 3 m/s. The bubble itself presents some oscillations due to Rayleigh instability .

A train of smaller bubbles can also be observed as shown Figure 12. This train may move along the vortex but keep a wavelength (distance between bubbles) quite constant between the bubbles. This may be due to local pressure minima. Work is in progress in order to characterise this instability.

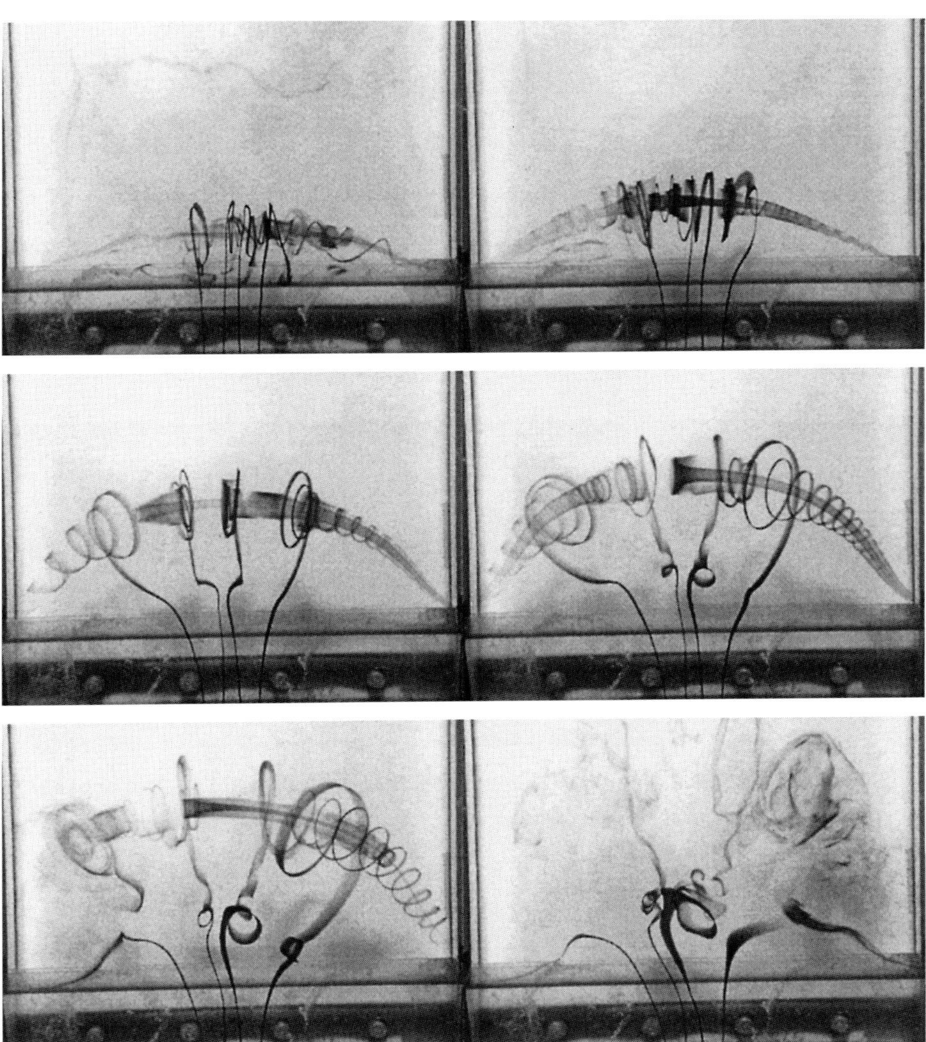

Fig. 8. Sequence of the explosion of a vortex. The flowrate downstream is of order 2 l/min, and the stretching is 5.7 s-1 (corresponding to a flow-rate of 5 l/min at each suction). The time interval between two consecutive images is 0.5 s.

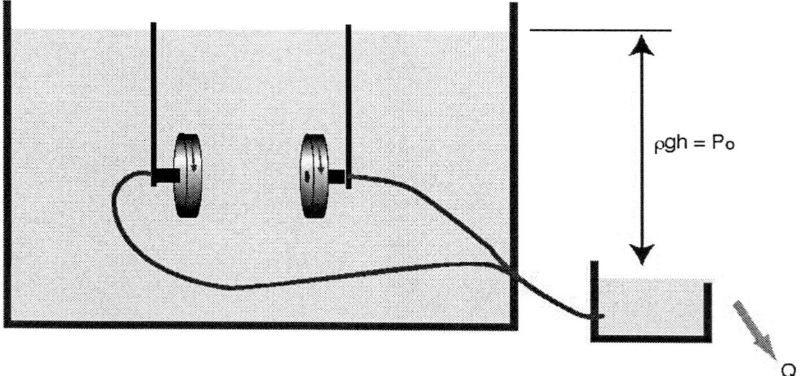

Fig. 9. Experimental set-up.

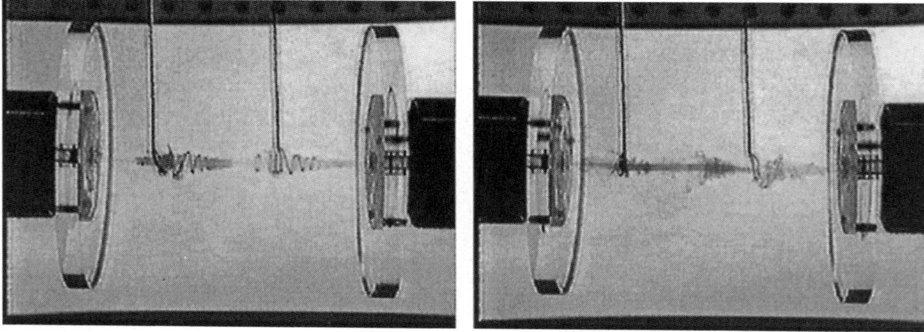

Fig. 10. Example of a view of a vortex generated between two co-rotating discs.

2 Conclusion

The experiments described in this paper allow to isolate a stretched vortex which models the filaments of vorticity of turbulent flows. Indeed, it is believed that these structures play an important role in the intermittency. The complexity of their dynamics is pointed out through visualisations, although lots of care should be taken in their interpretations: the pathlines are not the streamlines. Nevertheless, several instabilities can be observed with dye injection such as the precession of the vortex, vortex breakdown (or bulbs), instability of the stagnation point, destruction as turbulent burst,... Work is in progress on the two experimental set-ups described in this note in order to characterise, to quantify, and to model these instabilities.

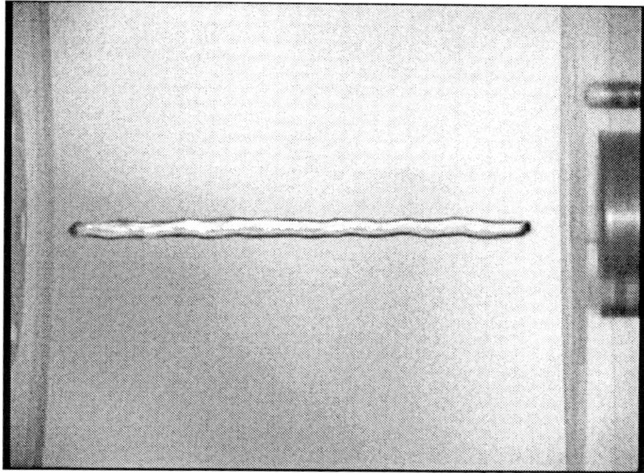

Fig. 11. View of a long air bubble trapped in the vortex core. The deformation of the bubble is due to a Rayleigh instability.

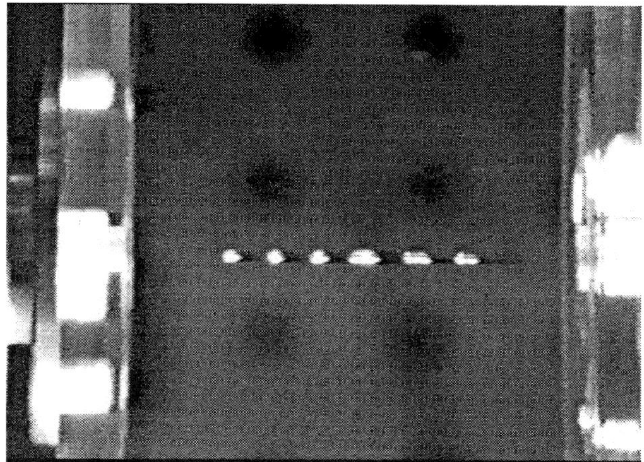

Fig. 12. Train of smaller bubbles trapped in the vortex core. The wavelength is quite constant for few minutes although the whole train oscillates along the vortex axis.

References

1. Cadot O.,Douady D. & Couder Y., *Phys. of Fluids*, **7**(3),630-646 (1995)
2. Siggia, E.D.,*J. Fluid Mech.*, **107**, 375-406 (1991)
3. Petitjeans P., Wesfreid J. E., & Attiach J.C., *Experiments in fluids*, **22**, 351-353 (1997)
4. Petitjeans P. & Wesfreid J. E., *Applied Scient.* (1997)
5. Manneville S., Robres J.-H., Maurel A., Petitjeans Ph. & Fink M., *Phys. Fluids*, **11**, 3380–3389 (1999)

6. Vassilicos J. C. & Brasseur J. G.,*Phys. Rev E*, **54**(1), 467-485 (1996)
7. Petitjeans P., Robres J.H., Wesfreid J.E. & Kevlahan N., *Europ. J. Mech.*, **17**, 549-560 (1998)

Three-Dimensional Stabilization of a Vortex Street in Viscoelastic Liquids

Olivier Cadot[1] and Satish Kumar[2]

[1] Université du Havre, Laboratoire de Mécanique, 25 rue Philippe Lebon BP540 76058 Le Havre CEDEX France
[2] Department of Chemical Engineering, University of Michigan, 2300 Hayward, 3074 H.H. Dow Building, Ann Arbor, MI 48105 USA

Abstract. The local injection of a viscoelastic liquid into the vortex street produced by a circular cylinder drastically modifies its two- and three-dimensional structure. At low injection rates, a stabilization of mode A (3D instability) is observed while the primary wavelength of the vortex street (2D instability) remains identical to that of the Newtonian case. There is also a flattening of the vortices in the street. At higher injection rates, the primary wavelength increases as well. These observations support the stabilization mechanism suggested by the numerical simulations of Kumar and Homsy (S. Kumar and G. M. Homsy: J. Non-Newtonian Fluid Mech. **83**, 251 (1999)).

1 Introduction

The addition of small amounts of macromolecules to a turbulent flow can drastically modify its macroscopic properties as evidenced by phenomena such as drag reduction [1] and inhibition of small scale intermittent structures [2] [3] [4]. The macromolecules give the flow viscoelastic properties and appear capable of modifying its hydrodynamic instabilities. In order to better understand these effects, we investigate a simpler flow: the flow around a circular cylinder at Reynolds numbers in the range 150-250. This flow displays generic two- and three-dimensional instabilities that have been well studied [5], and which are relevant to processes occurring in turbulent flows. The effects of polyethyleneoxide (PEO) additives on the 2D instability have been observed in a channel flow [6] and in a quasi-2D flow [7]. These consist of an increase in the wavelength of the vortex street along with a large delay in the roll-up process of the vortices. In the experiment of Cadot and Lebey [6], the viscoelastic liquid is injected locally through the cylinder and is assumed to introduce a surface tension in the unstable shear layers on each side of the cylinder. It then produces a short wavelength stabilization of the Kelvin-Helmholtz instability [8] which gives rise to the vortex street. The aim of this work is to study the effects of viscoelasticity on the 3D instability of the wake . In the case of a free shear layer, recent numerical simulations of Kumar and Homsy [9] suggest that 3D disturbances are more difficult to excite in viscoelastic liquids than in Newtonian ones. Here, we present experimental results concerning the 3D structure of the vortex street in the case of PEO additives. We will limit our study to the disturbance known as mode A [5] [10] [11].

2 Experimental setup

The experimental setup is similar to that of Cadot and Lebey [6]. A cylinder of 4 mm diameter is pierced with two parallel rows of 67 holes, each having 0.5 mm diameter. These rows are oriented upstream and symmetrically with respect to the flow, which descends vertically at fixed velocity U. In Figs. 1 brought into each end of the cylinder. The injection is controlled with a system containing a peristaltic pump which ensures a continuous and smooth injection [12]. The rate of injection is quantified with an injection parameter we define as $\beta = u/U$, where u is the velocity of the viscoelastic fluid at the exit of the cylinder holes. The injection parameter β is simply the fraction of injected fluid in the vicinity of the cylinder. In the experiments, β varies from 0.01 to 0.6; for larger values, the injected fluid tends to accumulate around the cylinder. Visualization is performed by video recording the flow with a CCD camera. For the views in Figs. 1(a)-(c) and Fig. 2, the flow is illuminated from behind and visualized from the front. For the views in Figs. 1(d)-(f), the flow is illuminated by a light sheet and the injected liquids are coloured with a fluorescent dye.

The viscoelastic liquids are semi-dilute aqueous solutions of PEO Water Solute Resin 303 from Union Carbide.The molecular weight of the WSR 303 is typically $9.10^6 g/mol$. We use 3 concentrations: 250, 600 and 1000 weight parts per million (wppm).All of these solutions show a strong shear thinning behavior and exhibit elastic properties which manifest themselves in a non vanishing first normal stress difference. The overlap concentration is about 300 wppm, hence, our solutions have to be considered as semi-diluted.We have to emphasize that their mechanical properties such as either their elasticity or the degradation are related to the architecture of the polymeric net in the solution rather to the mechanical properties of only one molecule.The consequence is that the relaxation time of the solution is much larger than either the Rouse [15] or the Zimm [16] time, both modeling diluted solutions (see Vlassopoulos and Scholwalter [13] for a detailed comparison of these time scales). From shear viscosity measurements [13], [12], the estimated relaxation time is typically around $0.5s$ at 500 wppm and $1s$ at 1000 wppm.

3 Results

Figure 1(a) and Figure 2 show the 3D instability of the wake at Re = 170 in the case of water injection.

This instability, called mode A , consists of an undulation of the primary Karman rolls just behind the cylinder. This undulation then leads to the formation of sheets which are more or less regularly spaced by a distance Λ. These sheets have their own dynamics and split themselves into two counter-rotating filaments which form loops. In the case of water injection, the amplitude of the injection parameter, β, does not change either the primary instability (i.e., the wavelength of the Karman street) or the undulation wavelength, Λ, of mode A. The injected water thus acts as a passive scalar in the wake. In the case of PEO

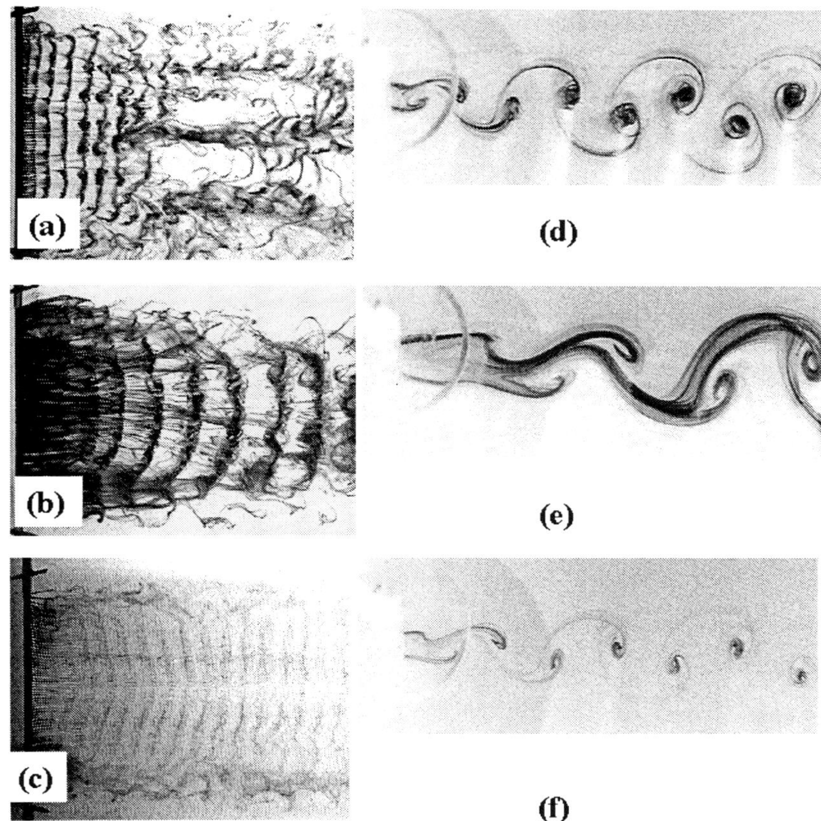

Fig. 1. Visualization of the wake at Re = 170 in the case of water injection and PEO injection at 1000 wppm for different values of the injection parameter β. For pictures (a), (b), (c) the cylinder is vertical on the left and the flow comes from left to right. For pictures (d), (e), (f) the cylinder is on the right and the flow comes from the right to left (a) Water injection and (b) PEO injection at $\beta = 0.5$. (c) PEO injection at $\beta = 0.05$. The close section views (d), (e) and (f) are related to another experiment at Re = 150 and correspond qualitatively to same injection parameters of (a), (b), and (c) respectively.

injection, the structure of the wake becomes significantly different. In particular, mode A is mostly suppressed for all β (Figs. 1(b)-(c)). At large β, this suppression is accompanied with a large wavelength (λ) increase, while at small β the wavelength is the same as in the Newtonian case.

The other pictures of the wake section are related to other experiments at Re = 150. For Figs. 1 (d)-(f), the injection parameters are similar to those of Figs. 1(a)-(c), respectively. We observe clearly that the wavelength increases as β increases and that there is a change in the eddy shape. The eddies are circular in the case of water injection, while they are flattened in the viscoelastic wake (Figs. 1(e)-(f)). Another observation is related to the large amount of solution that remains in the braid regions. This shows that the stretching flow between the vortices is weaker than that of the Newtonian case because it is less efficient at collecting the dye into the vortex core.

Figures 2(b) and (c) show rare cases where mode A was observable during PEO injection. They are compared to the case of water injection under the same experimental conditions (Fig. 2(a)). The wavelength of mode A adapts itself to the wavelength of the primary instability such that the ratio Λ/λ remains constant.

This was checked for 2 different solution concentrations. At 600 wppm the primary wavelength is twice the Newtonian value (Fig. 2(b)), while at 1000 ppm the primary wavelength is three times the Newtonian value (Fig. 2(c)). We did not see any significant effects for concentrations lower than 250 wppm on either the 2D or the 3D instability over the range of injection parameters we examined. We also stress the fact that Cadot and Lebey [6] showed that the observed effects are not related to the higher zero-shear viscosities of the PEO solutions, but are in fact due to their elasticity.

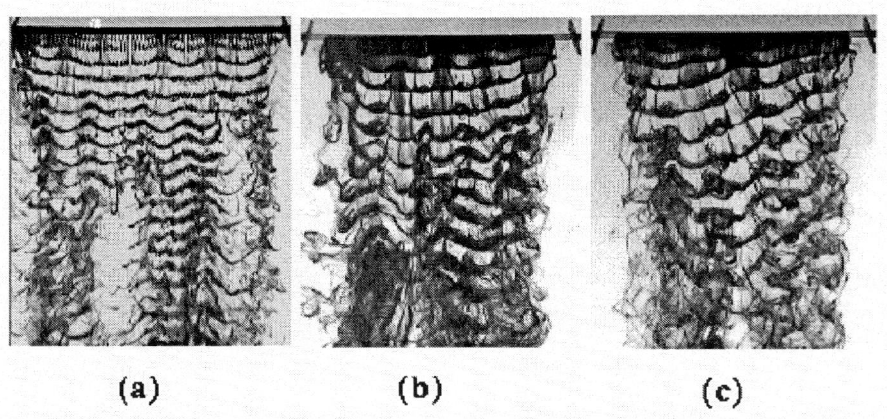

Fig. 2. Visualization of the wake in the case of (a) water injection, and PEO injection at (b) 600 wppm and (c) 1000 wppm. The injection parameter is $\beta = 0.5$ and the Reynolds number is Re $= 170$ for all three cases.

4 Discussion

The first result we discuss is the suppression of mode A by PEO injection. A possible mechanism for this may be related to the modification of the two-dimensional flow since the vortices in the Karman street are flattened and the stretching between them is weakened. These observations are consistent with the simulations of Kumar and Homsy [9]. Their work shows that polymers can modify the 2D flow in free shear layers by resisting the collection of vorticity into concentrated regions (such regions being similar to the vortices of the Karman street.) Their work also suggests that the resulting eddies are flattened and more stable to 3D perturbations.

The second result we discuss concerns the rare appearance of mode A in the viscoelastic case. For all the wavelength increases in the vortex street that we have observed, the ratio between the wavelengths of the 2D and 3D instabilities remains constant. Thus, the characteristic length scale which sets the wavelength of mode A is the separating distance between the vortices of the Karman street . The result brings some evidence to the idea that mode A is due to an instability of the primary rolls rather than to the presence of the cylinder [14].

5 Conclusion

Injection of viscoelastic fluid stabilizes the 3D disturbances in the wake past a circular cylinder. The mechanism may be related to a flattening of the eddies in the vortex street, which would be consistent with numerical simulations of viscoelastic free shear layers [9]. This scenario offers an interesting perspective concerning the inhibition of high vorticity structures (e.g., filaments) in turbulence by polymeric additives. If such structures form through instabilities similar to mode A, then our study suggests that their formation may be inhibited through interference with the associated 2D instability.

References

1. B. A Toms :'Observation on the flow of linear polymer solutions through straight tubes at large Reynolds numbers'. In: *Proceedings of the First International Rheological Congress* (North Holland, Amsterdam 1948) p. II-135
2. O. Cadot, S. Douady, Y. Couder: Phys. Fluids **7**, 630 (1995)
3. D. Bonn, Y Couder,P. H. J. van Dam, S. Douady: Phys. Rev. E **47**, R28 (1993)
4. O. Cadot, D. Bonn, S. Douady: Phys. Fluids **10**, 426 (1998)
5. C. H. K. Williamson: Ann. Rev. Fluid Mech.**28**, 477 (1996)
6. O. Cadot, M. Lebey: Phys. Fluids **11**, 494 (1999)
7. W. I. Goldburg: private communication
8. J. Azaiez, G. M. Homsy: J. Fluid Mech. **268**, 37 (1994)
9. S. Kumar, G. M. Homsy: J. Non-Newtonian Fluid Mech. **83**, 251 (1999)
10. H.-Q.Zang, U. Fey, B. R. Noack, M. Konig, H. Eckelmann: Phys. Fluids **7**, 779 (1995)
11. M. Brede, H. Eckelmann, D. Rockwell: Phys. Fluids **8**, 2117 (1996)
12. O. Cadot, S. Kumar: J. Fluid Mech., submitted
13. D. Vlassopoulos, W. R. Schowalter: J. Rheol. **38**, 1427 (1994)
14. T. Leweke, C. H. K. Williamson: J. Fluid Mech. **360**, 85 (1998)
15. P.E. Rouse: J. Chem. Phys. **21**, 1272 (1953)
16. B. A. Zimm : J. Chem. Phys. **24**, 269 (1956)

Bursting of a Swirling Jet Stemming from a Localized Perturbation

Ivan Delbende[1] and Jean-Marc Chomaz[2]

[1] Université de Paris VI, LIMSI–CNRS, B.P. 133, Bâtiment 508, F-91403 Orsay Cedex, France
[2] LadHyX, École Polytechnique–CNRS, F-91128 Palaiseau Cedex, France

Abstract. The present study is aimed at understanding the mechanisms which lead coherent vortex structures to burst, as observed in fully turbulent flows [2]. A single structure is considered here, modeled by a Batchelor-vortex flow. The linear and non-linear evolution of an initially localized perturbation is followed by DNS. As already known [5], the flow is linearly unstable towards helicoidal disturbances. Vortical helices are found to arise from the saturation of this primary instability. At the same time, the vortex bursts in the neighborhood of the initial perturbation location, where numerous linear instability modes grow in place, the interaction of which is likely to cause rapid vortex dissipation.

1 Introduction

Coherent structures may emerge from fully turbulent flows as long, rather straight, intense vortices. Before these structures dissipate, they may part in several co-rotating helical filaments, as shown experimentally by Cadot *et al.* [2] Getting some insight into the mechanism responsible for this phenomenon is a difficult task, as information on both vortex and surrounding medium is lacking. Cadot *et al.* [2] have measured the pressure signature induced by a passing-by structure. They have shown that the azimuthal component V_θ of velocity is essentially the same as in the Burgers vortex:

$$V_\theta \propto \frac{1 - e^{-r^2}}{r} \; . \tag{1}$$

By contrast, very few studies focus on the axial component V_x of velocity prevailing along such structures. Numerical simulations performed by She *et al.* [6] indicate that V_x *"is significant"* along the vortex. By analogy with atmospheric tornadoes, one might suspect that the axial flow is concentrated within the vortex core, thereby forming a swirling-jet flow, very different from a Burgers vortex (for which the axial flow is uniform in each plane perpendicular to the axis). The axial shear is likely to strongly affect the instability of coherent structures.

The present contribution is aimed at understanding the bursting of an axisymmetric vortex when a jet flow is superimposed along the axis. The basic flow and the numerical method are presented in Sect. 2. The impulse response is described in Sect. 3, and its structure is elucidated in Sects. 4 and 5, respectively for the linear and nonlinear regimes of evolution. In Sect. 6, the space- and time-scales predicted in the present numerical study are tentatively confronted to experimental results by Cadot *et al.* [2]. Concluding remarks are given in Sect. 7.

2 Basic flow profile and numerical method

The Batchelor vortex has been selected as basic flow. Let $R^\star$ denote the vortex radius and $\Omega_x^\star$ its centerline axial vorticity, where * stands for a dimensional variable. In the following, variables are presented in their dimensionless form: radius $R^\star$ and the typical azimuthal velocity $\Omega_x^\star R^\star$ have been chosen as reference length and velocity scales respectively. The axial and radial velocity components V_x and V_r, and the Burgers-like azimuthal velocity component V_θ are then given by:

$$V_x = u\,\mathrm{e}^{-r^2} \qquad (u \geq 0)\ , \tag{2a}$$

$$V_r = 0\ , \tag{2b}$$

$$V_\theta = \frac{1-\mathrm{e}^{-r^2}}{r}\ . \tag{2c}$$

In system (2a–2c, u is the maximum of the axial velocity component V_x. The situation $u = 0$ corresponds to a pure vortex without axial stream. The above flow (2a–2c) is known [5] to be significantly unstable as soon as u exceeds a critical value $u_\mathrm{c} \approx 0.7$. In the following, the maximum axial velocity is set to the value $u = 1.25$, in the vicinity of which maximal linear amplification is reached. However, it should be emphasized that the linear stability properties of the Batchelor vortex do not evolve qualitatively much when u is increased from 1 to about 10. The present study is thus thought to provide a generic picture of impulse responses provided the maximum axial velocity u is of unit order.

The numerical code [7,3] implements the time integration of incompressible Navier–Stokes equations perturbed around the basic flow (2a–2c). The velocity and vorticity perturbations are developed on $1024 \times 64 \times 64$ Fourier modes pertaining to the three Cartesian directions. The viscous diffusion of the basic flow is assumed to be exactly compensated by a body force, which allows one to connect the linear instability properties to the subsequent nonlinear development of perturbations in the same basic-flow conditions. The Reynolds number defined by

$$Re = \frac{\Omega_x^\star\, R^{\star 2}}{\nu^\star}, \tag{3}$$

where $\nu*$ stands for the kinematic viscosity, is set to 530.

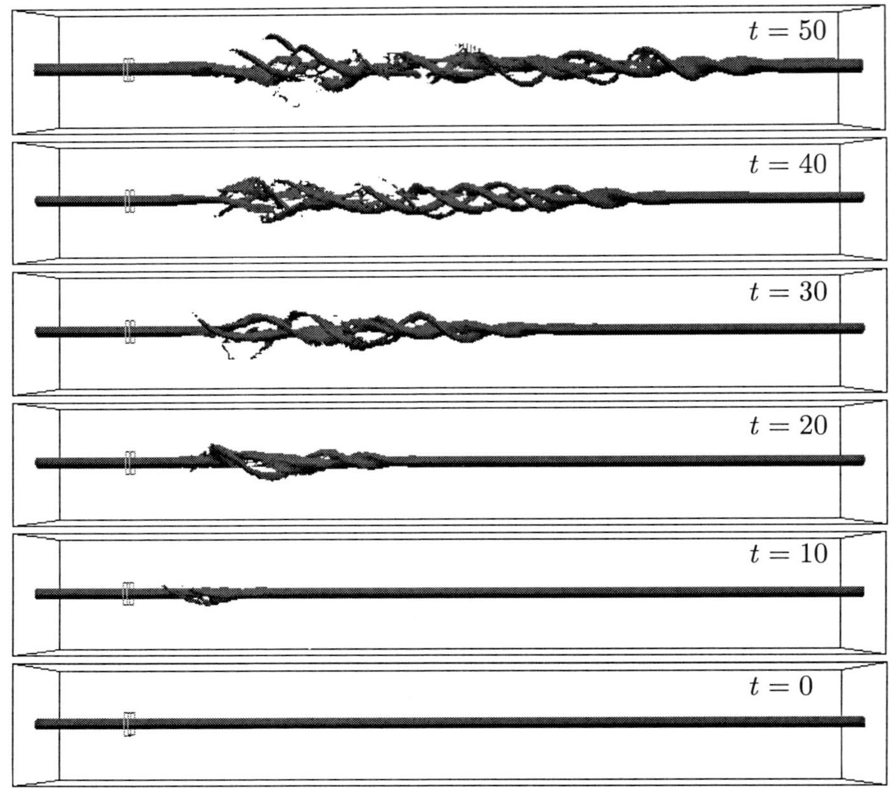

Fig. 1. Nonlinear impulse response of a swirling jet: isosurfaces of the total axial vorticity at several times. The isosurface level is chosen to be about 20% the maximum axial vorticity for the basic flow. Fluid flows from left to right. The domain, only partially represented, is meshed by a $1024 \times 64 \times 64$ Cartesian grid. The small frame indicates the plane of the initial impulse. The simulation parameters are $Re = 530$ and $u = 1.25$

3 Impulse-response phenomenology

At $t = 0$, the flow is locally perturbed in the plane $x = 0$. As a result of the flow instability, a wavepacket grows for $t > 0$ along the vortex core. On Fig. 1, the total (i.e. basic flow + perturbation) axial vorticity component is represented. Time increases from top to bottom by regular steps. The initial impulse has been selected in such a way that the perturbation amplitude is of unit order near $t = 10$. In other words, the exponential growth prevailing in the linear regime takes place from $t = 0$ up to $t = 10$, where nonlinearities become significant.

At later times ($t > 10$), the perturbation expands along the $x-$axis (nonlinear regime). At immediate proximity of the initial impact (symbolized by the small frame on Fig. 1), the flow widens and becomes strongly disorganized. At the wavepacket downstream end (on the right), the initial vortex parts into several vortex helices (visible at $t = 40$).

Both regimes are described in detail in the following two sections.

4 Linear regime

At the early stage of evolution ($t < 10$), the perturbation can be viewed as a linear combination of many "elementary" wavepackets, pertaining to distinct spatio-temporal branches of the Batchelor vortex linear instability. In particular, azimuthal modes m are found to be unstable for $-12 \leq m \leq -1$ for the parameter set $u = 1.25$ and $Re = 530$. Each $m-$wavepacket grows exponentially with growth rates $\sigma_m(V)$ along the spatio-temporal rays $x = Vt$, where the variable V denotes a ray velocity. By definition, the two particular values V_m^- and V_m^+ for which $\sigma_m(V_m^-) = \sigma_m(V_m^+) = 0$ denote the velocities of the $m-$wavepacket trailing and leading edges. At a given fixed time, the DNS-calculated perturbation may be decomposed by a Hilbert-transform technique (see Delbende *et al.* [4] for the detailed procedure). At $u = 1.25$ and $Re = 530$, 12 elementary unstable wavepackets pertaining to modes $m = -1, -2, \cdots, -12$ arise from the decomposition (Fig. 2a). When applied at two distinct times, the above decomposition allows one to recover the growth rate $\sigma_m(V)$ of the spatio-temporal instability mode m along each ray $x = Vt$ for each value of m (Fig. 2b).

On Fig. 2b, two regions may readily be distinguished: the leading-edge region, denoted as (I), rapidly moving away from the initial impact location, and the trailing-edge region, denoted as (II), situated in the neighborhood of the impact location. In region (I), isolated $m-$modes prevail on distinct ranges of ray velocity V, i.e. their growth rate $\sigma_m(V)$ is clearly higher than those for other modes (the dominant modes are either $m = -1, -2$ or -3 depending on V, as indicated by the arrows on the right part of Fig. 2b). In region (II), numerous azimuthal modes ($-7 \leq m \leq -2$) coexist with comparable growth rates. These modes thus grow with comparable amplitudes along the same spatio-temporal rays.

The above linear instability considerations deeply affect the evolution of the wavepacket in the nonlinear regime, as described below.

5 Nonlinear regime

Nonlinear effects become significant as soon as the perturbation variables reach values of unit order (here for $t > 10$). In region (I), each dominant $m-$mode undergoes a saturation process on the spatio-temporal rays along which it used to dominate the linear wavepacket dynamics. This leads to the formation of helical vortices with $|m|$ filaments, which connect to the filaments of neighboring zones of region (I), where a different value of m prevails. Such a phenomenon is clearly visible on Fig. 1 at $t = 40$: in region (I), three co-rotating helices connect to two helices and eventually to a single helical vortex (when you go from left to right). Similar patches of helices have been observed by Abid & Brachet [1] on

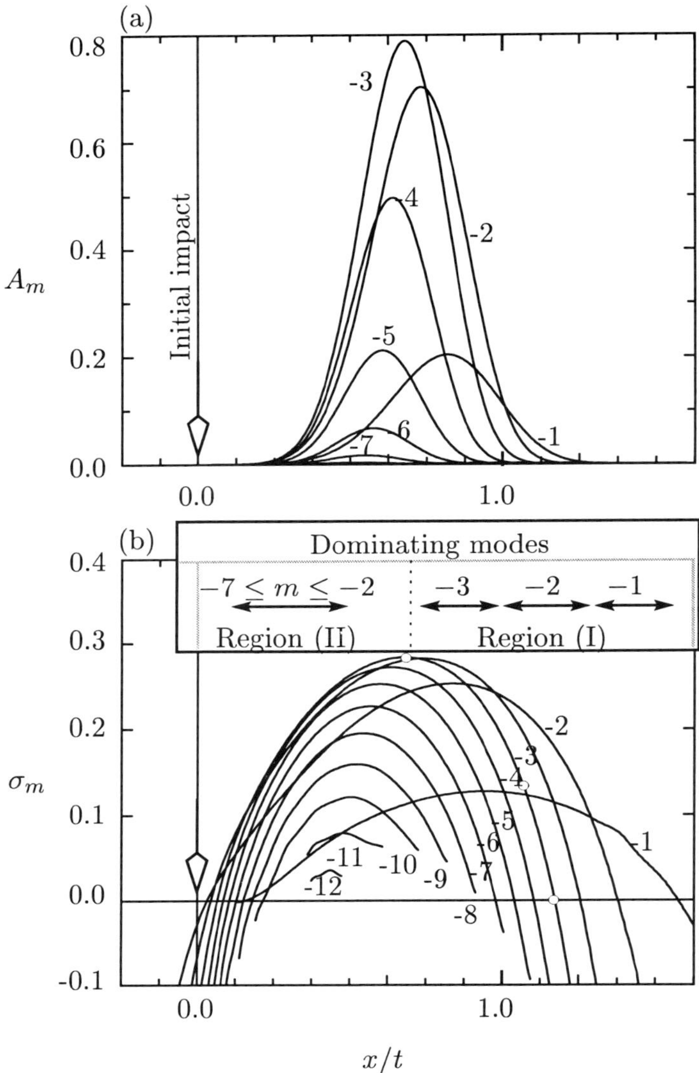

Fig. 2. (**a**) Amplitude A_m and (**b**) growth rate σ_m evaluated along the spatio-temporal rays $x/t = V$, for each elementary m−wavepacket developing in the swirling jet during the linear regime. The parameters are $u = 1.25$ and $Re = 530$.

temporal numerical simulations initiated by white noise or by a superposition of eigenmodes. The present study unveils the *spatio-temporal* organization of the filaments, in particular the connected zones with 1, 2, and 3 helices.

In region (II), situated close to the initial impulse location (on the left of Fig. 1), the evolution is far more intricate as a result of the numerous linear unstable modes which grow simultaneously. Throughout the region, no dominant azimuthal symmetry m is clearly selected. The nonlinear interaction of these modes may cause the vortex to burst. In this case also, the phenomenon originates in the spatio-temporal organization of the linear instability modes.

Note that the wavepacket-expansion velocity along the vortex axis is in both regimes – linear and nonlinear – of order u.

6 Space- and time-scale of the phenomenon

Before proposing the present mechanism as a candidate for the vortex bursting observed experimentally by Cadot *et al.* [2], one may compare the space- and time-scales of the phenomenon in both instances. For a specific event (shown as a representative sample on Fig. 2 in their article) the bursting duration $T^\star \sim 0.05$ s separating the appearance of wiggles and the explosion, normalized by the detected azimuthal velocity $V_\theta^\star = 250$ cm·s^{-1} and the vortex radius $R^\star = 0.25$ cm, is given by:

$$T = \frac{V_\theta^\star T^\star}{R^\star} \sim \frac{250 \times 0.05}{0.25} = 50 \; .$$

The above value is the very same order of magnitude as in the present numerical study. The typical 'wavepacket' (comprising the exploded zone and the helices) length $L^\star = 10$ cm observed experimentally, normalized by the vortex radius $R^\star$ is given by:

$$L = \frac{L^\star}{R^\star} \sim \frac{10}{0.25} = 40 \; ,$$

in good agreement with the wavepacket length predicted here (around 40 for $u = 1.25$). It should be however noticed that the latter agreement holds because of the following assumption: the basic axial velocities are of same order of magnitude as their azimuthal counterparts (i.e. u is of unit order), which remains to be checked experimentally.

7 Concluding remarks

A vortex-bursting process is investigated by DNS of the nonlinear response of a swirling jet to an initially localized perturbation. The dynamics of the nonlinear wavepacket is described in the light of the linear spatio-temporal instability properties of the basic flow. This leads to propose two simultaneous mechanisms for vortex dissipation. Due to axial shear acting along the initial vortex, helical structures (region I) may develop and propagate along the vortex core at a speed comparable to the maximum axial velocity. These structures take the form of helical vortices with 1, 2 and 3 filaments. Because of the perturbation formulation adopted here, it has not been possible to investigate whether they may be subject to secondary instabilities, which would possibly enhance dissipation. The vortex may also burst locally under the influence of numerous instability modes growing at the same rate *and* at the same place (in region II), which favors complex interactions and, eventually, dissipation.

Numerical simulations have been performed on the C90 Cray at IDRIS–CNRS under grant #980173/CP2, which is greatly acknowledged.

References

1. M. Abid, M.-E. Brachet: Phys. Fluids **10**(2), 469 (1998).
2. O. Cadot, S. Douady, Y. Couder: Phys. Fluids **7**(3), 630 (1995).
3. I. Delbende, J.-M. Chomaz: Phys. Fluids **10**(11), 2724 (1998).
4. I. Delbende, J.-M. Chomaz, P. Huerre: J. Fluid Mech. **355**, 229 (1998).
5. M. Lessen, P.J. Singh, F. Paillet: J. Fluid Mech. **63**, 753 (1974).
6. Z.S. She, E. Jackson, S.A. Orszag: Nature **344**, 226 (1990).
7. A. Vincent, M. Meneguzzi: J. Fluid Mech **225**, 1 (1991).

Computation of Equilibria Between Two Corotating Nonuniform Vortices

Uwe Ehrenstein[1] and Maurice Rossi[2]

[1] Laboratoire J.-A. Dieudonné, Université Nice-Sophia Antipolis, Parc Valrose, F-06108 Nice Cedex, France

[2] Laboratoire de Modélisation en Mécanique, Université de Paris VI, 4, Place Jussieu, F-75252 Paris Cedex 05, France

1 Introduction

Understanding decaying two-dimensional turbulence has important consequences for geophysics, meteorology and astrophysics. Various aspects appear in a recurrent way: from an initially unorganized two-dimensional flow, vorticity rapidly concentrates in localized regions called coherent structures and also in vorticity filaments . In this framework, the computation of a nonlinear equilibrium between the vortices and its stability becomes an essential ingredient. Focusing on vorticity patches, equilibria of two unequal-area vortices have been computed for instance by Dritschel[1]. However, vortices of two-dimensional turbulence are not vorticity patches. Our purpose is thus to verify how previous results obtained in the context of uniform vorticity are modified when nonuniform vortex pairs are considered.

2 Governing equations and solution procedure

To compute the equilibria between two corotating vortices with different circulation and size, Green's function integrals

$$\psi_i(\mathbf{r}) = -\frac{1}{4\pi}\int\int_{A_i} \log\left(|\mathbf{r}-\mathbf{r}_i'|^2\right)\omega_i(\mathbf{r}_i')dS_i', \tag{1}$$

are considered. In the above formula $\mathbf{r}$, $\mathbf{r}_i'$ are taken from the center of vortex i, $i = 1, 2$. In the reference frame (x, y) rotating with Ω, the total stream functions read

$$\psi_{t,1}(\mathbf{r}_1) = \psi_1(\mathbf{r}_1) + \psi_2(\mathbf{L}+\mathbf{r}_1) + \frac{1}{2}\Omega\,|\mathbf{a}+\mathbf{r}_1|^2 \tag{2}$$

and

$$\psi_{t,2}(\mathbf{r}_2) = \psi_2(\mathbf{r}_2) + \psi_1(-\mathbf{L}+\mathbf{r}_2) + \frac{1}{2}\Omega\,|-\mathbf{L}+\mathbf{a}+\mathbf{r}_2|^2\,. \tag{3}$$

The vectors $\mathbf{a} = \overrightarrow{OO_1} = (a, 0)$, $\mathbf{L} = \overrightarrow{O_2O_1} = (L, 0)$ and $\mathbf{r}_i$ are the coordinate vectors with origin at O_i, $i = 1, 2$ (see Fig. 1). For an equilibrium solution of the

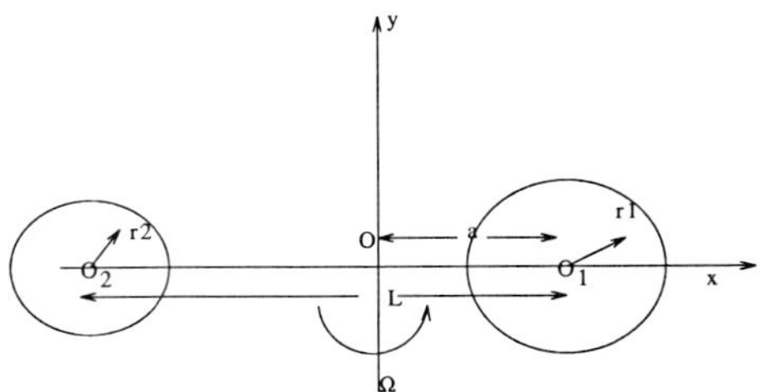

Fig. 1. Geometry of equilibrium configuration.

Euler's equations the isovorticity contours are streamlines. When vortices are far apart the vortices are supposed to be axisymmetric and the radii of the circular streamlines $r_{0,i}(\omega_i), i = 1, 2$, in each vortex are function of vorticity. When the vortices are at a finite distance from each other, the streamlines are deformed. Introducing the functions $f_i(\omega_i, \theta), i = 1, 2$, the same constant vorticity ω_i now lies on the perturbed streamlines

$$r_i(\omega_i, \theta) = \sqrt{r_{0,i}^2(\omega_i) + f_i(\omega_i, \theta)}, 0 \leq \theta \leq 2\pi, i = 1, 2. \tag{4}$$

Using this relation inside each vortex, Green's function integrals (1) are expressed as vorticity integrals:

$$\psi_i(\mathbf{r}) = -\frac{1}{4\pi} \int_{\omega_{i,0}}^{\omega_{i,1}} \int_0^{2\pi} \log\left(|\mathbf{r} - \mathbf{r}_i(\omega_i', \theta')|^2\right) \frac{1}{2}\omega_i' \left(-\frac{\partial r_i^2}{\partial \omega_i}(\omega_i', \theta')\right) d\omega_i' d\theta', \tag{5}$$

the vorticity ranging from its minimum value $\omega_{i,0}$ to its maximum value $\omega_{i,1}$ for each vortex region $i = 1, 2$. Note that $\partial r_{0,i}^2/\partial \omega_i < 0$; when the vortices get closer (at least for the equilibrium solutions we are looking for) we suppose that two distinct streamlines never cross each other and hence $\partial r_i^2/\partial \omega_i < 0$ also for the disturbed streamlines.

The vectors $\mathbf{r}_i = (r_i \cos(\theta)\ , r_i \sin(\theta))$ can be written in terms of the vorticity ω_i and the angle θ using (4). Consequently, the condition that the total stream function (2) (respectively (3)) is a constant on each vorticity level inside vortex 1 (respectively vortex 2) can now be written as

$$\psi_{t,i}(\omega_i, \theta) = C_i(\omega_i), 0 \leq \theta \leq 2\pi, i = 1, 2, \tag{6}$$

the integrals being evaluated using (5). Equation (6) can be viewed as nonlinear constraints for the unknowns $f_i(\omega_i, \theta), i = 1, 2$. The family of solution is supposed to be stationary in the frame of reference rotating with Ω. The rotation rate as well as the distance a between the center of rotation and the center of vortex 1 should be determined as function of the distance between the vortex centers.

This is achieved by letting the vortex centers to be stagnation points in the rotating reference frame which gives two extra-conditions

$$\lim_{r_i \to 0} \frac{\partial \psi_{t,i}}{\partial r_i}(r_i, \theta = 0) = 0, i = 1, 2$$

(the derivatives of the total stream functions with respect to θ being zero at $\theta = 0$, due to the symmetry with respect to the x-axis). The vorticity integrals are discretized using Fourier-collocation in θ and Chebyshev-collocation in ω, together with quadrature rules for the integrals [2]. By decreasing the distance L betwen the initially axisymmetric vortices, the increasing deformation of the streamlines is characterized by the unknown functions $f_i(\omega_i, \theta), i = 1, 2$, which are solution of a large nonlinear system, resulting from the discretization of the vorticity integrals.

3 Computations for Lamb-like vortices

Jimenez *et al.*[3] have shown that vorticity structures in decaying two-dimensional turbulence are well described by perturbed Lamb vortices. Starting with axisymmetric Lamb-like vortices

$$\omega_i = \omega_{\max,i} \exp\left(-\alpha r^2 / r^2_{\max,i}\right), 0 \le r \le r_{\max,i}, i = 1, 2, \tag{7}$$

equilibrium configurations may be computed. The parameter α measures the ratio between the vorticity maximum at the center and the minimum vorticity located at the boundary for each vortex. The system is made dimensionless in such a way that vortex 1 has a circulation $\Gamma_1 = \pi$ and an area $A_1 = \pi$. This implies that $r^*_{\max,1}$ is the reference length and $\omega^*_{\max,1}\beta$ with $\beta = (1 - e^{-\alpha})/\alpha$ the reference vorticity. Accordingly, the dimensionless vorticity distribution writes for each undisturbed vortex

$$\begin{aligned} \omega_1(r) &= \frac{1}{\beta} \exp(-\alpha r^2), 0 \le r \le 1, \\ \omega_2(r) &= \frac{\alpha_\omega}{\beta} \exp(-\alpha r^2 \alpha_r^2), 0 \le r \le 1/\alpha_r, \end{aligned} \tag{8}$$

with $A_1 = \pi, \Gamma_1 = \pi$ and $\alpha_\omega = \omega^*_{\max,2}/\omega^*_{\max,1}, \alpha_r = r^*_{\max,1}/r^*_{\max,2}$, hence $A_2 = \pi/\alpha_r^2$ whereas $\Gamma_2 = \alpha_\omega \pi / \alpha_r^2$.

Starting for a sufficiently large L value with the asymptotic rotation rate $\Omega = (\Gamma_1 + \Gamma_2)/2\pi L^2$, distance $a = L\Gamma_2/(\Gamma_1 + \Gamma_2)$ and with the axisymmetric vortex configuration (8), a first deviation (4) from the axisymmetric case is computed by solving the nonlinear system. The whole family of equilibrium states is then computed step by step by decreasing the continuation parameter L. The results are interpreted in terms of the distance between the centroids l which may easily be computed once an equilibrium state obtained. Different cases are considered according to the circulation ratio α_ω, area ratio α_r as well as the quantity α (see (7),(8)) which characterizes the non-uniformity of the vortex profiles. In Dritschel[1] the normalization is such that the total circulation is equal to π. For comparison, in all subsequent results the distance l is multiplied by the scale factor $\sqrt{\pi/(\Gamma_1+\Gamma_2)}$, where $\Gamma_1+\Gamma_2$ is the total circulation in our computations. Computing the spectrum of the Jacobian matrix for each family of equilibrium states points of exchange of stability may be detected. One result is depicted in Fig. 2 where δ (the gap between the vortices) is shown as function of l. For these computations two equal vortices have been considered, ranging from the almost uniform case $\alpha = 0.001$ to nonuniform vortices ($\alpha = 1$). The

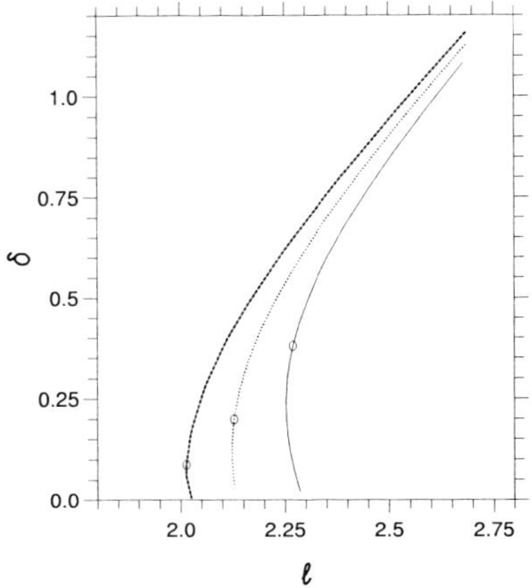

Fig. 2. Gap δ as function of distance between centroids l for equilibrium solutions with equal vortices; —— : vortex patches ($\alpha = 0.001$); ..... : non-uniform vortices ($\alpha = 1$); – – – : non-uniform vortices ($\alpha = 2.25$); ∘ : points of exchange of stability.

solid line corresponds to the almost uniform case and the circles indicate the parameter values for the appearance of the zero eigenvalue. To be more specific the instability point is located at $l = 2.26$ and $\delta = 0.34$, which agrees with the values of Fig. 4 in Dritschel[1]. The value given in Saffman[4] (p. 180) for a highly accurate computation is $l = 2.243$. The difference is due to the small but finite α value and the different algorithms used. In the extreme configuration where the vortices almost touch Saffman gives the same l-value as that for loss of stability and in our case this is what we approximately find (see Fig. 2).

The streamlines inside the pair of nonuniform vortices with $\alpha = 2.25$ are depicted in Fig. 3. The solution corresponds to the point of exchange of stability marked as a circle on the broken line of Fig. 2. Whereas the inner streamlines are almost elliptical the outer ones are deformed and a cusp forms at the outer vortex boundary for this vanishing δ-value. Similar computations have been performed by considering two unequal vortices. One example for two unequal vortices with $\Gamma_2/\Gamma_1 = 0.5$ is shown in Fig. 4 where streamlines for the entire flow field are depicted. This is numerically performed using Green's function integrals. The equilibrium state is close to marginal stability and the streamlines exhibit a flow stagnation point. Fixing $\alpha = 2.25$ and $\alpha_\omega = 1$ in (7), (8), the angular impulse J and the excess energy E has been computed for equal and unequal

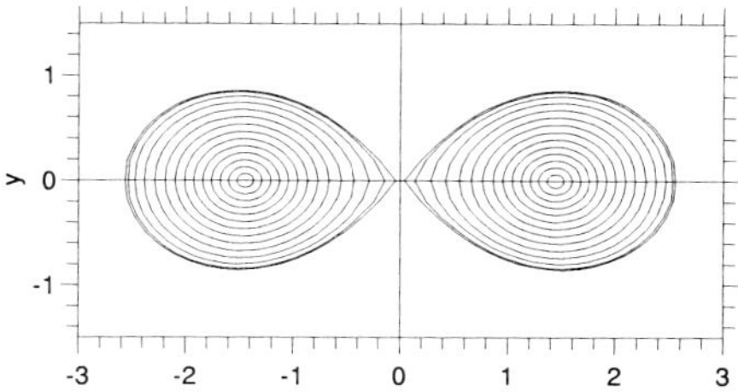

Fig. 3. Iso-vorticity contours for non-uniform equilibrium solution ($\alpha = 2.25$) at the point of exchange of stability $\circ$ shown in Fig. 2.

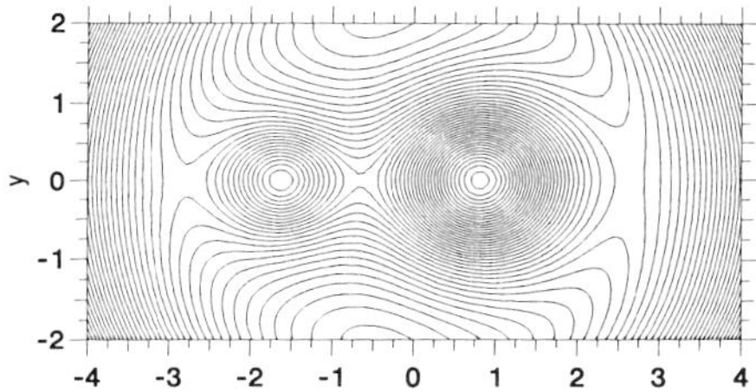

Fig. 4. Streamlines generated by the equilibrium state with $\Gamma_2/\Gamma_1 = 0.5$,($\alpha = 2.25$), for gap $\delta = 0.26$.

vortices. Again, the quantities have been normalized such that the total area and circulation is equal to π. Once an equilibrium state determined, the normalized angular impulse and excess energy are using the scale factor $s = \sqrt{(\Gamma_1 + \Gamma_2)/\pi}$,

$$J = \frac{1}{s^4} \int \omega(\mathbf{r})|\mathbf{r}|^2 dS, \quad E = \frac{1}{2s^4} \int \omega(\mathbf{r})\psi(\mathbf{r}) dS$$

with the stream function

$$\psi(\mathbf{r}) = -\frac{1}{4\pi} \int \int \log \left(\frac{|\mathbf{r} - \mathbf{r}'|^2}{s^2} \right) \omega(\mathbf{r}') dS'.$$

The above integrals are again expressed inside the vortex regions as vorticity integrals and are hence easily computed as a by-product of our solution procedure. The results for different circulation ratios are shown in Fig 5. For the family of unequal vortex pairs J exhibits a local minimum where the energy E goes through a local maximum, as function of the gap δ. These points coincide with exchange of stability in agreement with the theory of Saffman[4]. For equal vortices exchange of stability occurs for vanishing δ and indeed J and E reach an optimum for $\delta = 0$. The limiting case of the interaction between a finite vortex and a point vortex may be approximated by use of our solution procedure.

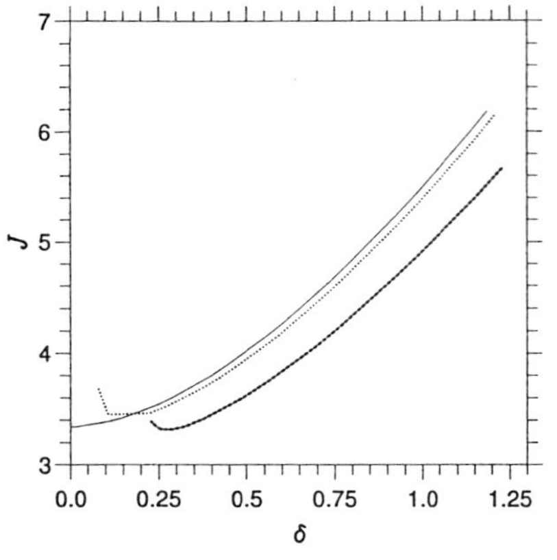

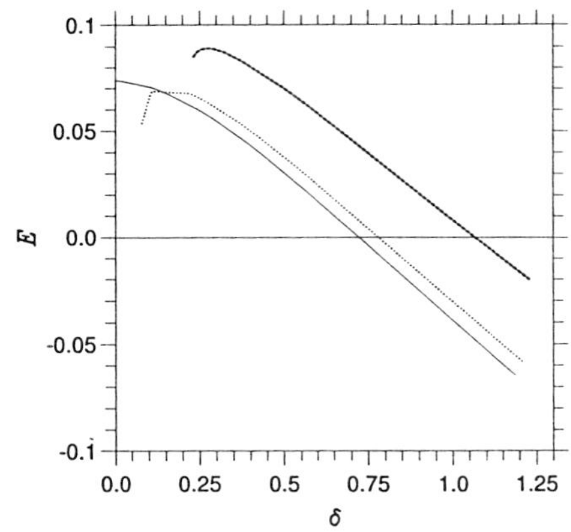

Fig. 5. (a) Angular impulse J and (b) excess energy E as function of the gap between centroids, for nonumiform vortices with —: $\Gamma_2/\Gamma_1 = 1$; .....: $\Gamma_2/\Gamma_1 = 0.75$; – – –: $\Gamma_2/\Gamma_1 = 0.5 (\alpha = 2.25.)$

Indeed, choosing for instance the parameter $\alpha_r = 10$ in (8) the area $A_2 = \pi/100$ becomes negligible and the second vortex is now characterized by the circulation $\Gamma_2 = \alpha_\omega \pi/100$. One marginally stable equilibrium state in this case is shown in Fig. 6.

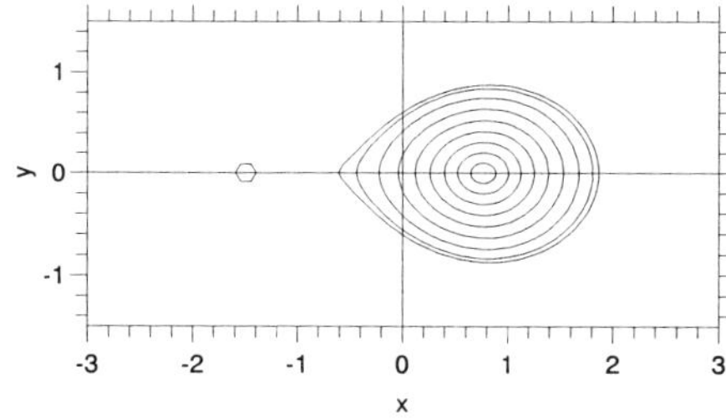

Fig. 6. Marginally stable equilibrium state for a finite vortex with area and circulation Γ_1 interacting with a point vortex with Γ_2, for $\Gamma_1/\Gamma_2 = 0.5$ $(\alpha = 2.25)$

4 Remarks

While we only considered the case of vortex pairs solution of Lamb-type profiles, the algorithm can *a priori* compute equilibrium states starting from general vortex profiles. More generally, the outlined solution procedure may provide a general tool for the computation of interacting vortices, solution of Euler's equations.

References

1. D. G. Dritschel, "A general theory for two-dimensional vortex interactions," J. Fluid Mech. **293**, 269 (1995).
2. U. Ehrenstein, M. Rossi,"Equilibria of corotating nonuniform vortices," Phys. Fluids **11**, 3416 (1999).
3. J. Jimenez, H. K. Moffat, C. Vasco, "The structure of the vortices in freely decaying two-dimensional turbulence," J. Fluid Mech. **313**, 209 (1996).
4. P. G. Saffman, *Vortex Dynamics* (Cambridge University Press, 1992).

Experimental Observation of the Multipolar Instability Inside a Deformed Elastic Cylinder

Christophe Eloy, Patrice Le Gal and Stéphane Le Dizès

Institut de Recherche sur les Phénomènes Hors Équilibre, UMR 6594.
12 Avenue Général Leclerc, F-13003 Marseille, France.

Abstract. The instability of a vortex subject to a stationary n-fold constraint is studied experimentally. Such a vortex is obtained by the rotation of a deformable cylinder wall on which two or three rolls are pressed. The two control parameters of the flow are the aspect ratio and the Reynolds number. Different regimes of instability are observed in the parameter space. The wavelength and frequency of the resulting three-dimensional unstable mode are measured by visualization and compared to theoretical predictions.

1 Introduction

Experimental and numerical studies show that filaments of intense vorticity are present in turbulent flows [12,3]. A fine observation of these vortical structures [1] suggests that some of their dynamical properties can be explain by the presence of Kelvin modes on the filament [8]. Besides, a vortex filament in a turbulent field is subject to an external strain field that deforms its streamlines into ellipses at first order [15]. This deformation leads to the elliptic instability whose natural modes are just the Kelvin modes [15,20]. The elliptic instability mechanism can be generalized to a n-fold azimuthal symmetry and gives rise to what we shall call a *multipolar instability* for $n = 2$, 3 or 4 [5]. The aim of this contribution is to study experimentally the multipolar instability, for elliptic ($n = 2$) and triangular ($n = 3$) deformations, in order to validate theoretical linear predictions [5] and to have an insight into its nonlinear regime. We argue that our result could explain the later stages of vortex dynamics in large Reynolds number flows which lead to the destruction of the filaments.

Few experiments have been designed to study the elliptic instability. Chernous'ko [4,6] studied the flow inside a rotating cylinder of elliptic cross-section sharply stopped. He gave an instability diagram showing the number of structures as a function of aspect ratio and eccentricity. Vladimirov & Tarasov [19] used the same set up with a small orifice drilled at the bottom of the container to allow a modification of the aspect ratio during the experiment. These two sets up had two main inconveniences: the instability occurred during the transient decay of the flow and Taylor-Görtler instability developed on the wall. These defects were avoided by Malkus [12,13] who used the boundary to drive the flow inside a deformable cylinder[1]. However only very qualitative results on this ex-

[1] The set-up used for in present paper is very similar to Malkus' one and is described below.

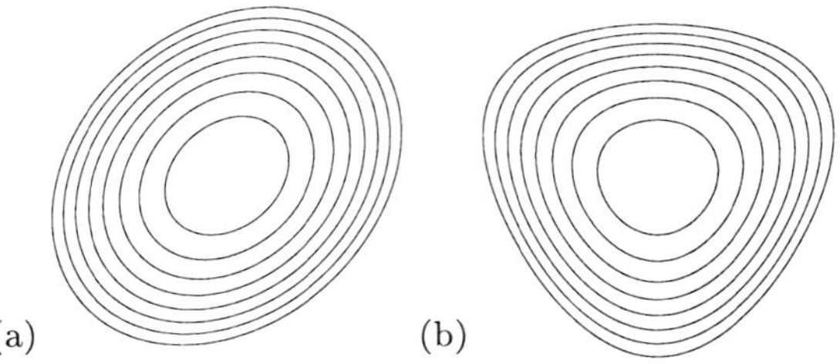

Fig. 1. Streamlines of the flow given by (1), with $\varepsilon = 0.25$ and (a) $n = 2$, (b) $n = 3$. The flow is rotating counterclockwise, the axis $\theta = 0$ is horizontal.

periment have been published so far and, to our knowledge, no experiment have been carried out for other azimuthal geometries.

2 Theoretical results

The elliptic instability has been studied theoretically by many authors in the past years. They used global mode techniques [15,20] and geometrical optics methods [11] (see also [2,17]). A generalization of these studies have been done for flows with higher azimuthal symmetry [5,10]. Results can be summarized as follow.

The basic flow whose stability is addressed here is given in cylindrical coordinates (r, θ, z) by the streamfunction:

$$\psi = -\frac{1}{2}r^2 + \frac{\varepsilon}{n}r^n \sin(n\theta), \tag{1}$$

where ε measures the deformation of the streamlines (ε is the eccentricity of the ellipses for $n = 2$) and n is the degree of azimuthal symmetry of the flow. The outer streamline of the flow is given by $\psi = -\frac{1}{2}$ (see Fig. 1).

For axisymmetric flow ($\varepsilon = 0$), Kelvin modes can be superimposed linearly to the basic flow. Their velocity field can be written as:

$$\mathbf{U}(r, \theta, z, t) = \mathbf{u}(r) \exp\left(\mathrm{i}(kz + m\theta - \omega t)\right) + \text{c.c.}, \tag{2}$$

with k the axial wavenumber, m the azimuthal wavenumber and ω the frequency. For an infinite cylinder, Kelvin modes are travelling waves with a full three-dimensional structure. However, for a given aspect ratio, one has to add two symmetric Kelvin modes (k, m, ω) and $(k, -m, -\omega)$ to form standing waves that satisfy the boundary conditions. These Kelvin modes are marginally stable for an axisymmetric and inviscid flow and satisfy the dispersion relation $D(k, m, \omega) = 0$ (few branches of D are drawn on Fig. 2). In the limit of small ε, Kelvin modes still exist. Moreover, two Kelvin modes can resonate if they have same k, same ω and have azimuthal wavenumbers m separated by n. On Fig. 2, intersection points of the dispersion relations for $m_1 = 10$ and $m_2 = 13$ thus

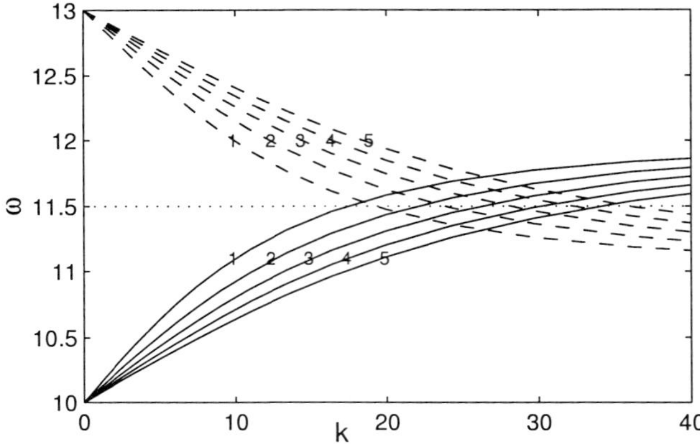

Fig. 2. Dispersion relation of the Kelvin modes inside a cylinder. Five branches are shown for $m = 10$ (solid lines) and $m = 13$ (dashed lines).

correspond to points of resonance for a triangular vortex ($n = 3$). Through theoretical analysis, it appears that combinations of Kelvin modes corresponding to branches of the dispersion relations with same label (see Fig. 2) are significantly more amplified than other combinations. These particular combinations are named *principal modes* and are noted (m_1, m_2, i) where m_1 and m_2 are the azimuthal wavenumbers of the two Kelvin modes satisfying $m_2 - m_1 = n$ and i is the common label of the dispersion curves. Note that the axial wavenumber k is an increasing function of i.

The theoretical results can be interpreted using the two control parameters of the experiment: the aspect ratio H/R, (with H the height of the cylinder and R its mean radius) and the Reynolds number $Re = \frac{\Omega R^2}{\nu}$, where Ω is the angular speed and ν the kinematic viscosity. On Fig. 3, are plotted the marginal stability curves of all principal modes in the plane (k, Re) for elliptic deformation ($n = 2$). It shows that, for a given deformation $\varepsilon = 0.1$, all modes can be stabilized by viscosity for low Reynolds numbers. Above a critical Reynolds number, $Re_c \approx 435$, the first principal mode $(-1, 1, 1)$ becomes unstable. For larger Reynolds numbers, other modes can also be observed with different critical wavenumbers k. For finite deformation ε, modes can be unstable over an interval of wavenumbers of bandwidth $\Delta k = O(\varepsilon)$. However, when the aspect ratio H/R is fixed, the axial wavenumber must satisfy $k = l\pi R/H$, with l an integer. This additional constraint may discard some modes and select particular ones.

Linear stability analysis predict which principal mode should be observed for a given couple $(H/R, Re)$. The mechanism of selection by viscosity and aspect ratio is similar for a triangular deformation ($n = 3$).

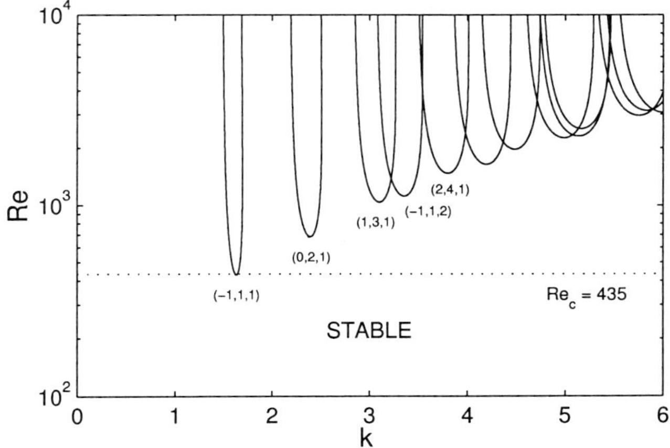

Fig. 3. Marginal stability curves for $n = 2$ and $\varepsilon = 0.1$.

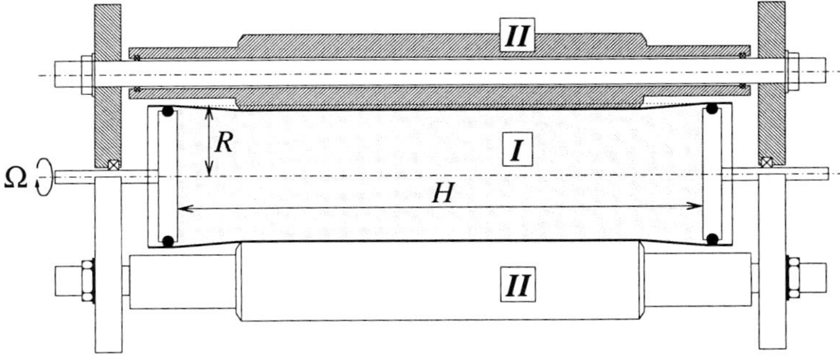

Fig. 4. Experimental set-up: (I) elastic cylinder filled with water; (II) rolls.

3 Experimental set-up

The core of the experiment is a transparent plastic extruded cylinder of $R = 2.75$ cm radius and variable height $H = 19$–22 cm. Its small thickness (0.5 mm) allows to deform its wall with two or three rolls parallel to its axis as shown on Fig. 4. The flow produced by this set-up is well described by (1). Therefore, theoretical results summarized above apply. The position of the rolls in the experiment is such that the deformation of the streamlines is $\varepsilon \approx 0.10$ for $n = 2$ and $\varepsilon \approx 0.12$ for $n = 3$. The cylinder is rotated by a 300 W variable speed motor at angular speed $\Omega = 0.5$–$10\,\mathrm{rad.s}^{-1}$ (i.e. Re= 400–8000). The cylinder is filled with water and anisotropic particles (Kalliroscope) are added for visualization.

A laser sheet is formed in the plane $\theta = 0$ (corresponding to the axis of maximal stretching for $n = 2$). Visualizations are made with a usual video camera. Wavelength and frequency of the mode appearing in the cylinder when it is rotated can be measured by image analysis.

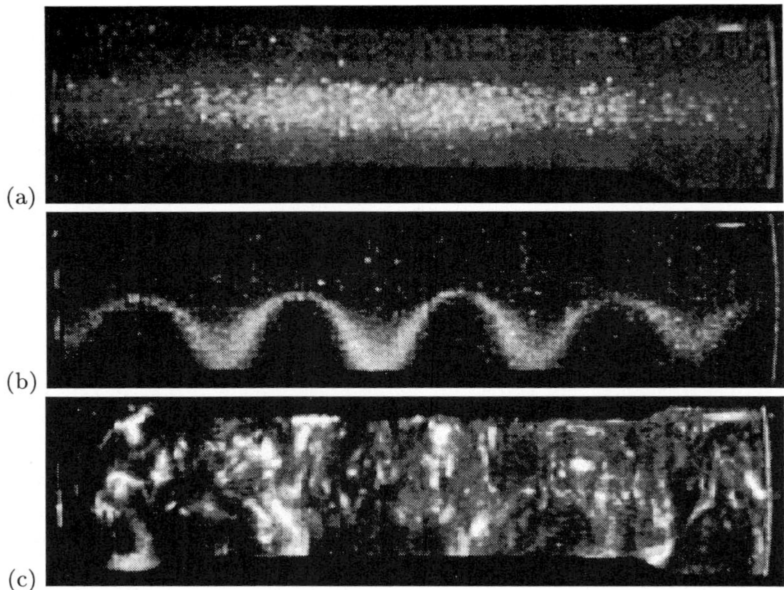

Fig. 5. Three successive pictures of the triangular flow ($n = 3$) for $\Omega = 6\,\mathrm{rad.s^{-1}}$, $H = 21.3\,\mathrm{cm}$ and (a) $t = 25\,\mathrm{s}$, (b) $t = 60\,\mathrm{s}$, (c) $t = 90\,\mathrm{s}$.

4 Results

In both cases ($n = 2$ and $n = 3$), when the elastic cylinder is rotated, the fluid is first spun-up (see Fig. 5a). Solid body rotation is reached after about 50 rotations. If the Reynolds number Re is sufficiently large, the multipolar instability takes place and a selected principal mode grows in the cylinder. This is illustrated for the triangular deformation case on Fig. 5b. For large Re, the developing mode may grow until it breaks down into small scales as it is shown on Fig. 5c. This very disordered state is not maintained by the rotation of the cylinder. It eventually evolves back to solid body rotation by a relaminarization process through viscous dissipation of the small scales. The re-establishment of solid body rotation allows again the development of the multipolar instability and so on. A cycle of instability–disorder–relaminarization then takes place. This scenario occurs both for elliptic ($n = 2$) and triangular ($n = 3$) deformations. However, contrarily to what has been claimed by Malkus [13], this scenario is not valid anymore for sufficiently small Reynolds numbers. Indeed, close to the threshold of instability, the unstable mode grows and reaches an amplitude of saturation which has been observed to maintain over long periods of time (over 1000 rotations).

The effect of the finite length of the cylinder has also been studied. As it was expected from theoretical analysis, the aspect ratio of the cylinder permits a mode selection. This selection is illustrated on Fig. 6 for $n = 2$. Depending on

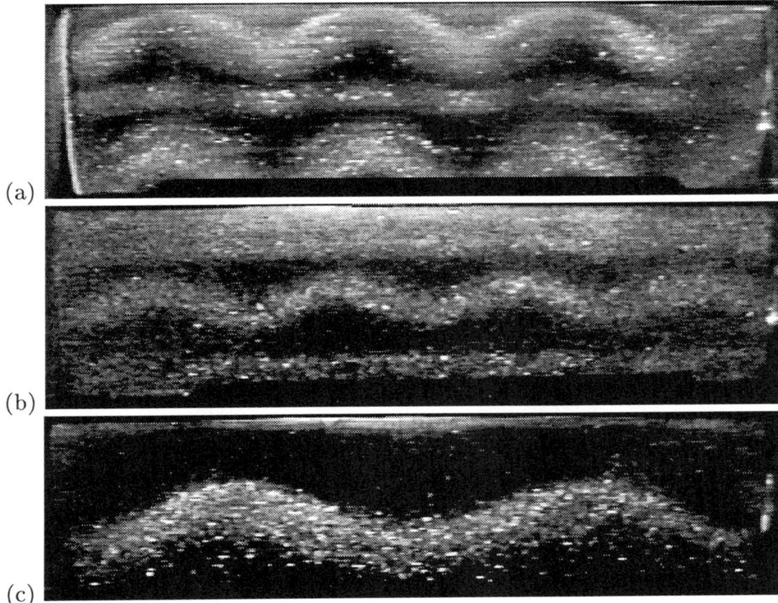

Fig. 6. Visualizations of the flow for $\Omega = 3\,\mathrm{rad.s}^{-1}$ and $n = 2$ at (a) $H = 19.6$ cm, (b) $H = 20.9$ cm, (c) $H = 21.9$ cm.

the length of the cylinder, three different principal modes have been observed: the mode $(1, 3, 1)$ for $H = 19.6$ cm, $(-1, 1, 2)$ for $H = 20.9$ cm and $(-1, 1, 1)$ for $H = 21.9$ cm. Their wavelength λ is such that $H = 3.5\lambda$, 4λ and 2λ respectively that is $\lambda = 1.99R$, $1.85R$ and $3.71R$. These wavelengths are in good agreement with the theoretical predictions of the most unstable wavelength (see Fig. 3): respectively $2.07R$, $1.91R$ and $3.98R$.

For $n = 2$, the principal modes $(-1, 1, i)$ are stationary in the laboratory frame. These modes produce undulation of the vortex core in a plane $\theta =$ constant (Fig. 6b,c). This undulation is in qualitative good agreement with theoretical predictions as shown on Fig. 7. Indeed, on this figure is plotted the axis of rotation of the vortex flow deformed by the mode $(-1, 1, 1)$. For other principal modes (m_1, m_2, i), when saturation is reached, the flow is time periodic. The theoretical analysis predicts a frequency $\omega \approx \Omega(m_1 + m_2)/2$. On Fig. 8, is shown a spatio-temporal diagram of the mode $(-1, 2, 1)$ over 45 rotations of the cylinder. The axis horizontal video line of Fig. 5 have been extracted on every image of the movie to construct this diagram. On this figure, the mode frequency shows a good agreement with its theoretical prediction since one can count 20.5 periods of the unstable mode ($\omega \approx \Omega/2$). Moreover, it is observed that this frequency decreases as mode amplitude increases. This effect may be due to nonlinearity. On this figure, one can also see that the instability of the flow in the cylinder is constituted by a standing wave made by the interaction

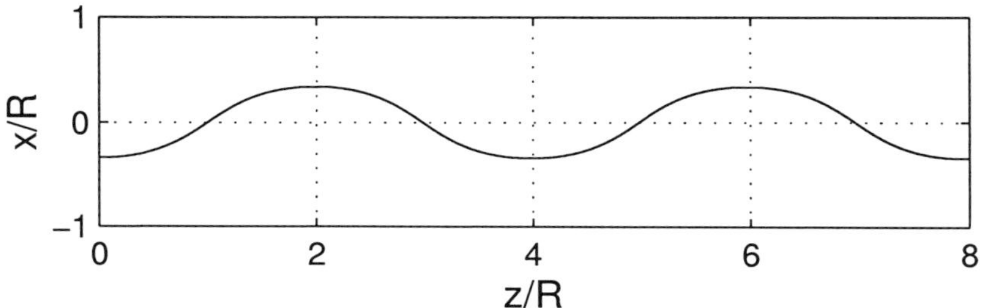

Fig. 7. Theoretical prediction of the position of the rotation axis displaced by the principal mode $(-1, 1, 1)$. To be compared to Fig. 6c.

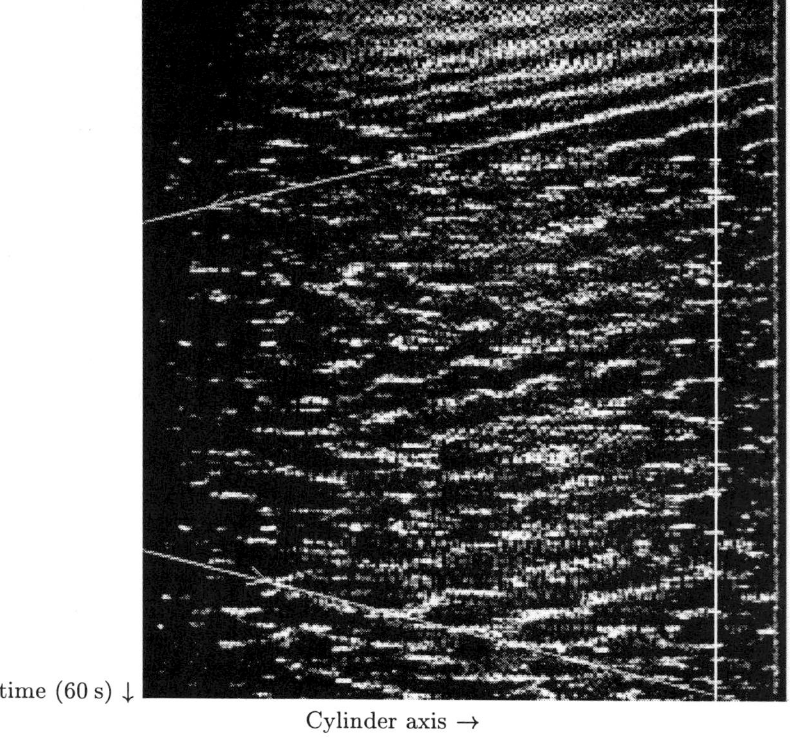

Fig. 8. Spatio-temporal diagram of the triangular flow ($n = 3$). The angular speed is $\Omega = 5\,\mathrm{rad.s}^{-1}$ and $H = 21.3\,\mathrm{cm}$. Time is vertical, the diagram shows 45 rotations.

of two propagative waves (on Fig. 5, diagonal lines have been added to highlight these propagations).

5 Discussion

In this contribution, the multipolar instability have been evidenced experimentally. The Reynolds number and the aspect ratio of the experiment have been varied to select some particular modes. Wavelengths and frequencies of those modes have been shown to be in good agreement with theoretical predictions.

In the nonlinear regime, a stationary saturated state has been observed slightly above threshold. However, this state has been observed to break down into small scales for sufficiently large Reynolds numbers. This "explosion" could be related to the dynamics of vortices observed in turbulent flows. Indeed, experimental and numerical observations [1,12] show that these filaments experience deformation into strands before breaking down. This observation is very similar to what has been observed in our toy experiment. This nonlinear regime could be explained by the appearance of a secondary instability [9,14] but so far, no experimental proof of this statement is available.

References

1. S. Arendt, D. C. Fritts, and Ø . Andreassen: Eur. J Mech. *B*/Fluids **17**,595 (1998)
2. B. J. Bayly: Phys. Rev. Lett. **57**, 2160 (1986)
3. O. Cadot, S. Douady, and Y. Couder: Phys. Fluids **7**, 630 (1995)
4. Y. L. Chernous'ko: Izv. Akad. nauk SSSR FAO **14**, 151 (1978)
5. C. Eloy and S. Le Dizès, *Stability of Rankine vortex in a multipolar strain field*, submitted to Phys. Fluid (1999)
6. E. B. Gledzer and V. M. Ponomarev: J. Fluid Mech. **240**, 1 (1992)
7. J. Jiménez and A. A. Wray: J. Fluid Mech. **373**, 255 (1998)
8. Lord Kelvin: Phil. Mag. **10**, 155 (1880)
9. R. R. Kerswell: J. Fluid Mech. **382**, 283 (1999)
10. S. Le Dizès and C. Eloy, Phys. Fluids **11**, 500 (1999)
11. A. Lifschitz and E. Hameiri: Phys. Fluids *A* **3**, 2644 (1991)
12. W. V. R. Malkus: Geophys. Astrophys. Fluid Dynamics **48**, 123 (1989)
13. W. V. R. Malkus and F. A. Waleffe: 'Transition from order to disorder in elliptical flow: a direct path to shear flow turbulence'. In: *Advances in turbulence III*, 1991, ed. by A. V. Johansson and P. H. Alfredsson, Springer-Verlag.
14. D. M. Mason and R. R. Kerswell: J. Fluid Mech. **396**, 73 (1999)
15. H. K. Moffatt, S. Kida, and K. Ohkitani: J. Fluid Mech. **259**, 241 (1994)
16. D. W. Moore and P. G. Saffman: Proc. Roy. Soc. Lond. *A* **346**, 413 (1975)
17. R. T. Pierrehumbert: Phys. Rev. Lett. **57**, 2157 (1986)
18. C.-Y. Tsai and S. E. Widnall: J. Fluid Mech. **73**, 721 (1976)
19. V.A. Vladimirov and V. F. Tarasov: 'Resonance instability of the flows with closed streamlines'. In: *Laminar–turbulent transition, IUTAM Symposium Novosibirsk (1984)*. ed. by V. V. Kozlov V. V., Springer-Verlag (1985)

Absolute/Convective Instabilities and Spatial Growth in a Vortex Pair

David Fabre[1], Carlo Cossu[1,2] and Laurent Jacquin[1]

[1] ONERA, 29 Av. de la Division Leclerc, BP 72, F-92322 Châtillon Cedex, France
[2] LadHyX, CNRS–École polytechnique, F-91128 Palaiseau Cedex, France

1 Introduction

Airplane trailing vortices have a destabilizing effect on ensuing aircrafts. Security spacings, related to the trailing vortices "lifetime" are actually enforced between take-offs and landings. This spacing limits the maximum take-off and landing frequency in saturated airports [19]. A number of studies have been devoted to the understanding of vortex wake dynamics, usually modeled by a pair of counter-rotating vortices. Two types of vortex pair three-dimensional instabilities have been identified in the past: a long-wave instability (of the order of the spacing b between the two vortices) and a short-wave instability (of the order of the vortex core radius a) have been first considered respectively by Crow [1] and by Moore & Saffman [15] and Tsai & Widnall [20]. These two mechanisms, which are thought to participate in the vortex wake dissipation, have been observed in recent experiments [12]. One possible technique to accelerate the dissipation of aircraft wakes is to force these instabilities by on-board control devices [2,21,22]. Until now only *temporal* vortex pair instability analyses are available. If one wants to force these instabilities, however, it would be more appropriate to analyze their *spatial* stability in the airplane reference frame. As the spatial analysis makes sense only when the instabilities are convective, an absolute/convective stability analysis is required. The results presented consist of the absolute/convective and spatial stability analyses of both long– and short–wave instabilities.

2 Base Flow

The vortex wake is modeled by a pair of counter-rotating vortices of circulation $\pm\Gamma$, core radius a and separated by a distance b. The vortices are moving in a direction perpendicular to their axes with a self-induced velocity approximately given by $W_0 = \Gamma/2\pi b$ [see figure 1–(b)]. In the reference frame of the airplane, in addition to the self-induced velocity, the pair of vortices is subject to the advection velocity U_∞ which has an U_0 component along the vortex axes [see figure 1–(a)]. The spatio-temporal stability properties of the vortex pair will depend upon the ratio a/b, but also upon the velocity ratio W_0/U_0. In order to give a simple geometrical interpretation of this velocity ratio we introduce the angle θ defined by

$$\tan\theta = \frac{W_0}{U_0}, \tag{1}$$

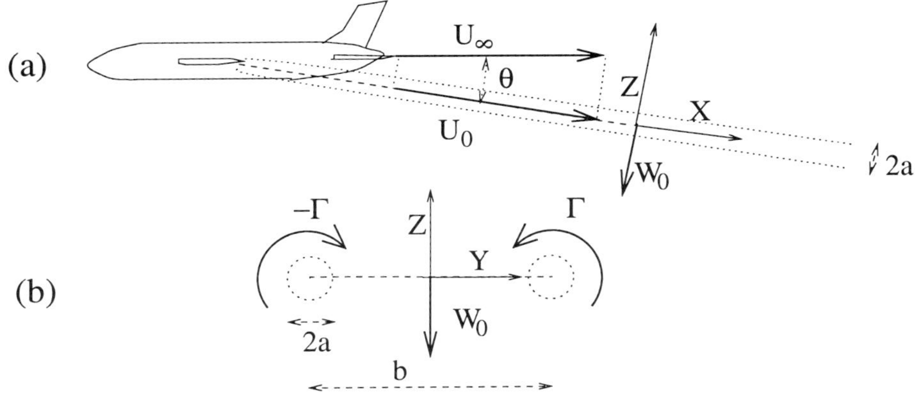

Fig. 1. Vortex pair base flow: (a) Side view, (b) view in a plane orthogonal to the axes of the vortices.

which measures the inclination of the vortex pair with respect to the airplane flight axis, as shown in figure 1.

Aside from inducing a downstream translation velocity W_0, the effect of one vortex on the other is to produce in its neighborhood a strain field of intensity $\Gamma/2\pi b^2$, and principal strain axes inclined of 45 with respect to the vortex wake plane [17]. This strain field is destabilizing for the vortex pair and is responsible for the two instabilities we consider here.

3 Temporal and Spatial Instability

The stability analysis is performed by linearizing the Euler equations around the vortex pair base flow described above. The perturbations are decomposed into normal modes proportional to $e^{i(kx-\omega t)}$. Thus the equations reduce to a dispersion relation

$$D(k, \omega; a/b, \theta) = 0, \tag{2}$$

which relates the wavenumber $k = k_r + ik_i$ (where $i = \sqrt{-1}$) and the pulsation $\omega = \omega_r + i\omega_i$ of each normal mode and depends upon the parameters of the problem and upon the particular frame of reference. It is usually very difficult to obtain explicitly the "full" dispersion relation derived from the linearized Euler equation so that one is lead to consider some physically relevant asymptotic limit of (2). In the following two asymptotic limits will be considered: the long– and the short–wave limits obtained by assuming $(a/b)^2 \ll 1$ and, respectively, $|kb| = O(1)$ and $|ka| = O(1)$. Temporal stability analyses of long– and short–wave instabilities [1,15,20] have been performed in a frame attached to the vortex pair. In that case, one considers the development of perturbations from an initial condition, for example resulting from the atmospheric turbulence

[5]. In these studies the spatial origin of the wake was not considered, and the angle θ therefore was not a relevant parameter of the problem. The temporal stability analysis is performed by solving dispersion relation (2) as $\omega^T(k;a/b)$ where k is real and ω complex. One may in particular find the wavenumber k^T_{max} corresponding to a maximal growth rate $\omega^T_{i,max}$.

In the frame of reference of the aircraft, the linearized equations lead to a similar dispersion relation, with simply a Doppler shift in the frequency. Referring to figure 1 this amounts to replacing ω with $\omega - kU_0$ in the dispersion relation obtained previously. Depending on the values of the parameters, two situations may occur [1,3,9,10]:

- If all the unstable perturbations are convected downstream and decay in the frame of the aircraft, the instability is said to be *convective*. In this case one may perform a spatial stability analysis, i.e. solve the dispersion relation with real ω and complex k. One generally obtains two families of spatial branches, referred as k^+ and k^-. These respectively correspond to the causal downstream ($x \to +\infty$) and upstream ($x \to -\infty$) spatial growth or damping of perturbations.
- If at least one perturbation can grow in place in the frame of the aircraft, the instability is said to be *absolute*. In this case, perturbations grow temporally in the frame of the aircraft, and the spatial approach does not make sense.

Before performing a spatial stability analysis, one must first determine the absolute or convective nature of the instability. For this purpose one has to consider the behavior of perturbations with zero group velocity in the aircraft reference frame. These perturbations correspond to *saddle points* of the dispersion relation, and their wavenumbers $k_0 = k_{0,r} + ik_{0,i}$ satisfy

$$D(k_0, \omega_0; a/b, \theta) = 0, \tag{3}$$

$$\frac{d\omega}{dk}(k_0, \omega_0; a/b, \theta) = 0. \tag{4}$$

The instability is absolute if one can find a saddle point (k_0, ω_0) satisfying the Briggs-Bers criterion [1,3,9,10], i.e. such that $\omega_{0,i} > 0$ and if this saddle point results from a pinching between two spatial branches $k^+(\omega)$ and $k^-(\omega)$ (this latter condition ensures that the corresponding perturbation verifies causality).

There is generally no simple relation between the spatial and temporal stability results. However, in the particular case $\theta \ll 1$ (i.e. $U_0 \gg W_0$), the spatial growth rate can be related to the temporal growth rate by a Taylor expansion of the dispersion relation [10]. This result is formally equivalent to the Gaster relation [8] and takes the following form:

$$k^G(\omega) = \frac{\omega - i\omega^T_i(\omega/U_0)}{U_0}.$$

In particular, the maximum spatial growth rate, for $\theta \ll 1$, is:

$$k^G_{i,max} = -\frac{\omega^T_{i,max}}{U_0}. \tag{5}$$

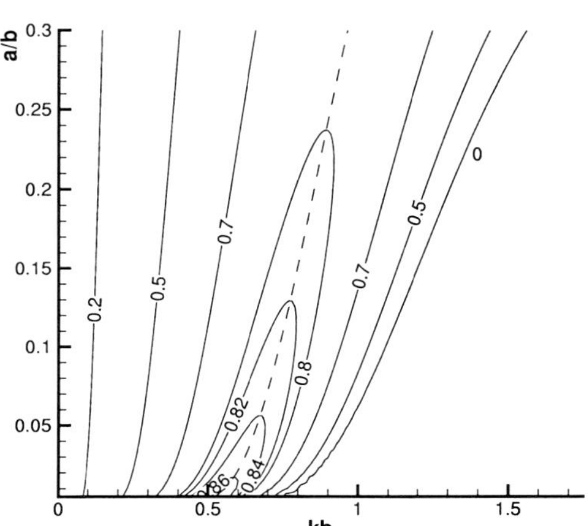

Fig. 2. Long-wave instability: temporal growth rate $(2\pi b^2/\Gamma)\omega_i^T$ as a function of (a/b) and of the nondimensional wavenumber kb. Dotted line: locus of the most amplified wavenumbers $k_{max}^T b$.

4 Long-Wave Instability

4.1 Dispersion Relation

Let us first consider the long-wave instability named after Crow [1], which results in symmetrical displacement of the vortex centerlines. This instability can sometimes be observed behind aircrafts in the presence of condensation in the cores of the vortices. Crow [1] modeled the vortex pair by two vortex filaments and used the Biot-Savart law and the cut-off approximation in order to calculate the vortex-induced velocities. This derivation assumes $(a/b)^2 \ll 1$ and $|kb| = O(1)$, i.e. small vortex core compared to vortex separation and perturbation wavelength of the order of the vortex spacing. In the reference frame moving with the vortex pair the dispersion relation for the symmetric modes (the antisymmetric ones are stable) takes the following form [1,17]:

$$D(k, \omega; a/b) = \left(\frac{2\pi b^2}{\Gamma}\right)^2 \omega^2 + (1 - \psi + \varpi)(1 + \chi - \varpi) = 0. \qquad (6)$$

Three different mechanisms enter the instability dynamics: the strain which induces the instability, a self-induced rotation with frequency ϖ, which re-stabilize each vortex for sufficiently large k and the mutual induction, characterized by

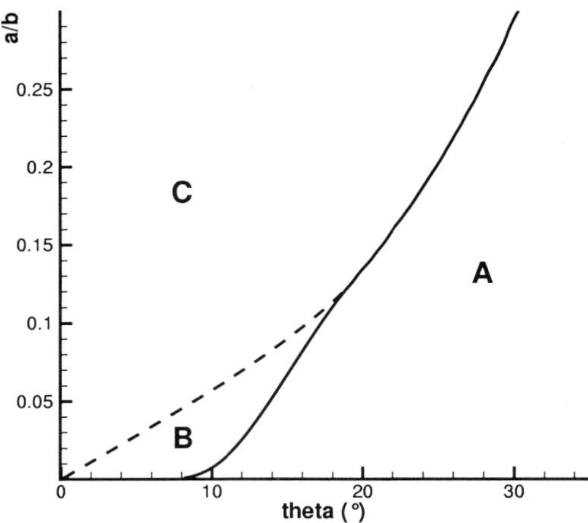

Fig. 3. Long-wave instability: convective/absolute regions. A: absolute instability. B and C: convective instability.

the functions χ and ψ. We extend these functions to complex values of k as

$$\begin{aligned}\varpi &= \frac{k^2b^2}{2}\left(\ln\frac{2}{ska} - \gamma + K\right),\\ \psi &= skbK_1(skb) + k^2b^2K_0(skb),\\ \chi &= skbK_1(skb),\end{aligned}$$

where K_0, K_1 are modified Bessel functions of the second kind, s is the sign of the real part of k, $\gamma \approx 0.577$ is the Euler constant, and the parameter K depends upon the vortex core structure [11,14,21]. For a constant vorticity core (Rankine vortex) $K = 1/4$, while for a Gaussian core vorticity distribution (Lamb-Oseen vortex) $K = 0.056$ [11,18]. In this section results are given for a Rankine vortex.

4.2 Temporal Instability

The temporal growth rate $\omega_i^T(k; a/b)$ given by equation (6) is plotted in figure 2. The maximum growth rate is $\omega_{i,max}^T \approx 0.8\Gamma/(2\pi b^2)$ and the corresponding wavelength varies with a/b. The more concentrated are the vortices, the longer is the most unstable wavelength. Temporal analysis [1] also predicts that the perturbations develop in planes tilted of an angle ϕ relative to the xy plane, with ϕ given by:

$$\tan\phi = \sqrt{\frac{1+\chi-\varpi}{1-\psi+\varpi}}. \tag{7}$$

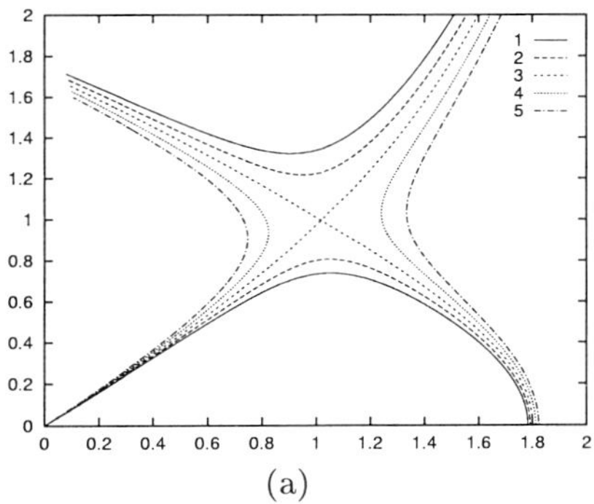

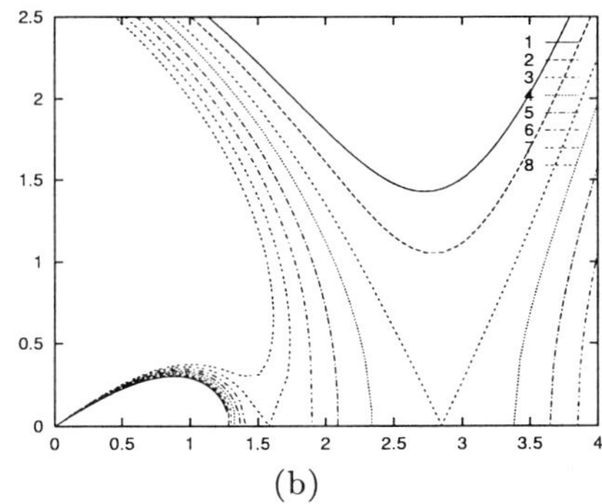

Fig. 4. Long-wave instability: details of the pinching between k^+ and k^- spatial branches in the complex plane $(k_r, -k_i)$ as θ rises across the convective/absolute threshold. (a) : $a/b = 0.3$. Curves labeled 1 5 : θ varies from 29, 5 to 31, 5. (b) : $a/b = 0.1$. Curves labeled 1 8 : θ varies from 15, 9 to 17, 7.

Inspection shows that for a/b varying from 0.01 to 0.3, the most unstable mode corresponds to $\phi \approx 48$. Finally, one notes that equation (6) predicts a second instability domain for larger values of k which is spurious [17] because these wavenumbers do not verify the assumption $|kb| = O(1)$.

4.3 Convective/Absolute and Spatial Stability Analyses

We now rewrite dispersion relation (6) in the frame of the aircraft:

$$D(k, \omega, a/b, \theta) = \left(\frac{2\pi b^2}{\Gamma}\right)^2 (\omega - kU_0)^2 + (1 - \psi + \varpi)(1 + \chi - \varpi) = 0. \qquad (8)$$

We numerically investigated dispersion relation (8) in order to determine the absolute and convective instability regions in the $(\theta, a/b)$ parameters plane. For the considered parameters it is found that this dispersion relation admits three saddle points in the complex half-plane $k_i > 0$. One of these roots is always real and cannot satisfy the absolute instability condition $\omega_{0,i} > 0$. Considering the two other roots, one may identify three regions in the plane of parameters, which are depicted in figure 3.

- Region A: two saddle points are complex conjugates, and one of them verifies $\omega_i > 0$. As will be confirmed by the inspection of spatial branches in the vicinity of the threshold, the instability is absolute.
- Region B: three saddle points are real. The instability is convective.
- Region C: two saddle points are complex conjugates and do not satisfy $\omega_i > 0$, and the third is real. The instability is convective.

Figure 4-(a) details the pinching between k^+ and k^- branches is the vicinity of the transition from region C to region A, for $a/b = 0.3$. Figure 4-(b) shows

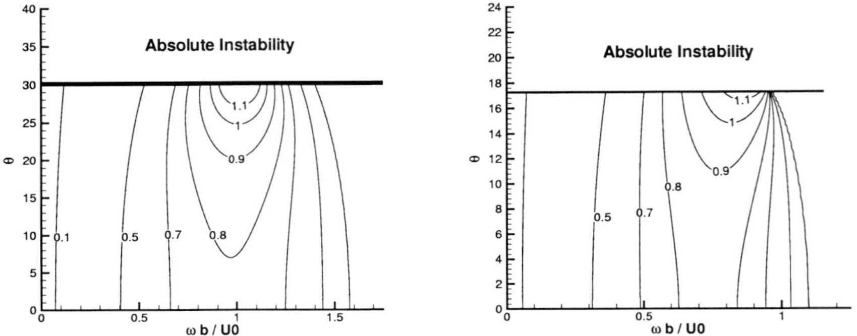

Fig. 5. Long-wave instability: spatial growth rate $\left(2\pi b^2/\Gamma\right) k_i U_0$ as function of the nondimensional excitation frequency and of θ. Left plot: $a/b = 0.3$. Right plot: $a/b = 0.1$.

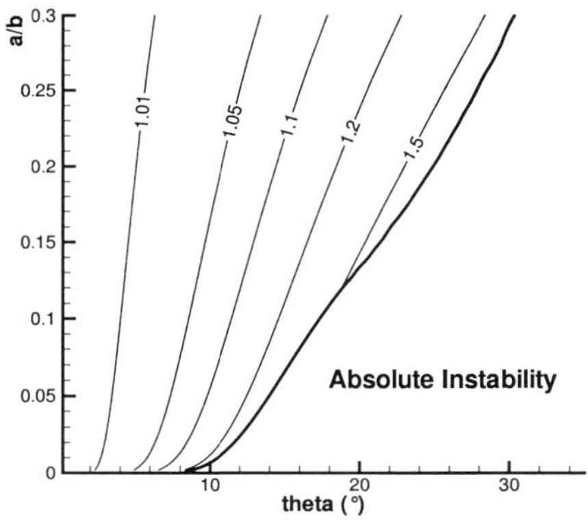

Fig. 6. Long-wave instability: ratio $k_{i,max}/k^G_{i,max}$ from the spatial growth rate to the corresponding temporal growth rate displayed in the region of convective instability.

the same detail for $a/b = 0.1$. In this case the transition involves a double pinching on the real axis. As θ is increased, the k^- spatial branch undergoes a first pinching with a neutral branch, and then a second pinching occurs with the k^+ branch. Between these two pinchings one observes the presence of three real saddle points. Figure 5 shows the spatial growth rates of the long wave instability for $a/b = 0.3$ and 0.1. In order to allow a comparison with temporal results, $k_i U_0$ has been represented as a function of θ and ω/U_0. For $\theta \ll 1$ the result is equivalent to the temporal result presented in figure 2. Increasing θ leads to higher values of the growth rates and one may see that the corresponding

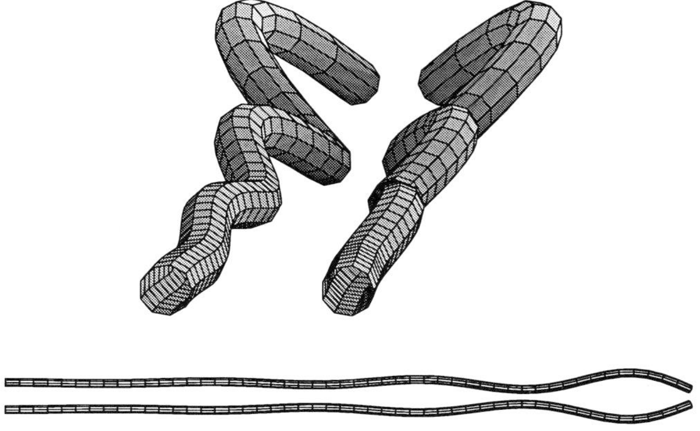

Fig. 7. Long-wave instability: example of spatial mode ($a/b = 0.15, \theta = 10$). Perspective and top views.

frequency slightly increases. Figure 6 shows the ratio $k_{i,max}/k^G_{i,max}$ between the computed spatial growth rate and the asymptotic value given by equation (5). When approaching the absolute instability threshold, differences may reach 50%. Finally, it may be remarked that, contrary to temporal modes, spatial modes do not lie in planes: as θ is increased, the progressively take an helical shape. As an illustration, we have represented in figure 7 the most instable spatial mode for the set of parameters $a/b = 0.15, \theta = 10$. The helical shape is slightly visible.

5 Short-wave instability

The second type of instability we consider is the short-wave one [20,15]. This instability arises in the neighborhood of wavenumber values k_c where a resonance condition occurs between two inertial (Kelvin) waves of one isolated vortex and the straining field induced by the other vortex. The dispersion relation is derived in references [15,17] by using an asymptotic method, assuming weak strain, i.e. $(a/b)^2 \ll 1$ and short waves i.e. $|ka| = O(1)$:

$$\left(\frac{2\pi a^2}{\Gamma}\right)^2 \omega^2 = Q^2(ka - k_c a)^2 - (a/b)^4 R^2. \tag{9}$$

Q and R have to be determined for each k_c and depend on the vortex core type. These constants have been computed by Tsai & Widnall [20] for a constant vorticity core (Rankine vortex), and by Eloy & Le Dizès [7] and Sipp [18] for a Gaussian vorticity core (Lamb-Oseen vortex). The temporal stability analysis [15,20] predicts the existence of a narrow band of unstable wavenumbers centered on each k_c, with a width $|ka - k_c a| < (a/b)^2 R/Q$ and a maximal growth rate $\omega^T_{i,max} = \Gamma/(2\pi b^2)R$. Writing dispersion relation (9) in the frame of the aircraft

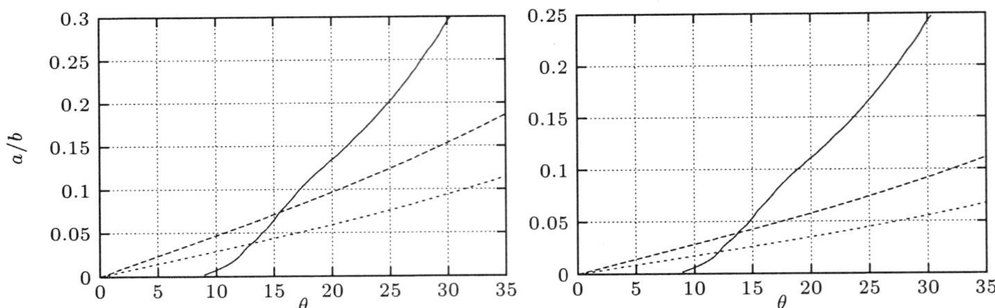

Fig. 8. Convective/absolute transition thresholds for a Rankine vortex (left), and a Lamb-Oseen vortex (right). Solid line: long-wave instability. Dashed lines: first two modes of short-wave instability. The curves separate the region of convective instability (up, left) from the region of absolute instability (bottom, right).

leads to:

$$D(k,\omega;a/b,\theta) = \left(\frac{2\pi b^2}{\Gamma}\right)^2 (\omega - kU_0)^2 - (b/a)^2 Q^2 (kb - k_c b)^2 + R^2 = 0. \quad (10)$$

This simple dispersion relation, which is quadratic in k and ω, constitutes a generic model for weakly interacting non-dispersive waves and its absolute/convective properties are well known [1,13]. The saddle points of this relation are solutions of

$$\left[1 - \left(\frac{b}{a}\right)^2 \left(\frac{W_0}{U_0}\right)^2 Q^2\right] (kb - k_c b)^2 = \left(\frac{a}{b}\right)^2 \frac{R^2}{Q^2}, \quad (11)$$

and it is easily seen that the condition for absolute instability is $W_0/U_0 > (a/b)Q^{-1}$. In the convective instability regime the maximum spatial growth rate is given by

$$k_{i,max} U_0 = \frac{-iR}{1 - Q^2 (b/a)^2 (W_0/U_0)^2} \left(\frac{\Gamma}{2\pi b^2}\right),$$

to be compared to the asymptotic result (5), for $\theta \ll 1$, given by $k^G_{i,max} U_0 = -iR\Gamma/(2\pi b^2)$. The absolute/convective instability thresholds for the first two unstable wavenumbers k_c are represented in figure (8) and compared to the result for the long-wave instability (for higher unstable wavenumbers k_c, the convective/absolute instability transition occurs for larger θ). Except for very small ratios a/b, the convective/absolute transition of the short-wave instability occurs for angles θ larger than for the long-wave instability.

6 Conclusions

In this study the convective/absolute and spatial stability analyses of the long– and short–wave instabilities developing in a pair of counter-rotating vortices have been performed. For this system, only the temporal instability properties were known [1,15,20]. Recent analyses have revealed the possibility of absolute instabilities in isolated vortices in presence of core axial flow [6,14,16]. The present results show that absolute instabilities may develop in a vortex pair, even in the absence of core axial flow.

Two parameters appear in the problem: the ratio a/b of the core radius of the vortices to their separation and the angle $\theta = \tan^{-1}(W_0/U_0)$ characterizing the ratio of the pair descent velocity to the axial advection velocity. The absolute and convective instability regions have been determined in the plane of these two parameters, and the spatial growth rates have been evaluated in the convective region. When $\theta \to 0$, the spatial stability analysis is fully equivalent to the temporal stability analysis. When $\theta \ll 1$ both analyses give similar results. For sufficiently large θ, instabilities may become absolute.

References

1. A. Bers. Linear waves and instabilities. In C. De Witt and J. Peyraud, editors, *Physique des Plasmas*, 117–225. Gordon & Breach, New York, 1975.
2. A. J. Bilanin and S. E. Widnall. Aircraft wake dissipation by sinusoidal instability and vortex breakdown. *AIAA-73-107*, 1973.
3. R. J. Briggs. *Electron-Stream Interaction with Plasmas.* MIT Press, Cambridge, Mass., 1964.
4. S. C. Crow. Stability theory for a pair of trailing vortices. *AIAA J.*, 8:2172–2179, 1970.
5. S. C. Crow and E.R. Bate. Lifespan of trailing vortices in a turbulent athmosphere. *J. Aircraft*, 13(7):476–482, 1976.
6. I. Delbende, J.M. Chomaz, and P. Huerre. Absolute/convective instabilities in the Batchelor vortex: a numerical study of the linear impulse response. *J. Fluid Mech.*, 355:229–254, 1998.
7. C. Eloy and S. Le Dizs. Instability of the Burgers and Lamb-Oseen vortices in a strain field. *J. Fluid. Mech.*, 378:145–166, 1999.
8. M. Gaster. Growth of disturbances in both space and time. *Phys. Fluids*, 11:723–27, 1968.
9. P. Huerre and P. A. Monkewitz. Local and global instabilities in spatially developing flows. *Annu. Rev. Fluid Mech.*, 22:473–537, 1990.
10. P. Huerre and M. Rossi. Hydrodynamic instabilities in open flows. In C. Godrèche and P. Manneville, editors, *Hydrodynamic and Nonlinear Instabilities*, 81–294. Cambridge University Press, Cambridge (UK), 1998.
11. S. Leibovitch, S.N. Brown, and Y. Patel. Bending waves on inviscid columnar vortices. *J. Fluid Mech.*, 173:595–624, 1986.
12. T. Leweke and C. H. K. Williamson. Cooperative elliptic instability of a vortex pair. *J. Fluid Mech.*, 360:85–119, 1998.
13. E. M. Lifshitz and L. P. Pitaevskii. *Physical Kinetics.* Pergamon Press, Oxford, 1981.
14. T. Loiseleux, J.M. Chomaz, and P. Huerre. The effect of swirl on jets and wakes: Linear instability of the Rankine vortex with axial flow. *Phys. Fluids*, 10(5):1120–1134, 1998.

15. D. W. Moore and P. G. Saffman. The instability of a straight vortex filament in a strain field. *Proc. R. Soc. Lond. A*, 346:413–425, 1975.
16. C. Olendraru, A. Sellier, M. Rossi, and P. Huerre. Inviscid instability of the batchelor vortex: absolute-convective transitions and spatial branches. *Phys. Fluids*, 11(7):1805–1820, 1999.
17. P. G. Saffman. *Vortex Dynamics.* Cambridge Univ. Press, Cambridge, UK, 1992.
18. D. Sipp. Instabilits dans les coulements tourbillonnaires. PhD Thesis, École polytechnique, Palaiseau, France, 1999
19. P.R. Spalart. Airplane trailing vortices. *Annu. Rev. Fluid Mech.*, 30:107–138, 1998.
20. C.-Y. Tsai and S. E. Widnall. The stability of short waves on a straight vortex filament in a weak externally imposed strain field. *J. Fluid Mech.*, 73:721–733, 1976.
21. S. E. Widnall, D. Bliss, and A. Zalay. Theoretical and experimental study of the stability of a vortex pair. In *Aircraft Wake Turbulence.* Plenum Press, 1971.
22. C. H. K. Williamson, T. Leweke, and G. D. Miller. Fundamental instabilities in spatially-developing wing wakes and temporally-developing vortex pairs. In *Proceedings of FEDSM '98*, 1998. FEDSM98-4993.

Ultrasound Scattering by Forced Laminar Wakes

Rodrigo H. Hernández[1] and Christophe Baudet[1]

[1] LEAF-NL, Depto. Ingeniería Mecánica
- Universidad de Chile Casilla 2777, Santiago, Chile
[2] Laboratoire de Physique (UMR 5672),
Ecole Normale Supérieure de Lyon
46 Allée d'Italie 69364 Lyon, France.

1 Introduction

Most of flows found in nature belong to the class of spatially developing open flows. In this hydrodynamic class, fluid particles enter and leave the flow boundaries of the observation domain continuously. It is the case of boundary layers [1], jets [3,4] and notably wakes [5].

Actually, concepts of absolute and convective instability have proved to be useful to classify the different types of open flows according to their local dynamic behavior. A flow system like the wake, created by the presence of a thin flat plate, fall into the class of noise amplifiers [6,7]. The system become very sensitive to external noise which it is therefore amplified. It is well known that every portion of the wake is convective (stable or unstable); the system presents an extrinsic dynamics, what means that the spatial evolution of the flow is essentially determined, either by the external noise entering the system or by a coherent particular applied forcing. A wake can also display an intrinsic dynamics either by some adequate hydrodynamic resonance or by the onset of localized regions of absolute instability. Vortical structures that are formed become however very insensitive to the incoming external noise and display a very definite oscillation frequency [9].

A renewed interest on wakes comes from the idea of disorder and, perhaps more properly, transition to turbulence. The wake of a thin flat plate does not present spontaneous self–sustained oscillations. Initial disturbances can grow and then saturate but are continuously advected by the mean flow. However its sensitiveness to coherent external forcing makes possible to recover a very rich family of spatially evolving vortical structures. The onset of dynamical vortical structures at large and small scales will determine any further behavior of the wake.

This paper is concerned with the pattern forming structures on the wake of a flat plate under forcing. This choice was made mainly because the laminar velocity profile downstream the plate shows two well defined symmetric sheets of vorticity (of opposite sign). If the plate is submitted to a controlled forcing, one will find its prints in the wake under the form of a spatio–temporal modulation of velocity profiles and thereafter the onset of spatial modes of vorticity. These modes make the modulated wake a benchmark for an ultrasound scattering experiment.

Scattering of sound waves of high frequency in air by laminar and turbulent vortex flows has been the subject of some recent experimental [10–12] as well as theoretical works [13–15]. The presence of vorticity and the further non linear coupling between the sound wave and the base vortex flow guarantees a coherent sound emission (scattering) which gives considerable information of its spatial structure as well as about its temporal dynamics.

Theoretical works have clearly concluded that the process of sound scattering is well adapted for an acoustic flow diagnostic. In the first Born approximation, the sound scattered pressure (or density) is found to be proportional to the Fourier transform in space and time of the vorticity of the base flow $\boldsymbol{\Omega}(\mathbf{k}, \nu)$, where $\mathbf{k}$ is the wave vector and ν is the frequency. This constitutes a spectral probe of the vortex flow structure.

Well controlled and coherent vortical structures are obtained if we remain in the category of noise amplifiers. Therefore the flat plate must be very thin, i.e., the thickness–to–length ratio must be small, $e/b \ll 1$. Thus, if the wake is stationary, one can apply an adequate forcing without triggering eigenmodes caused by any instability onset. On the contrary, the system falls into the class of oscillators [7,16] presenting an intrinsic dynamics. Even if fluid particles leave the domain of interest, they remain enough time inside so that infinitesimal disturbances can grow, giving place to the onset of global modes of the system. Comprehensive examples are the flow around a circular [8,9] or square (varying aspect ratio) [17,16,18] cylinders.

At low values of the plate's Reynolds number, $Re_t = U_o e/\underline{\nu}$, (where $U_o, e, \underline{\nu}$ are the free stream mean velocity, plate thickness and kinematic viscosity respectively), and if the boundary layer remain laminar without separation, the near wake remain stationary. Even though two small recirculation regions attached to the trailing edge of the plate always exists [17]. Recirculation regions are unavoidable, even if the cross–stream plate dimension (thickness e) is small. However the smaller the thickness e the shorter the turnover time will be, so avoiding the onset of instabilities of the wake originated on these recirculating regions.

Modulation of the velocity profile downstream from the plate is performed through a harmonic forcing. A flat plate performing rotary oscillation about the leading edge, creates necessarily a vorticity modulated wake. This is accompanied by the onset of well defined spatial modes of vorticity because the forcing is a priori known.

Forcing consists in making vary the angle of attack of the plate through a periodic oscillation of the plate support (see Figure 1). A very broad range of spatial patterns of vorticity $\boldsymbol{\Omega}(\mathbf{r}, t)$ can be obtained through simple settings of the wave form amplitude ζ_m, the uniform upstream velocity U_o, and the wave form frequency f_m.

If the oscillation amplitude ζ_m is too high (and therefore the angle of attack α) there will be strong longitudinal pressure gradients along the plates's faces. Boundary layer separation will be enhanced and further but undesired vortex shedding sets up, perturbing the initially modulated wake pattern [19]. In

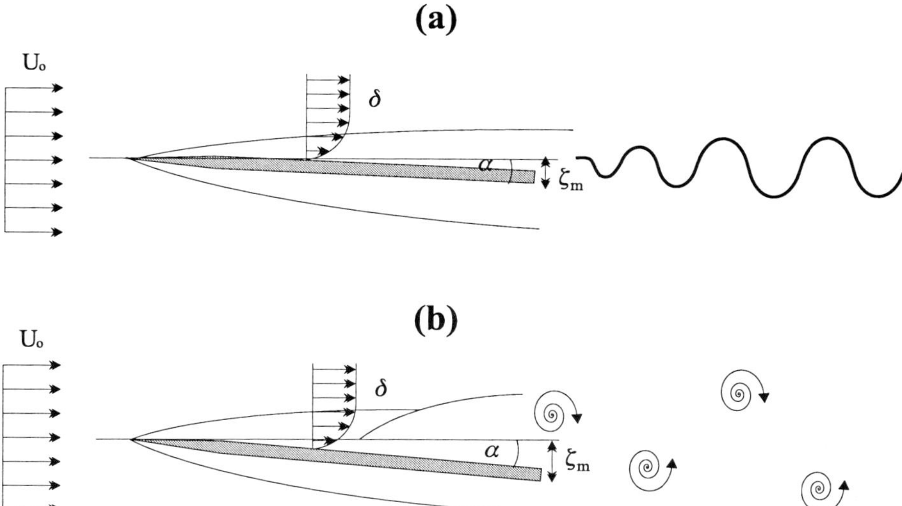

Fig. 1. Expected periodic patterns on a modulated wake through a periodic forced oscillation of the trailing edge of a thin flat plate. (a) Wavy wake for small forcing amplitude ζ_m, (b) Large scale vortex shedding for higher forcing amplitude ζ_m.

this case, we will get nevertheless a vorticity–modulated wake, but over spatial scales of the order of the length of the plate. An obvious very interesting case but inadequate for an ultrasound scattering experiment in air, where the sound wavelength is too small with respect to those spatial scales. Therefore here the modulation amplitude, ζ_m, will be kept of the order of e.

This situation is doubly advantageous. First the transverse (cross stream) spatial scale of vorticity is small enough and therefore better adapted to ultrasound sound wavelengths. On the other hand a linear forcing becomes feasible at even higher frequencies.

Even using a very simple way of forcing like here, the wake's spectral dynamics can be very complicated due to non linear interactions. At first, we can expect at least two scenarios as a function of ζ_m and U_o according to the symmetry of the vorticity distribution that one gets (Figure 1).

This system presents two characteristic time–scales: one given by the forcing frequency, f_m^{-1} and the other by the ratio b/U_o both of them defining a characteristic dimensionless parameter $F = f_m b/U_o$ of the forced wake.

At low forcing frequencies, f_m for a $F < 1$ condition, we get a slowly varying streamwise sinusoidal wake pattern, where the wave number or spatial mode is given by $2\pi/\lambda_\Omega$ and $\lambda_\Omega = U_o/f_m$ is the resulting vorticity wavelength. At each forcing cycle, fluid particles under the boundary layer have already left the surface of the plate, therefore they 'feel' a non localized spectral forcing (frequency is undefined).

On the contrary, if we increase f_m approaching the $F > 1$ case, the vorticity distribution becomes more complicated, since fluid particles don't have left the plate before a cycle of oscillation takes place. So particles under the boundary layer are now submitted to a harmonic forcing in time. We can now trigger instabilities of the boundary layer itself [2].

In order to remain in the neighborhood of the $F \sim 1$ condition, but working with reasonable high forcing frequencies f_m, we chose a plate where the width b is small enough so that $F \sim 1$ but greater than e so that we don't loose the class of noise amplifier of the whole system for a wide range of flow velocities U_o.

2 Experimental Set Up

We use a metal (copper, small roughness) thin plate of length $l = 20$ cm, width $b = 4$ cm and thickness $e = 1.6$ mm. The plate is placed in the test section of an open low turbulence wind tunnel. (maximum free stream velocity $U_o = 2$ m/s, rate of turbulence 0.05 %). The leading and trailing edges of the plate are of semi–cylindrical and triangular shape respectively [20].

The plate is maintained in vertical position through two half–axes passing through the geometric center of the front end of the plate (at the middle of the test section). We use ball-bearings supporting axes to perform oscillatory forcing without too much friction. (Figure 2a).

Oscillatory forcing is performed with a B&K electromagnetic shaker coupled to a rigid aluminum arm system which is fixed on both sides of the plate. The whole system was mounted on a completely independent support, uncoupled from the wind tunnel in order to avoid vibrations associated to fans. Measures of wake velocity profiles were made with a hot wire probe TSI-1210-60 at a distance of $x/e = 79.4$ downstream the plate. It can be moved across the wake over $|y/e| < 11$ with the help of a computer controlled XY stepper motor system.

Data acquisition, periodic forcing and probe movements were computer controlled using a HP700 series work station and a 4 channel HP3565 spectrum analyzer. A sweep in forcing amplitude as well as in forcing frequency was possible using a HP33120A function generator through the HPIB bus. Forcing waveforms signals were previously amplified by a KEPCO BOP-50-4M amplifier before the B&K shaker.

The zero attack angle $\alpha = 0$ was found iteratively, through a cycle of measures of the symmetry of the velocity profiles downstream the plate. Successive corrections were done through a micrometer screw to change α.

2.1 Sound scattering

Ultrasound scattering by vortex flows appears as a consequence of non linear coupling between an incident sound wave and a target vorticity distribution. It can be found that the scattered acoustic pressure p_s by an arbitrary vorticity field $\boldsymbol{\Omega}(\mathbf{r}, t)$ is directly proportional to the spatio–temporal Fourier transform of vorticity, $\Omega(\mathbf{q}, \nu)$. In three dimensions,using the first Born approximation, p_s can be written as [15],

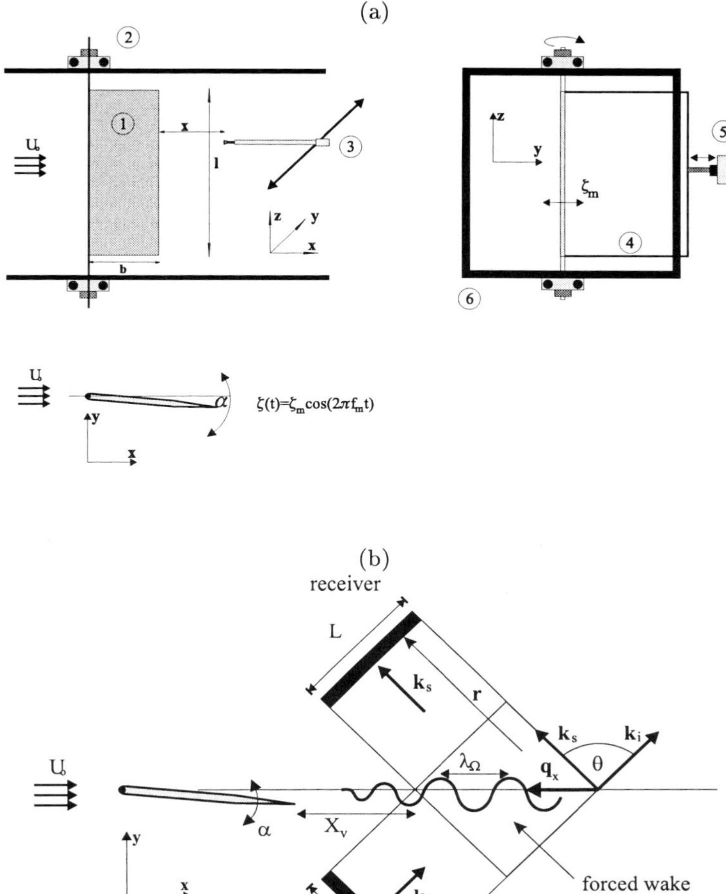

Fig. 2. (a) Experimental Set Up. (1) A vertical metallic flat plate of small roughness. (2) Supporting axes mounted on ball bearings. (3) Hot wire probe scanning the whole test section. (4) Aluminium supporting axes. (5) Shaker. (6) Test section. (b) Sound scattering experimental set up. Square (size $\Lambda = 15$ cm) Sell type ultrasound emitter and receiver under a symmetric configuration. The scattering angle between the incident (k_i) and the scattered (k_s) wave vector is $\theta = 30^o$. The beginning of the scattering volume, x_v, corresponds to a $x_v/b = 13$ ratio downstream the plate.

$$p_s(\mathbf{r},\nu) = p_o \frac{i\nu\pi^2}{c^2|\mathbf{r}|} \mathrm{e}^{i2\pi\nu\mathbf{r}/c} \frac{\cos\theta}{1-\cos\theta} \sin\theta \, \Omega_z(\mathbf{q},\nu-\nu_o) \tag{1}$$

Where p_o is the incident sound pressure, θ the scattering angle, $\mathbf{r}$ the distance from the target, c the speed of sound and ν_o the incoming sound frequency. The scattering wave vector is defined by $\mathbf{q} = \mathbf{k}_s - \mathbf{k}_i$, and can be written as $q \sim 4\pi(\nu_o/c)\sin(\theta/2)\hat{x}$ in our experimental scattering set up (because $\mathbf{q}\cdot\hat{y} = 0$).

Sound scattering experiments were performed under the symmetrical configuration shown by figure 2 b). Both the sound emitter and receiver are square Sell type transducers of size $\Lambda = 16$ cm having a flat response between 5 to 100 kHz within 10 dB. A detailed description can be found in previous work on sound scattering experiments [10,11].

The emitter and receiver (placed outside the test section) are focused toward the wake of the plate (Figure 2 b). The incident, $\mathbf{k}_i$, and scattered, $\mathbf{k}_s$, wave vectors forming an angle θ. With such a geometry we probe the vortex flow at length scales corresponding to the wave vector $\mathbf{q} = \mathbf{k}_s - \mathbf{k}_i$ whose components are $q_y = 0$ and $q_x = 4\pi(\nu_o/c)\sin(\theta/2)$. For the actual set–up, considering a constant angle $\theta = 30^o$, the wave vector $\mathbf{q}$ can be varied simply by changing the emitter frequency ν_o thus probing different length scales associated to the base vortex flow.

Diffraction effects coming from the limited size of the sound transducers were measured with a transmission diffraction grating of constant step 1.70 ± 0.01 cm. We found a measured diffraction angle of the order of 3° at 40 kHz.

We will see that the presence of the mean flow introduces a frequency shift on the acoustic signal, a Doppler effect [10], around the incoming frequency ν_o. Broadening of scattering peaks is due to diffraction effects which in turn determine our spectral resolution.

A heterodyne detection procedure (demodulation of the received acoustic signal) gives us an analytic signal [21] of low-frequency that can be easily sampled. The phase of this analytic signal is directly related to the Doppler shift. Working at constant angle θ the theoretical resolution on a single spectral component is given by, $\Delta q = (4\pi/\Lambda)cos\theta/2$ [11].

3 Results

3.1 Hot Wire Wake Measurements

To know accurately the wake response to different forcing regimes, we performed systematic measurements of wake velocity profiles using thermal anemometry. Scanning the cross stream coordinate with the hot wire probe gave us a complementary picture of the wake dynamics before to proceed with scattering experiments. Note that we use harmonic forcing in time, i.e., $\zeta(t) = \zeta_m \cos(2\pi t f_m)$ where the amplitude ζ_m remains always weak ($\sim e$). As we mention above, we will create periodic oscillations of the near wake which are advected downstream, forming spatial vorticity modes given approximately by a simple dispersion relation of the form $\lambda_\Omega \sim U_o/f_m$. Remember that if the system remains as a noise amplifier, we can obtain a very rich family of wake's spatial modes.

Once the zero attack angle is obtained, wake velocity recordings without forcing have the typical aspect shown in figure 3a). We can see also the cross spatial scale of vorticity computed approximately by $\sim \partial u/\partial y$. It is evident that vorticity has two terms $\Omega = \partial u_y/\partial x - \partial u_x/\partial y$, but the hot wire probe only measures absolute values of velocity. So, we get only an approximate of cross stream velocity gradients. Accurate vorticity measurements are performed in the scattering experiment.

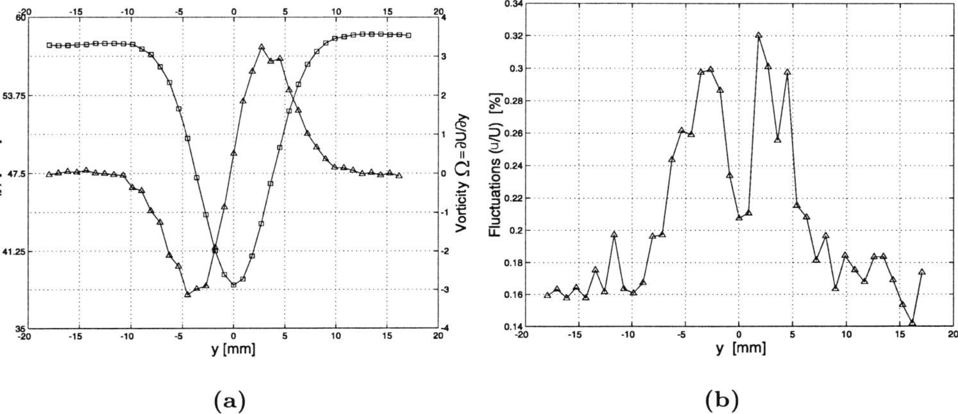

Fig. 3. Free wake behavior from hot wire measurements at $x/e = 79.4$ with $\alpha = 0°$ and $U_o = 57$ cm/s. (a) Mean velocity profile (⊔) . We show also an estimation of the wake's vorticity (△) using first order numerical derivatives. (b) Wake velocity fluctuations.

Even if the wake's velocity fluctuations are weak, $u'(y)/\overline{u}$ of the order of < 0.3 %, two maxima are found precisely where the velocity gradient is maximum (Figure 3b).

This figure gives us an idea of the spatial extent of vorticity, and in particular, of the angle $\beta \sim 2.26^o$ that determines the (weakly) divergent behavior of the wake. It is an effect of the sharpness of the plate and its width. The Reynolds number based on the plate's width, Re_b allow to estimate the boundary layer thickness at the plate's end; $\delta = b/\sqrt{(Re_b)} \sim 1.7$ mm ($\delta/e \sim 1$). Figure 3a) shows roughly that the cross spatial scale of vorticity is of the order of $3.8e$.

In general, even at very weak forcing amplitudes, the wake velocity fluctuations are very high. In some cases there is a factor of 10 between them and the natural wake velocity fluctuations (without forcing). On figure 4 we show, as an example, two temporal series of wake velocity under harmonic forcing of constant frequency $f_m = 10$ Hz. We see both the fundamental mode far from the plate $y \sim -8e$, and the first harmonic close to the plate.

A fundamental difference between a natural and a forced wake is the cross stream increased scale of the later. Figure 5 shows how the wake is enlarged under an harmonic forcing and that this effect is more important at $f_m = 5$ Hz than $f_m = 14$ Hz.

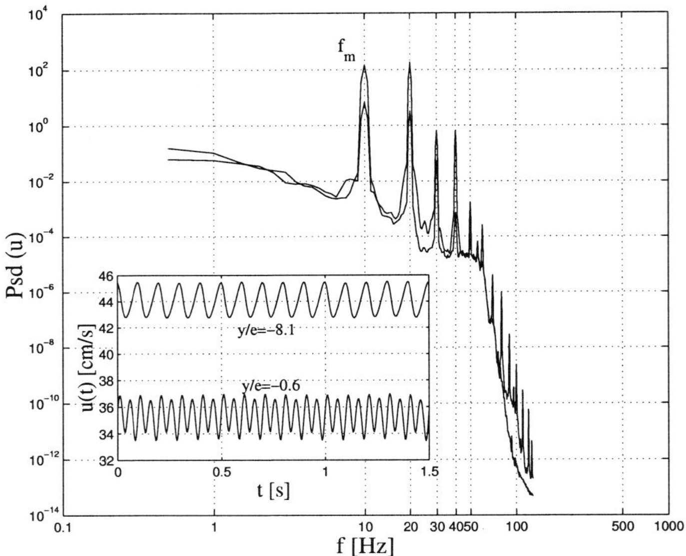

Fig. 4. Forced wake behavior from hot wire measurements. Forcing with amplitude $\zeta_m = 0.3$ mm at $f_m = 10$ Hz and Reynolds number $Re_b = 1185$. Power spectral density computed at two cross stream sites, $y/e = -0.6, -8.1$. The subplot shows typical velocity recordings at those positions.

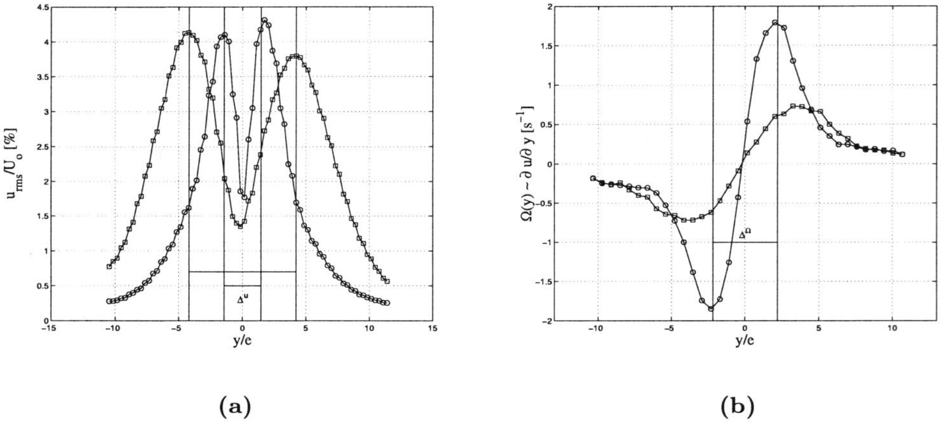

Fig. 5. Wake forcing at two different frequencies $f_m = 5$ (∘) and $f_m = 14$ (⊔) Hz with same amplitude $\zeta_m = 0.3$ mm and Reynolds number $Re_b = 1185$. (a) The velocity fluctuations $u'(y)/U_o$ $(x/e = 79.4)$ and (b) An estimation of the wake vorticity $\Omega(y)$. We indicate the increased cross stream scale of wake velocity fluctuations (Δ^u) and vorticity (Δ^Ω) with the forcing frequency.

A comprehensive picture of these effects is found on figure 6. The shape of mean velocity profiles $u(y) = \overline{u(y,t)}$ and their local fluctuations $u'(y) = u_{rms}$ as a function of the forcing frequency f_m (with constant forcing amplitude $\zeta_m = 0.3$ mm) is compared with the natural situation (without forcing). We note an increase of both local mean velocity and local fluctuations behind the plate as the forcing frequency decreases. On figure 7 we show the power spectrum of velocity fluctuations at different increasing forcing frequencies. This frequency and its harmonics appears clearly on each plot.

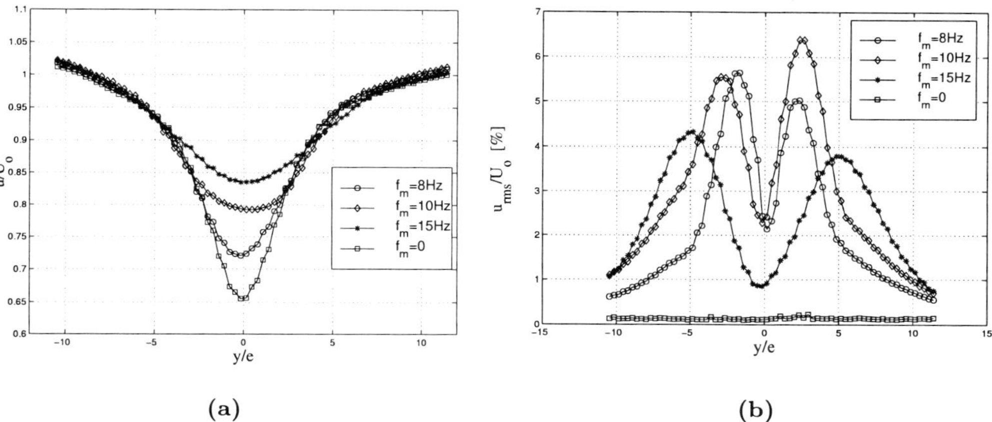

Fig. 6. Systematic effect of forcing frequency on the wake behavior ($x/e = 79.4$) at constant amplitude $\zeta_m = 0.3$ mm and Reynolds number $Re_b = 1185$. (a) Mean velocity profiles $\overline{u}/U_o$ and (b) velocity fluctuations u'/U_o. Note that the free wake behavior is indicated by a $\sqcup$ ($f_m = 0$ Hz).

3.2 Sound Scattering and Spectral Wake Dynamics

As we have shown above, a kind of harmonic forcing like $\zeta(t) = \zeta_m \cos(2\pi f_m t)$ traduces into an oscillatory wake pattern. Any wake modulation traduces into a vorticity modulation at the forcing frequency f_m.

If we compute the spatial and temporal Fourier transform of the vorticity equation in 2D, neglecting diffusion terms, we obtain a dispersion relation relating frequency and wave number of the form; $2\pi f = \mathbf{q} \cdot \mathbf{u}$ where $\mathbf{u}$ corresponds to the advection velocity of vorticity. This relationship will explain why we find a Doppler effect associated to scattered pressure signals [11]. We will prove that a scattering peak is found when the scattering wave vector q matches the wave length of the spatial vorticity pattern λ_Ω related to the forcing frequency by $\lambda_\Omega \sim U_o/f_m$.

A first agreement with theory (Eq.1) is the fact that the scattering pressure signal, p_s, shows a Doppler frequency shift $(\nu_o - f_m)$, where its sign is given by

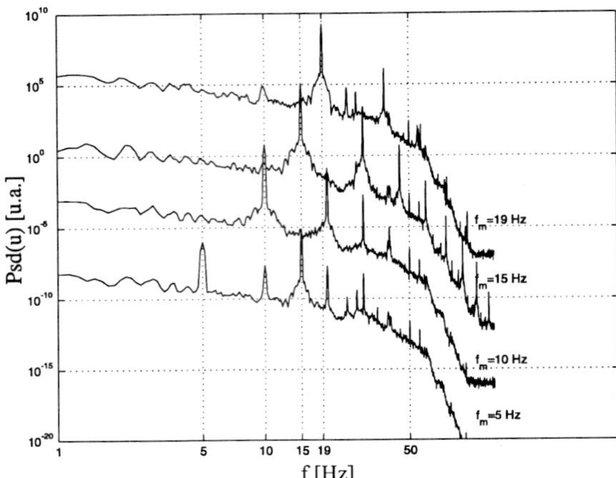

Fig. 7. Power spectra of wake velocity fluctuations at different forcing frequencies. Hot wire measurements performed at $y/e = -8.1$, $x/e = 59.4$ for a constant forcing amplitude $\zeta_m = 0.3$ mm and Reynolds number of $Re_b = 10^3$.

the sense of the scattering vector q. After a heterodyne demodulation, the power spectrum of the scattered pressure signal is centered on ν_o and one measures with good accuracy ($d\nu = 50$ mHz) the spectral line associated to p_s, as shown by the subplot of figure 8. The scattering peak appears exactly at $\nu - \nu_o = f_m$ as predicts the equation 1.

If we change the incoming sound frequency ν_o, for a constant scattering angle $\theta/2$, we are exploring a wide band of spectral components of the wake pattern. Remember that the scattering vector is given by $q = 4\pi(\nu_o/c)\sin(\theta/2)$.

After a frequency sweep of the incoming sound frequency, ν_o, we record the evolution of the Doppler spectral line at f_m as a function of the scattering vector q. We found, as expected, a resonant behavior for the scattering amplitude p_s as shown in figure 8. A resonant scattering vector q_r associated to a particular sound frequency, corresponds to the fundamental wave vector of the modulated wake pattern under forcing, as given by the relationship $2\pi/\lambda_\Omega$. The second peak in figure 8 (indicated as q_r^Λ), originates by diffraction effects due to the finite size window of the sound emitter and receiver. The spectral window is theoretically given by a sin_c profile for square windows. Note that the relative position of the peak $q_r^\Lambda - q_r$ does not change with the mean wake velocity as it was already proved in other experiments [11].

For a constant forcing frequency f_m and varying upstream flow velocity U_o, the spectral wake components will change, obeying to the simple relationship, $\lambda_\Omega \sim U_o/f_m$. If we consider that the sound wave is modulated by the vorticity pattern, the corresponding scattering spectral line obtained at different flow velocities must not change, even though the resonant wave vector q_r must be different. This is confirmed on figure 9. The Doppler peak on the power spectrum (Figure 9a) is independent of the flow velocity. However the corresponding

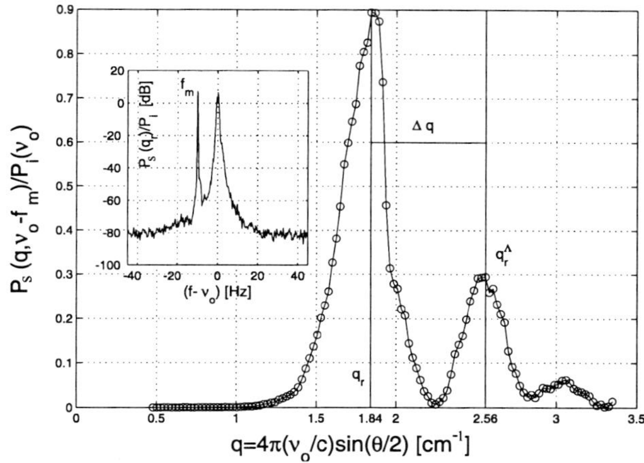

Fig. 8. Ultrasound scattering by the forced wake (forcing amplitude, frequency and Reynolds number are: $\zeta_m = 0.3$, mm, $f_m = 10$ Hz, $Re_b = 10^3$ respectively). We display the normalized amplitude of the scattered pressure as a function of the scattering wave vector q. The spatial resonance is obtained at $q = q_r$ ($\nu_o = 19.25$ kHz). The subplot shows the corresponding power spectrum of scattered pressure, where a Doppler peak is found at $f - \nu_o = f_m$.

resonant scattering vectors occur for very different wake pattern wave lengths (Figure 9b).

So we confirm the validity of the dispersion relation. The scalar product $\mathbf{q}{\cdot}\mathbf{u}$ is exactly conserved using the resonant scattering wave vectors. This picture allow us to understand the sound scattering mechanism, where the incoming sound wave is linearly modulated in frequency, which explains the Doppler shift on scattered pressure signals. This is the reason why we use a heterodyne detection technique, largely used in the radio frequency domain. The carrier wave at ν_o is modulated at f_m which corresponds to the scattered pressure by flow vorticity.

If we take a look at the scattering set up (Figure 2b), we see that scattering pressure comes from the volume defined by the intersected acoustic beams (incident and scattered beams) downstream the plate. Therefore the advection velocity for the vorticity field is not exactly the free stream velocity U_o. An effort was made to determine a characteristic advection velocity from the velocity profiles measured downstream the plate with the hot wire probe.

The advection velocity is simply defined here as the spatial average (in the cross stream direction) of $u(y)$, $\langle u \rangle$,

$$\langle u \rangle = \frac{1}{Y} \int_{-Y/2}^{Y/2} u(y)dy$$

where Y is the characteristic scale of wake velocity profiles at the beginning of the scattering volume. We can see on figure 10b) that it is a very reasonable definition of advection velocity, because if we compute the dimensionless number

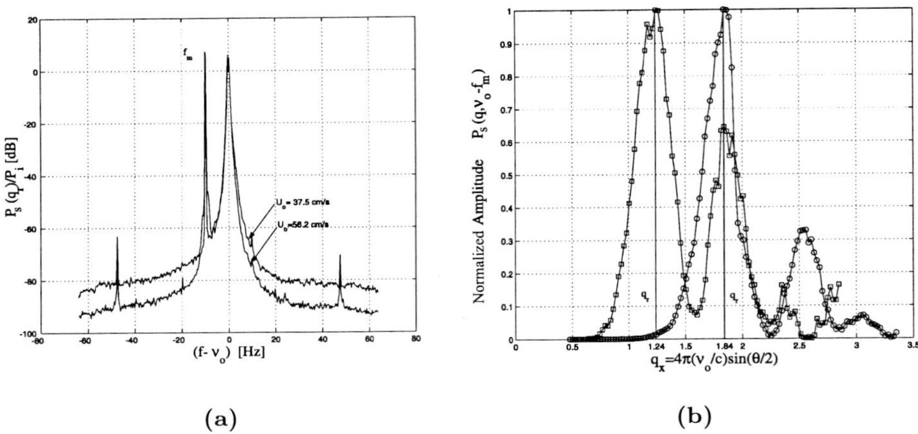

Fig. 9. Ultrasound scattering by the forced wake at two different free stream velocities $Re_b = 10^3$ (⊔), $Re_b = 1.5 \times 10^3$ (∘), where the forcing amplitude and frequency are: $\zeta_m = 0.3$, mm and $f_m = 10$ Hz, respectively. (a) The power spectra of scattered pressure display the same Doppler peak at the forcing frequency f_m. (b) A sweep of the scattering wave vector $q = 4\pi(\nu_o/c)\sin(\theta/2)$ shows that two clearly different spatial resonances are found at very definite length scales.

F using this velocity $F = f_m b/\langle u\rangle$, we found that the resonant wave length λ_r/b evolves linearly with F as one should expect. Moreover, we found that at $F = 1$ both $\lambda_r/b \sim 1$ and the resonant scattered pressure amplitude is maximum.

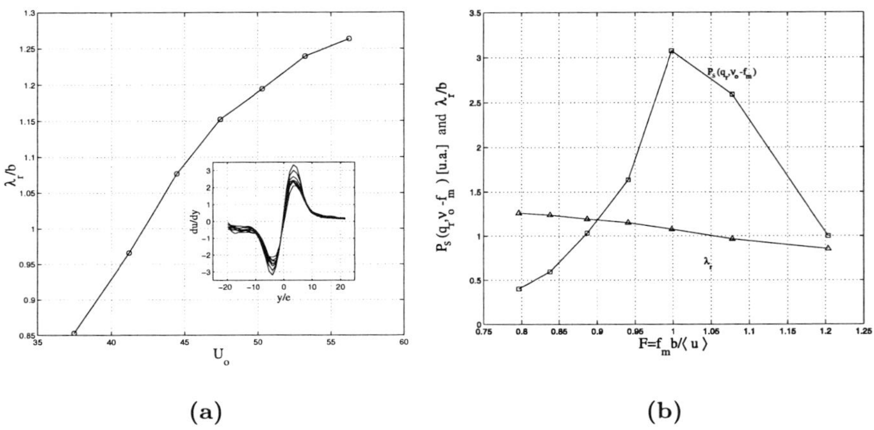

Fig. 10. (a) Evolution of the normalized resonant wavelength $\lambda_r = 2\pi/q_r$ with the free stream velocity U_o for a forced wake ($\zeta_m = 0.3$ mm, $f_m = 10$ Hz). (b) Evolution of the normalized resonant scattered pressure amplitudes (⊔) and resonant wavelengths (△) with the dimensionless parameter $F = f_m b/\langle u\rangle$.

In the vicinity of $F = 1$ the path of fluid particles is exactly a half of the plate's width, $b/2$ on every half forcing period, so the spatial wake pattern has a wave length of order b^{-1}. At this scale the associated vorticity of that mode reaches a maximum, i.e., a vorticity–resonance as a function of the forcing frequency. This behavior was unexpected. But it can be explained as follows: Fluid particles inside the boundary layer whose path is exactly $b/2$ acquire a complete momentum transfer given by the half–forcing cycle, they pick up an increased cross velocity and so increase the wake velocity gradient and therefore vorticity.

At very weak upstream velocities, U_o, the wake can be considered as linear. But as U_o is increased we expect the wake to become non linear. Two effects can be attributed to non linear terms. First, the linear growth of natural as well as artificial instabilities is actually limited or saturated by non linear terms, and second the production of higher harmonics which is the richness of a non linear oscillator system.

Measurements of a non linear wake with hot wire probes make non sense because of the intrinsic non linear character of the hot wire itself (production of harmonics of a pure sinusoidal signal). Therefore it is a well controlled scattering experiment which shed light on such a non linear behavior, due to the fundamental linear relationship between scattering pressure and vorticity (See equation 1).

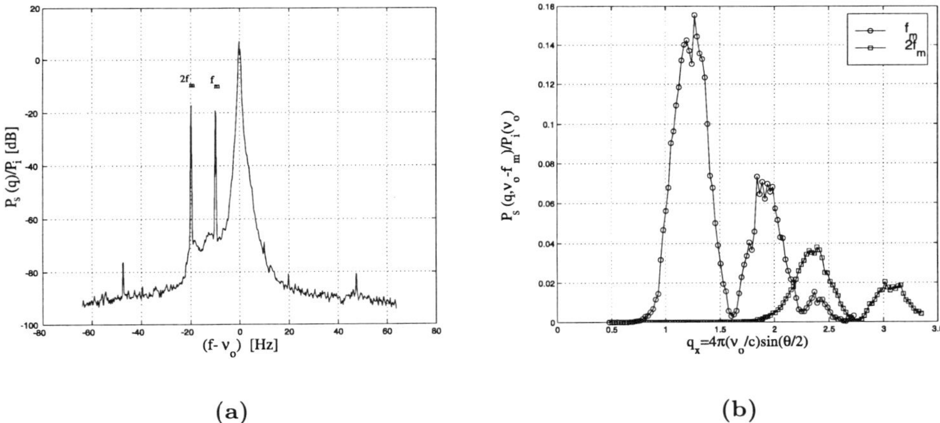

(a) (b)

Fig. 11. Example of higher harmonics production by a non linear wake. Forcing amplitude, frequency and Reynolds number are: $\zeta_m = 0.3$, mm, $f_m = 10$ Hz, $Re_b = 1.4 \times 10^3$ respectively. (a) Power spectrum of scattered pressure showing two peaks associated to the fundamental (f_m) and first harmonic ($2f_m$). (b) Evolution of the peak amplitude of the fundamental (∘) and first harmonic (⊔) as a function of the scattering wave vector $q = 4\pi(\nu_o/c)\sin(\theta/2)$.

Figures 11 and 12 show such a non linear process. At constant forcing frequency $f_m = 10$ Hz, we increase the upstream velocity or the Reynolds number

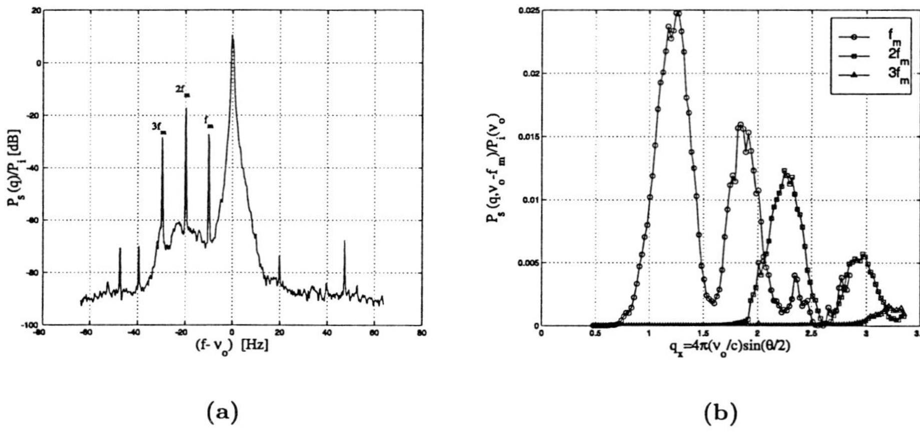

(a) (b)

Fig. 12. Another example of higher harmonics production by a non linear wake. Forcing amplitude, frequency and Reynolds number are: $\zeta_m = 0.3$, mm, $f_m = 10$ Hz, $Re_b = 1.5 \times 10^3$ respectively. (a) Power spectrum of scattered pressure showing three Doppler peaks associated to the fundamental, first and second harmonics. (b) Spatial scales associated to the fundamental (○), first (⊔) and second harmonics (△).

Re_b. The power spectra in figures 11a) and 12a) show clearly the onset of the first and second harmonics $2f_m$ and $3f_m$ respectively.

The amplitude of the fundamental, first and second harmonics was obtained as before, with a frequency sweep of the incoming sound wave (See figures 11b and 12b), to resolve accurately each spatial mode.

As we see we are able to resolve the complete temporal and spatial dynamics of a non linear oscillating wake. With this result in mind we pursued a more ambitious forcing. At constant upstream velocity, the forcing wave form is the sum of two waves of incommensurate frequencies.

$$\zeta(t) = \zeta_m (\mathrm{e}^{i2\pi f_m^1} + \mathrm{e}^{i2\pi f_m^2})$$

We digitally synthesized the sum of two pure sinusoidal waves at frequencies of $f_m^1 = 8$ and $f_m^2 = 10$ Hz with the same amplitude ζ_m recorded on the volatile state memory of the function generator. This forcing serves also to inspect non linearities of the wake. We should find spectral lines at frequencies $\delta f_m = f_m^1 \pm f_m^2$ and successive combinations of the original waves with δf_m.

On figure 13 we see that the power spectrum of scattered pressure at $Re_b = 10^3$ shows four individual spectral lines (peaks) at the following frequencies: $-6, -8, -10, -12$ Hz. They are found at the negative frequency axis because of the sign of the corresponding Doppler effect $(\mathbf{q} \cdot \mathbf{u})$. The spectral line at $\delta f_m = 2$ Hz can not be seen due to the broadening of the spectral line of the incoming sound wave. However exact combinations between the original waves and the δf_m wave are demonstrated by the peaks at 6 and 12 Hz.

Again a sweep in wave number q allow us to find the corresponding spatial modes associated to each peak in the power spectrum. We verify that at each spectral line corresponds a resonant spatial wave vector as indicated on figure 13b), $q_{r1}, q_{r2}, q_{r3}, q_{r4}$ On the subplot of figure 13a) we show the experimental dispersion relation $\omega(q)$ for those vorticity modes.

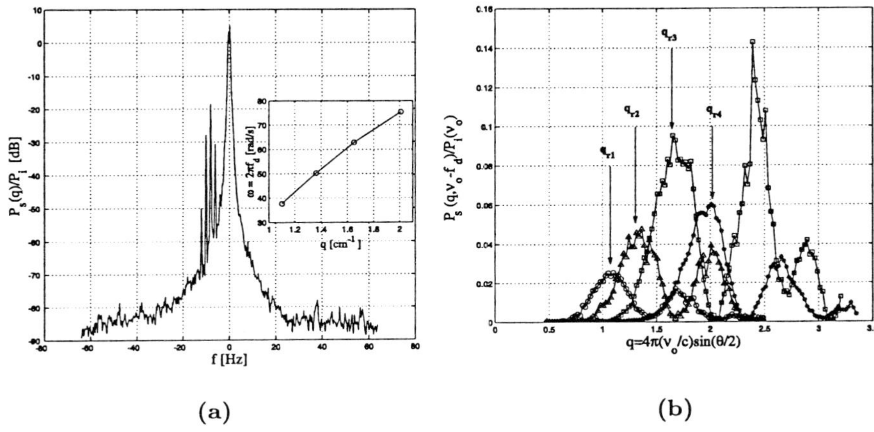

Fig. 13. Wake modulation at two forcing frequencies $f_m^1 = 8, f_m^2 = 10$ Hz. ($\zeta_m =$ 0.3mm, $Re_b = 10^3$). (a) Power spectrum of scattered pressure showing four Doppler peaks (modes). The subplot showing their dispersion relation $\omega(q)$. (b) Spatial resonances at each Doppler peak, showing the presence of very different spatial modes in the wake. Doppler shifts are -6 Hz ($\circ$), -8 Hz ($\triangle$), -10 Hz ($\sqcup$) et -12 Hz ($*$).

It is evident that if increase the number of forcing modes we should speak of wave packets. The propagating properties of the wake can then be investigated. This experiment immediately suggests that one can study the evolution of wave packets introduced into a laminar wake. We can then follow accurately their evolution in time using ultrasound scattering and try to determine possible energy transfers between different modes. A work of this kind will permit to better understand actual ideas at the origin of energy transfers in turbulent open flows. The mechanical energy being injected at integral length scales is then transferred toward the smaller scales trough non linear coupling, being finally dissipated as heat by viscosity [10].

4 Conclusion

The objective of this experimental work was to study the dynamical behavior of a laminar wake under a linear harmonic forcing. Forcing was introduced by small amplitude rotary oscillations of the trailing edge of a flat plate.

A throughout investigation of the wake behavior and vorticity modulation was accomplished using classical hot wire anemometry and ultrasound scattering methods respectively.

We confirmed the scattering mechanism of sound waves by a target of vorticity. We put forward and confirmed the presence of a Doppler effect in agreement with recent theoretical findings, in particular that the Doppler shift is accurately described by the scalar product $\mathbf{q} \cdot \mathbf{u}$.

Important findings on shedding vorticity were put forward using sound scattering. In particular, the resonant behavior of vorticity with the forcing frequency.

The fundamental difference between this experiment and those of sound scattering by von Kármán vortices [11] is that here λ_Ω and Re are not coupled parameters, since we can vary λ_Ω independently of the free stream velocity and to probe a wide band of spectral modes associated to the wake pattern.

Experiments on mixed harmonic forcing (the sum of two harmonic waves) can be viewed as an introduction to the propagation of wave packets through laminar wakes. It is the spatial resolution of the spectral measurements, in the limit of diffraction effects, that must permit to work with a vorticity distribution including several spectral modes.

Acknowledgment

We are grateful to S.Fauve for very fruitful discussions. R.H.Hernández acknowledges support from ECOS (action C94E02)

References

1. H. Schlichting, Boundary Layer Theory, McGraw Hill, New York, 1968.
2. J. J. Healey, A new boundary layer resonance enhanced by wave modulation: theory and experiment, J. Fluid Mech., **304**, 231, 1995.
3. A. Michalke, Survey on jet instability theory, Prog. Aerospace Sci., **21**, 159, 1984.
4. B. L. Smith and A. Glezer, The formation and evolution of synthetic jets, Phys. Fluids, **10**, (9), 2281, 1998.
5. M. Schumm, E. Berger and P. A. Monkewitz, Self–excited oscillations in the wake of two–dimensional bluff bodies and their control, J.Fluid Mech., **271**, 17, 1994.
6. A. Couairon and J. M. Chomaz, Global instability in fully nonlinear systems, Phys.Rev.Lett, **77**, 4015, 1996.
7. P. Huerre and P. Monkewitz, Local and Global instabilities in spatially developing flows, Annu.Rev.Fluid Mech., **22**, 473, 1990.
8. C. Mathis, M. Provansal and L. Boyer, The Bénard–von Kármán instability: an experimental study near the threshold, J.Physique Lett., **45**, L-483, 1984.
9. M. Provansal, C. Mathis and L. Boyer, Bénard-von Kármán instability: transient and forced regimes, J.Fluid Mech., **182**, 1, 1987.
10. C. Baudet, S. Ciliberto and J.-F. Pinton, Spectral analysis of the von Kármán flow using ultrasound scattering, Phys.Rev.Lett., **67**, (2), 193, 1991.
11. J.-F. Pinton and C. Baudet, Measurements of vorticity using ultrasound scattering, in *Turbulence in spatially extended systems*, Nova Science Publishers, 1993.

12. J.-F. Pinton, C. Laroche, S. Fauve and C. Baudet, Ultrasound scattering by buoyancy driven flows, J.Phys. II France, **3**, 767, 1993.
13. A. L. Fabrikant, Sound scattering by vortex flows, Sov.Phys.Acoust., **29**, (2), 152, 1983.
14. L. M. Lyamshev and A. T. Skvortsov, Sound scattering by a vortex soliton in axisymmetrical shear flow, Sov.Phys.Acoust. **35**, (3), 279, 1989.
15. F. Lund and C. Rojas, Ultrasound as a probe of turbulence, Physica D, **37**, 508, 1989.
16. K.Hannemann and H.Oertel Jr., Numerical simulation of the absolutely and convectively unstable wake, J.Fluid Mech., **199**, 55, 1989.
17. H. Oertel Jr., Wakes behind blunt bodies, Annu.Rev.Fluid Mech., **22**, 539, 1990.
18. A. Okajima, Strouhal numbers of rectangular cylinders, J.Fluid Mech., **123**, 379, 1982.
19. S. Taneda, Visual obserbations of the flow past a circular cylinder performing a rotatory oscillation, Journal of the Physical Society of Japan, **45**, (3), 1038, 1978.
20. S. Candel, Mcanique des fluides, Dunod, Paris, 1990.
21. A. Papoulis, Probability, Random Variables and Stochastic Processes, McGraw Hill, New York, 1965.

A Numerical Study of the Elliptic Instability of a Vortex Pair

Florent Laporte and Alexandre Corjon

CERFACS, European Centre for Research and Advanced Training in Scientific Computation, 42, Av. G. Coriolis, 31057 Toulouse Cedex, France

Abstract. The elliptic instability developing in a laminar vortex pair with a high ratio of the vortex core size to the separation distance between the vortices is studied by direct numerical simulations. The computations allows the prediction of both the linear regime of the instability development and the non-linear regime including the transition to a locally turbulent flow.

1 Introduction

Vortex stability is a subject of intense research for many reasons. As far as theoretical studies are concerned, the precise understanding of the vortex dynamics plays a major role in the global understanding of turbulence [1]. Some industrial applications are also directly related to this topic. Among them, the wake vortex problem [2] is of primary interest and has motived the present study. The energetic counter-rotating vortex pair that exists far downstream any civil transport aircraft can be hazardous for a following aircraft. The increase of the air traffic, hence the reduction of the separation distances between aircraft, is partly limited by the potential danger of wake vortices . The dynamics of these vortical flows needs being exhaustively understood in order to suggest realistic solutions to this problem. The flows under consideration for aircraft wakes are turbulent (see *e.g.* [3]) because the Reynolds number based on the circulation generated by the aircraft is of order of 10 to 100 millions, and also because the surrounding atmosphere is composed of turbulent eddies. The action of the atmosphere on the vortices is double: on the one hand the turbulent diffusion acts directly on the velocity to enhance the vortex decay, and on the other hand the unsteadiness of the atmosphere brings perturbations to the vortex system, triggering instabilities.

Instability phenomena developing in wake vortices have first been revealed by the existence of a long-wavelength three-dimensional instability (Crow instability [4], 1970). Many theoretical studies (for instance, refer to Widnall *et al.* [5], 1971), as well as experimental and numerical studies made it possible to analyze its characteristics. It has also been shown that the development of this instability quickly leads to the formation of vortex rings that may last for long times.

The first experiments of Sarpkaya & Suthon [6] (1991), Thomas & Auerbach [7] (1994), followed by the series of experiments and the exhaustive analysis of Leweke & Williamson [8] (1998), have shown that a pair of counter-rotating vortices also develops a short-wavelength three-dimensional instability. This instability has been theoretically investigated by several approaches. One can refer to the studies of Widnall, Bliss and Tsai [9] (1974), Moore and Saffman [10] (1975), Tsai and Widnall [11] (1976), Robinson and Saffman [12] (1984), Pierrehumbert [13] (1986), Bayly [14] (1986), Landman and Saffman [15] (1987), Waleffe [16] (1990) and the recent work of Eloy and Le Dizès [17] (1999). In the case of an isolated vortex in an infinitesimal uniform externally imposed strain field , the instability phenomenon is due to the ellipticity of the streamlines in the vicinity of the vortex cores, which depends itself on the strain rate and the vorticity. As far as single vortex flows are concerned, the transition to turbulence in an elliptic vortex (in a box) has been numerically studied by Lundgren and Mansour [18] (1996). In the case of a vortex pair, the ellipticity of the streamlines is linked to the vorticity of the considered vortex and to the strain rate induced on this vortex by the other of the pair. The first results of numerical simulations for a vortex pair are due to Orlandi, Carnevale, Lele & Shariff [19] (1998), and to Billant, Brancher & Chomaz [20] (1998).

It has to be mentioned that these experimental and numerical studies are not representative of aircraft wakes flow conditions. As far as the flows developing in wakes are concerned, the Reynolds number is many orders of magnitudes larger than the Reynolds numbers considered in the experimental and the numerical studies, and the ratio of the vortex core radius to the separation distance between the vortices is usually smaller in real wakes. Nevertheless, the computation of these "simplified" flows is a first and necessary step. Simulating a more realistic configuration of an aircraft wake will be the next step. The series of simulations we present aims at reproducing the experiments of Leweke & Williamson [8] numerically. The comparison between the experimental and the numerical results obtained is consequently a validation of the numerical tool used. It also allows some discussion on the method used to obtain an initial condition for the simulation. Finally, these simulations made it possible to analyze precisely the dynamics of this instability. The mechanisms leading to a fast vortex decay will be controlled only when the several instability phenomena will be properly described in simple configurations, and their interactions will be understood.

We first present the numerical tool used to perform the DNS, and we describe the initial condition. The results are then presented. The qualitative features of the instability and their evolution up to turbulence are described. A spectral analysis completes the previous part, giving a quantitative counterpart. A global characterization of the flow evolution is then proposed with the definition of integral quantities. We finally conclude on the main results brought by this study and mention the short-term perspectives.

2 Numerical tool and initialization

2.1 The Navier-Stokes solver

The numerical tool used for the direct numerical simulations is called NTMIX3D. It has been developed by the French Research Center for Turbulent Combustion (C.R.C.T.) and by the French Petroleum Institute (I.F.P.). This code is a parallel, three-dimensional, finite difference compressible Navier-Stokes solver. A sixth-order compact space-discretization scheme with spectral-like resolution [21] is used for the calculation of both convective and viscous terms. The time integration is performed with a three-stage Runge-Kutta method. The grids used are Cartesian and regular.

2.2 The initial condition

The simulations are performed for a Reynolds number based on the circulation, $Re_\Gamma = 2400$. The initial condition is a linear superposition of two rectilinear counter-rotating circular Lamb-Oseen type vortices. The velocity and pressure fields are given in cylindrical coordinates by:

$$v_\theta(r) = V_{max}\alpha\frac{r_c}{r}\left(1 - e^{-\beta\frac{r^2}{r_c^2}}\right)$$

$$\frac{dP}{dr} = \frac{\rho v_\theta^2(r)}{r} ,$$

with $\alpha = 1.40$ and $\beta = 1.2544$. V_{max} is the maximum tangential velocity, reached for $r = r_c$. The circulation Γ of one vortex is $\Gamma = 2\pi\alpha r_c V_{max}$. The time is set to the non-dimensional form:

$$t^* = t\,\frac{\Gamma}{2\pi b^2} ,$$

where $\Gamma_0 = \Gamma(t^* = 0)$ and $\tau_b = \frac{2\pi b^2}{\Gamma}$ is the time scale based on the separation distance of the vortices.

The linear superposition is suggested by Leweke & Williamson [8]. The ratio of vortex radius r_c to the separation distance b between the vortices centers is initially $r_c/b = 0.2$. Each of the two Lamb-Oseen vortices composing the initial condition are stationary solutions of the Euler equations if considered separately. However, for the ratio $r_c/b = 0.2$, a linear superposition is not a stationary solution (in the frame of reference linked to the vortex pair) of the non-linear Euler equations. A transient phase is observed, during which the initial condition adapts to the Navier-Stokes equations and becomes a stationary solution of the Euler equations. The originally circular streamlines become approximately elliptic due to the mutual strain induction. The adaptation is a purely two-dimensional and viscous process (see Sipp, Jacquin and Cossu [22], 1999), that principally takes place before the instability starts being amplified. Since the viscous diffusion acts during the adaptation period, the Navier-Stokes solution considered as the "real" initial condition obtained just before the instability sets in is characterized by the ratio $r_c/b \simeq 0.23$. These properties are detailed, discussed and demonstrated in Laporte & Corjon [23].

In order to trigger the instability, a white noise is added to the three components of the velocity. For the set of simulations presented, the amplitude of the noise (denoted by A_r) lies in the interval

$$10^{-4}V_{max} \leq A_r \leq 5\ 10^{-3}V_{max} .$$

The coordinates system can be described as seen in figure 1. The axial direction of the rectilinear vortices defines the Oz axis. In a transverse plane, the vortices centers define the Ox axis (horizontal), and the orthogonal Oy axis is vertical. The discretization used in transverse xy-planes is the same for all the simulations. In these planes, the number of points per core radius r_c is 6. Several axial discretizations have been used. We denote by np_z the number of grid points in the axial direction per wavelength of the elliptic instability . Table 1 gathers the various parameters used for each simulation. The total number of grid points lies between 700,000 and 2,200,000 approximately for the set of simulations.

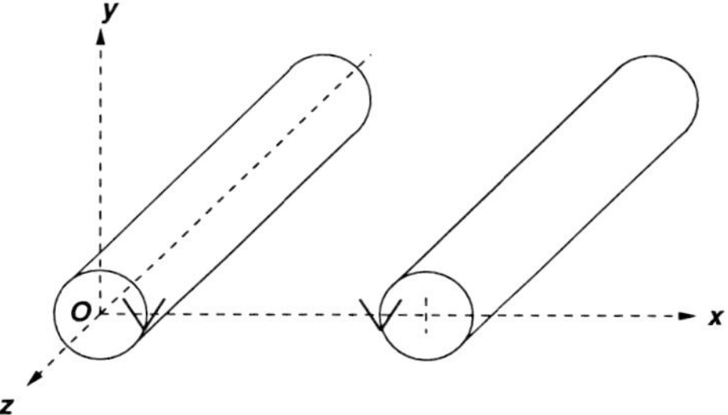

Fig. 1. The vortex pair and the coordinates system.

Table 1. Parameters of the simulations.

Run	L_x/b	L_y/b	L_z/b	A_r	np_z
1	9.0	8.0	1.8	5.10^{-3}	12
2	8.0	7.0	2.7	5.10^{-3}	8
3	4.5	3.5	1.8	5.10^{-3}	36
4	4.5	3.5	7.5	10^{-4}	8
5	4.5	3.5	13	5.10^{-3}	8
6	4.5	3.5	18	5.10^{-3}	8
7	5.5	4.0	4.5	5.10^{-3}	12
8	5.5	4.5	4.5	10^{-3}	8

Periodic boundary conditions are used in the three directions. The transverse dimensions of the simulation domains are kept constant within the series of computations. These dimensions are approximately $(L_x, L_y) = (5b, 4b)$. They are chosen such that the periodic replicas of the vortices have a weak influence on the considered pair. In the axial direction, the dimension varies from $1.8b$ to

18*b*. The corresponding simulations (Runs 2,3,4,5,6,7) give the opportunity to identify the influence of the axial dimension of the domain on the most amplified wavelength, and to release from it.

3 Results

These calculations simulate the development of the elliptic instability (linear and weakly non-linear phases), the non-linear phase, and the transition to turbulence. By linear phase, we mean the phase during which the perturbations remain very small compared to the base flow, and during which the linear stability analysis remains consequently valid. The development and the structure of the elliptic instability are revealed by visualizations of isolevels of a vorticity component (ω_x, ω_y or ω_z), or by isolevels of the vorticity magnitude $||\omega|| = (\omega_x^2 + \omega_y^2 + \omega_z^2)^{1/2}$. The levels are chosen in order to show particular and unsteady vortical structures existing in the flow, and are consequently adapted to each case. The main features of the elliptic instability are identified in this section. We recall that all the other unstable modes developing in the flow are also simulated, but their evolution is not analyzed here.

3.1 Linear regime

A dominant wavelength is amplified during the first part of the simulation (linear phase), after the transient corresponding to the adaptation period. As mentioned just above, this phase is properly defined referring to the linear stability analysis. The development of the instability can be characterized by the sinuous displacement of the vortices centerlines with respect to their unperturbed initial position, and by the displacement of the fluid in the vicinity of the vortex cores. Figure 2 shows the displacement of the centerlines and the axial modulation that exists in the surroundings. The experiments and the theoretical predictions of Leweke & Williamson [8], as the simulations of Orlandi *et al.* [19] reveal the same qualitative features. The corresponding mode of instability presents a particular structure. The centerlines of the two vortices are seen curved in phase if projected on a horizontal xz-plane, and out of phase if projected in a vertical yz-plane. Figure 3 shows the centerlines defined by the λ_2 criterion [24], and projected in two such planes. This strong co-operative development [8] of the instability in the two vortices is directly linked to the high ratio r_c/b that characterizes the considered flow, and which imposes a strong coupling on the component of the velocity that is normal to the median plane separating the vortices.

According to theoretical predictions [8], and under the assumption of a single vortex in a uniform externally imposed infinitesimal strain field, the angle of the perturbations (by perturbation we mean difference between the total flow and the unperturbed flow) identified by the centre of rotation should be aligned with the direction of the maximum stretching. The displacement of the centerlines should consequently lie on a line inclined 45 degrees with respect to the Ox axis.

Fig. 2. Linear phase ($t^* \simeq 11$). Three isosurfaces of vorticity magnitude $||\omega||$ are plotted.

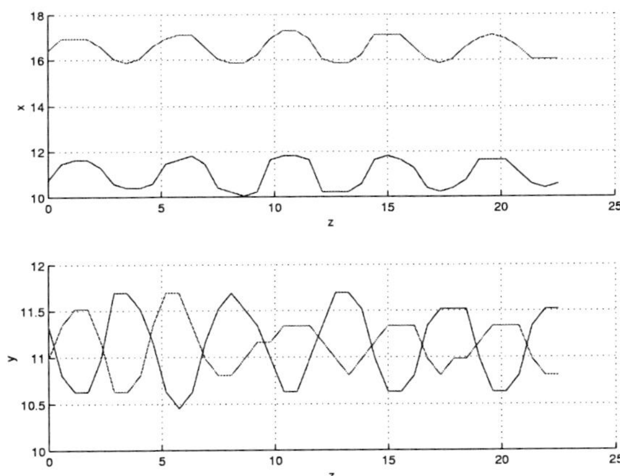

Fig. 3. Displacement of the vortices centerlines in the linear regime, top view (top) and side view (bottom).

The angle found here is very close to this prediction (40 degrees approximately) for all the simulations. Under the assumption recalled above, the theoretical calculation of the form of the perturbation on the velocity and the vorticity components has been carried out by Bayly [14], Pierrehumbert [13] and Waleffe [16] separately. The flow resulting from the superposition of a single vortex and a localized mode of perturbation is explicitly detailed by Leweke & Williamson [8]. An invariant tube of streamlines can be evidenced in the total flow. The layers of fluid inside and outside this tube are displaced in opposite radial directions. It can be demonstrated that an invariant tube of axial vorticity of larger diameter (40%) also exists in the considered flow containing the vortex pair (for details see Laporte & Corjon [23]). This tube is shown by the intermediate isosurface of vorticity magnitude of figure 2. No strictly invariant tube should exist for the vorticity magnitude, however it is found that the contribution of the axial component of vorticity in the vorticity magnitude is approximately $\omega_x/||\omega|| =$ 99% at the location of the tube, and at the instant shown in figure 2.

3.2 Non-linear regime

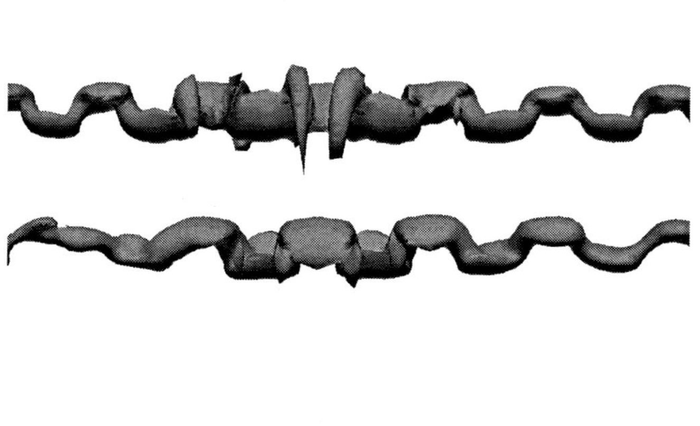

Fig. 4. Non-linear phase ($t^* \simeq 13$): development of shorter wavelengths than the elliptic instability wavelength. A transverse secondary vortex pair starts being formed on the top vortex. Isosurface of vorticity magnitude $||\omega||$, bottom view.

The weakly non-linear regime and the non-linear regime of development of the instability means the amplification of other wavelengths. In particular, the sub-harmonics of the fundamental mode corresponding to the elliptic instability wavelength are amplified. Their growth rate can be predicted in the framework of a weakly non-linear study, by establishing the amplitude equations of these eigenmodes [25]. These growth rates are different from those predicted by the linear stability analysis [20] for each mode taken separately, and are directly linked to the growth rate of the leading instability. Figure 4 shows the particular spatial structure of the vorticity magnitude concentration once the non-linear phase is reached. Once again, as far as the structure of the flow is concerned, the results obtained by Orlandi *et al.* [19] are very similar to the present results. The originally elliptic vortices transform their form during the growing of the instability. The vorticity concentrates then in crescent-like structures.

Due to the form of the displacement induced by the instability near the bottom stagnation point of the flow, fluid layers are extracted from one vortex and roll-up around the other of the pair (as observed in [8]). The azimuthal component of the instability mode acts on the extracted fluid layers and re-orient the vorticity in the transverse Ox direction (defined by the vortex centers). The above described combined effects results in the formation of a secondary counter-rotating transverse vortex pair per elliptic instability wavelength.

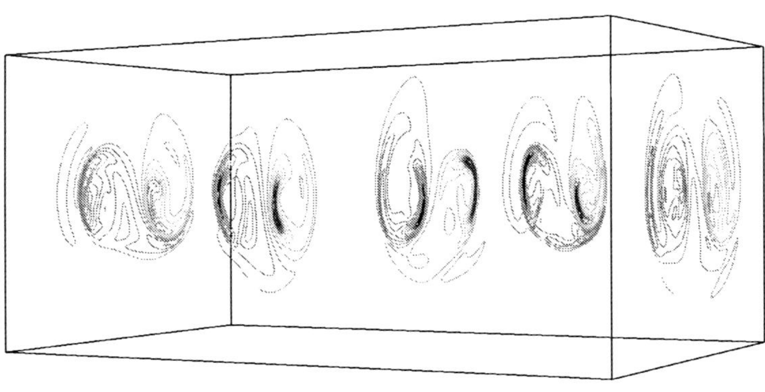

Fig. 5. Extraction of fluid layers during the phase of transition to turbulence. Isocontours of axial vorticity ω_z in several transverse cut planes.

Figure 4 shows the formation of a transverse vortex pair and figure 5 (axial vorticity contours) shows how the fluid is wrapped around the vortices. Due to the presence of a hyperbolic stagnation point near the location of formation of the secondary vortex pairs, these transverse structures experience stretching and compression. This results in the intensification of the secondary vortices which evolve towards fine 'bean-like' vortices located at the boundary of the primary vortex oval (see figure 6). These orthoradial secondary vortical structures lead to a fast transition to turbulence by the stretching they induce on the already strongly deformed primary vortices. This late-stage phenomenon is similar to the vortex decay simulated in a homogeneous isotropic turbulence field (Risso, Corjon & Stoessel [26], 1997), since the coherent eddies of the turbulence are organized in a similar way near the hyperbolic stagnation points. Note that a long-wavelength instability can be seen on figure 6.

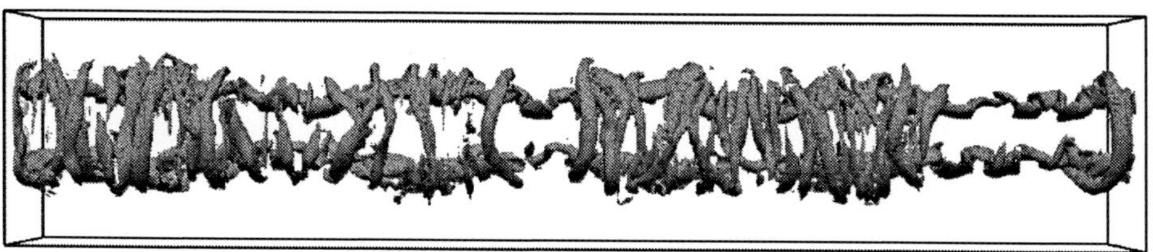

Fig. 6. Creation and development of transverse vortical structures between the main vortices in the non-linear regime (locally turbulent flow). The transverse secondary vortices are stretched and compressed near the hyperbolic stagnation point existing below the vortex pair. Vorticity magnitude isosurface $||\omega||$, bottom view.

3.3 Spectral Analysis

We focus on the analysis of a simulation that is representative of the set of results obtained (Run 8). Domain sizes are (L_x, L_y, L_z) and the corresponding discretization is (n_x, n_y, n_z). The grid points are defined by (x_i, y_j, z_m), $i \in \{1, n_x\}$, $j \in \{1, n_y\}$, $m \in \{1, n_z\}$. A Fourier transform in the axial direction makes it possible to access the temporal evolution of the kinetic energy of the perturbation existing in each mode of the simulation domain. We dispose of the kinetic energy field on the grid

$$Ec(x_i, y_j, z_m) = \frac{1}{2}(u^2 + v^2 + w^2)(x_i, y_j, z_m) \ .$$

The kinetic energy spectrum is obtained by performing discrete Fourier transforms (FFT) in the axial direction for each grid point (x_i, y_j), and then averaging over the transverse plane. The discrete Fourier coefficients are defined by

$$\widehat{Ec}_k(x_i, y_j) = \frac{1}{n_z} \sum_{m=1}^{n_z} Ec(x_i, y_j, z_m) e^{-2i\pi km/n_z} \ ,$$

for $k = 1, ..., n_z/2$. The average perturbation kinetic energy spectrum in the axial direction is consequently

$$Ec_k^* = \frac{1}{L_x L_y} \sum_{i=1}^{n_x} \sum_{j=1}^{n_y} \widehat{Ec}_k(x_i, y_j) \Delta_{x,i} \Delta_{y,j} \ ,$$

where $\Delta_{x,i} = x_{i+1} - x_i$ et $\Delta_{y,j} = y_{j+1} - y_j$. Ec_k^* represents the amount of perturbation kinetic energy contained in the mode number k of the simulation domain. The corresponding wavelength is defined by $\lambda_k = \frac{L_z}{k}$. The wavenumber $k_z = \frac{2k\pi}{L_z}$ is set to a non-dimensional form taking the vortex core radius at $t^* = 0$, $r_c(0)$, as reference length:

$$k_z^* = k_z \ r_c(0) = \frac{2k\pi}{L_z} \ r_c(0) \ .$$

Figure 7 shows the distribution of the kinetic energy in the axial direction over the simulated modes, and during the simulation period.

An initial damping of the kinetic energy can be observed on all the simulated modes, due to the viscous damping of the initial axial perturbation (white noise), and due to the adaptation of the initial condition to the flow. The decay of kinetic energy associated to small wavelengths (high wavenumbers) goes on decreasing as the elliptic instability (identified by the non-dimensional wavenumber $k_z^* \simeq 1.5$) starts being amplified. The first mode to be amplified is the elliptic one. As the weakly non-linear regime is reached, the perturbation energy is distributed over some selected modes, which are the sub-harmonics of the fundamental mode. As the non-linear phase arises, the elliptic mode is brought to saturation and all the simulated modes are amplified (see figure 7). This phase corresponds to the transition to a locally turbulent flow. We now focus on the elliptic instability mode.

The most amplified wavelength, denoted by λ_e, and corresponding to the elliptic mode is found to be

$$\frac{\lambda_e}{b} = 0.85 \pm 0.05 \ .$$

Another vortex radius, denoted by a, is defined in the case of a Lamb-Oseen vortex as $r_c = 1.12 \times a$. In our case, this non-dimensional vortex radius at the end of the adaptation period is (for details see Laporte & Corjon [23])

$$a/b \simeq 0.253 \ ,$$

which leads to the following non-dimensional most amplified wavenumber

$$k_z a = \frac{2\pi}{\lambda_e/b} \times a/b \simeq 1.87 \ .$$

This value is very close to the value $k_z a \simeq 1.6 \pm 0.2$ measured by Leweke & Williamson [8].

During the first part (transient period excepted) of the simulation corresponding to the linear and weakly non-linear regimes, the exponential growth of the dominant mode associated with the elliptic instability wavelength determined just above is observed. The growth rates are directly deduced from a spectral analysis, since the definition reads

$$\sigma = \frac{1}{2} \frac{d \ \ln(E)}{dt} \ .$$

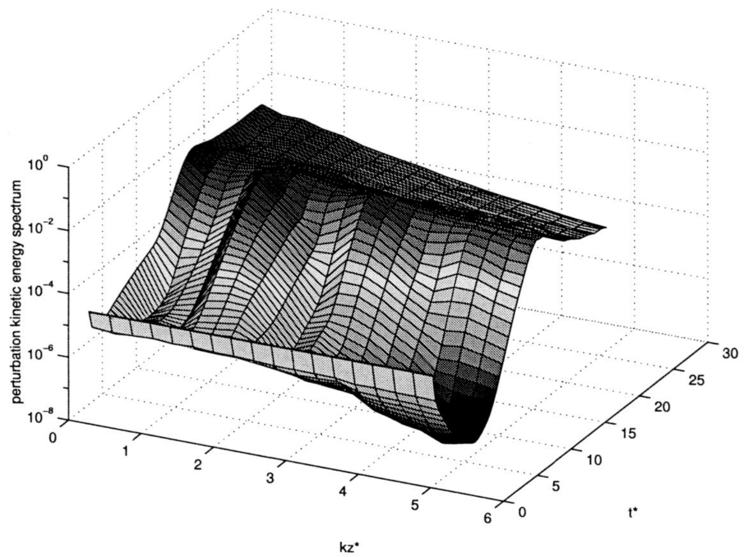

Fig. 7. Decomposition of the perturbation kinetic energy in Fourier modes during the simulation time.

Where E is the energy perturbation. Figure 8 shows the evolution of the elliptic instability perturbation mode ($E^{1/2}_{k^*_z=1.5}$) in the simulation denoted by run 8. The transient period can be clearly identified ($t^* \leq 3.1$). The amplification of the elliptic mode after the transient corresponds precisely to an exponential growth during the whole period $4.9 \leq t^* \leq 12.3$ as evidenced in figure 8. The corresponding non-dimensional growth rate reads

$$\frac{\sigma}{\Gamma_0/2\pi b^2} = 0.95 \pm 0.3 \ .$$

The growth rate measured in [8] is $\frac{\sigma}{\Gamma_0/2\pi b^2} = 0.94 \pm 0.12$, which is in excellent agreement with our results. The comparisons with the theoretical studies are detailed in [23]. Once the fundamental mode has come to complete saturation, each axial wavelength is amplified: this marks the transition to turbulence. The fully turbulent flow (but still with large scale coherent structures, as seen on figure 6) is characterized by a reduction to zero (or less) of all the simulated growth rates.

3.4 Evolution of the vorticity components

As mentioned by Orlandi *et al.* [19], the initial condition contains only one non negligible vorticity component, ω_z. The development of the elliptic instability generates vorticity on the two other components. We consider here the evolution

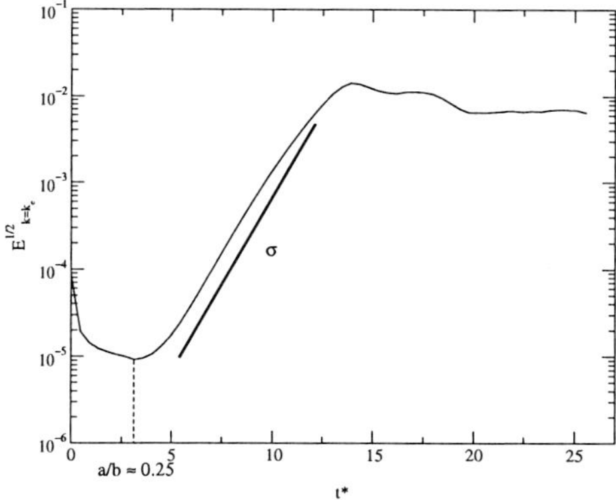

Fig. 8. Amplification of the perturbation mode $E^{1/2}_{k_z^*\simeq 1.5}$ of the elliptic instability.

of one of the two transverse vorticity components, ω_x. To this end, we define a global variable, Γ_x, associated to this vorticity component ω_x, and representative of its behavior in the whole simulation domain. We propose the following definition:

$$\Gamma_x = \frac{1}{L_x}\int_{L_x}\Big(\int_{L_y}\int_{L_z}|\omega_x(x,y,z)|dydz\Big)dx \ .$$

The discrete counterpart of this definition reads

$$\Gamma_x = \frac{1}{n_x}\sum_{i=1}^{n_x}\Big(\sum_{j=1}^{n_y}\sum_{m=1}^{n_z}|\omega_x(x_i,y_j,z_m)|\Delta_{y,j}\Delta_{z,m}\Big) \ ,$$

since we have regular grids. Γ_x is homogeneous to a circulation, and quantifies the average amount of vorticity in the x direction existing in the total flow. The amount of vorticity in each transverse plane is calculated and then averaged with respect to the axial direction. The variable Γ_z associated to the axial vorticity can be defined in a similar way. Figure 9 shows the evolution of these two values.

The axial vorticity Γ_z remains constant during the development of the elliptic instability, until the completely non-linear phase. On the contrary, an exponential growth on both transverse vorticity components is observed (only one component is shown here). The associated non-dimensional growth rate is

$$\frac{\sigma_{\Gamma_x}}{\Gamma/2\pi b^2} \simeq 0.73 \ .$$

The increase of Γ_x is a manifestation of the three-dimensional features of the originally two-dimensional flow. The transition to a fully turbulent flow becomes effective as soon as Γ_x reaches the order of magnitude of Γ_z. The fact that the different components follow the same evolution after the transition may indicate that the turbulent eddies are nearly homogeneous, therefore the locally turbulent

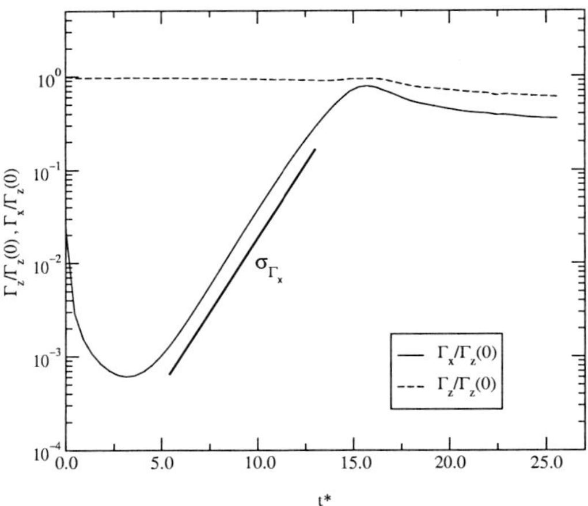

Fig. 9. Evolution of the averaged axial and transverse vorticity components.

mixing resulting from the elliptic instability saturation in the considered flow configuration seems to be efficient.

One of the most significant variables to quantify the vortex decay is the circulation of one vortex. First of all, we simply take inspiration from the method described by Orlandi *et al.* [19] to get an approximation of the circulation: the axial vorticity is integrated over the half simulation domain containing one vortex of the pair and averaged over the axial dimension, i.e.

$$\Gamma = \frac{1}{L_z} \int_{L_z} \left(\int_{L_x/2} \int_{L_y} \omega_z dx dy \right) dz \ ,$$

whose discrete counterpart is

$$\Gamma = \frac{1}{n_z} \sum_{m=1}^{n_z} \left(\sum_{i=1}^{n_x/2} \sum_{j=1}^{n_y} \omega_z(x_i, y_j, z_m) \Delta_{x,i} \Delta_{y,j} \right) .$$

Figure 10 shows the evolution of this circulation set to a non-dimensional form with respect the initial circulation.

At the beginning of the simulation (very small perturbations), the circulation remains constant, which is in agreement with the viscous diffusion of a single vortex. As the linear regime develops, a moderate decrease of the circulation appears. When the elliptic instability mode is brought to saturation, a fast and sudden decay occurs. This kind of evolution is observed by Leweke & Williamson [8] and Orlandi *et al.* [19]. The decay directly induced by the elliptic instability rises up to a 40% decrease of the initial circulation.

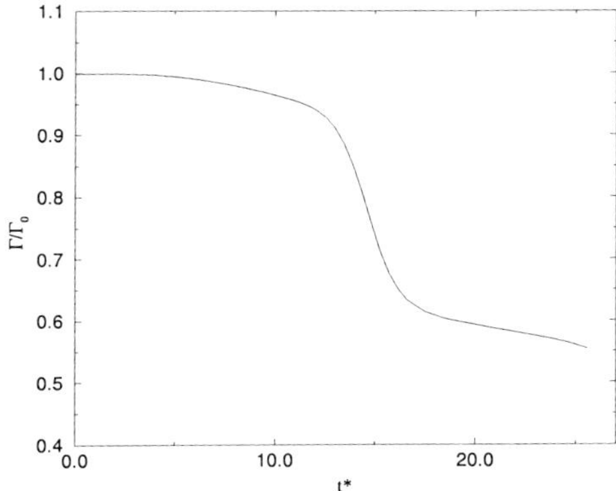

Fig. 10. Evolution of the average circulation of one vortex.

4 Conclusions

The three-dimensional short-wavelength elliptic instability developing in a counter-rotating vortex pair is simulated. The flow configuration is chosen to be close to the series of water-tank experiments of Leweke & Williamson [8]. The main qualitative and quantitative features of the simulated instability are in good agreement with both experimental measurements and theoretical predictions in the linear regime. The non-linear phase and the transition to turbulence are also predicted by the simulations, and discussed in details. The evolution of the vorticity components existing in the simulation domain and the circulation of one vortex during the simulations allows a global characterization of the flow evolution. The sudden decay of the circulation that occurs when the instability is brought to saturation is accompanied by an homogeneous spatial redistribution of the vorticity. This kind of behavior is searched for in the framework of wake vortices. The hypothetical onset of the elliptic instability in realistic aircraft configurations is consequently of interest, and is currently studied.

5 Acknowledgments

The authors are grateful to Thomas Leweke and Pierre Brancher for useful discussions.

References

1. B. J. Bayly, S. A. Orszag and T. Herbert, "Instability Mechanisms in Shear-Flow Transition," Ann. Rev. Fluid. Mech. 1988. 20 : 359-91.
2. P. R. Spalart, "Airplane trailing vortices," Ann. Rev. Fluid Mech. 1998. 30:107-38.
3. W. J. Devenport, M. C. Rife, S. I. Liapis and G. J. Follin, "The structure and development of a wing-tip vortex," J. Fluid Mech. **312**, 67-106 (1996).

4. S. C. Crow, "Stability theory for a pair of trailing vortices," AIAA J. **8**, 2172-2179 (1970).
5. S. E. Widnall, D. Bliss and A. Zalay, "Theoretical and experimental study of the stability of a vortex pair," In *Aircraft Wake Turbulence and its Detection*, Edited by J.H. Olsen et al., Plenum Press, New York, p. 305 (1971).
6. T. Sarpkaya and P. Suthon, "Interaction of a vortex couple with a free surface," Exps. Fluids **11**, 205-217, (1991).
7. P. J. Thomas and D. Auerbach, "The observation of the simultaneous development of a long and a short-wave instability mode on a vortex pair," J. Fluid Mech. **256**, 289-302 (1994).
8. T. Leweke and C. H. K. Williamson, "Cooperative elliptic instability of a vortex pair," J. Fluid Mech. **360**, 85-119 (1998).
9. S. E. Widnall, D. Bliss and C-Y Tsai, "The instability of short waves on a vortex ring," J. Fluid Mech. **66**, 35-47 (1974).
10. D. W. Moore and P. G. Saffman, "The instability of a straight vortex filament in a strain field," Proc. R. Soc. Lond. A **346**, 413-425 (1975).
11. C-Y Tsai and S. E. Widnall, "The stability of short waves on a straight vortex filament in a weak externally imposed strain field," J. Fluid Mech. **73**, 721-733 (1976).
12. A. C. Robinson and P. G. Saffman, "Three-dimensional stability of an elliptical vortex in a straining field," J. Fluid Mech. **142**, 451-466 (1984).
13. R. T. Pierrehumbert, "Universal short wave instability of two-dimensional eddies in an inviscid fluid," Phys. Rev. Lett. **57**, 2157 (1986).
14. B. J. Bayly, " Three-dimensional instability of elliptical flow," Phys. Rev. Lett. **57**, 2160-2163 (1986).
15. M. J. Landman and P. G. Saffman, "The three-dimensional instability of strained vortices in a viscous fluid," Phys. Fluids **30**, 2339-2342 (1987).
16. F. Waleffe, "On the three-dimensional instability of strained vortices," Phys. Fluids A **2**, 76-80 (1990).
17. C. Eloy and S. Le Dizès, "Instability of the Burgers and Lamb-Oseen vortices in a strain field," J. Fluid Mech. **378**, 145-166 (1999).
18. T. S. Lundgren and N. N. Mansour, "Transition to turbulence in an elliptic vortex," J. Fluid Mech. **307**, 43-62 (1996).
19. P. Orlandi, G. F. Carnevale, S. K. Lele and K. Shariff, "DNS study of stability of trailing vortices," Proceedings of the Summer Program 1998, Center for Turbulence Research, Stanford University.
20. P. Billant, P. Brancher and J-M. Chomaz, "Three-dimensional stability of a vortex pair," Phys. Fluids **11**, 2069-2077 (1999).
21. S. K. Lele, "Compact Finite Difference Schemes with Spectral-like Resolution," J. Comp. Phys. **103**, 16-42 (1992).
22. D. Sipp, L. Jacquin and C. Cossu, "Self-adaptation and viscous selection in concentrated 2D vortex dipoles" Submitted to Phys. Fluids (1999).
23. F. Laporte and A. Corjon "Direct Numerical Simulations of the elliptic instability of a vortex pair" Submitted to Phys. Fluids (1999), accepted.
24. J. Jeong and F. Hussain, "On the identification of a vortex," J. Fluid Mech. **285**, 69-94 (1995).
25. P. Brancher, "Etude numérique des instabilités secondaires de jets," Ph.D. Thesis, Ecole Polytechnique, France (1996).
26. F. Risso, A. Corjon and A. Stoessel, "Direct Numerical Simulations of Wake Vortices in Intense Homogeneous Turbulence," AIAA J. **35**, 1030-1040 (1997).

On the Stability of Vortices in an Ideal Gas

Stéphane Leblanc[1], Anne Le Duc[2] and Lionel Le Penven[3]

[1] Laboratoire d'Analyse Non Linéaire Appliquée, Université de Toulon et du Var, BP 132, F-83957 La Garde Cedex, France
[2] Lehrstuhl für Fluidmechanik, Technische Universität München, Boltzmannstrasse 15, D-85748 Garching, Germany
[3] Laboratoire de Mécanique des Fluides et d'Acoustique, Ecole Centrale de Lyon, BP 163, F-69131 Ecully Cedex, France

Abstract. Two distinct mechanisms of three-dimensional instability in compressible planar vortices in an ideal gas are presented. Both mechanisms have been obtained with the geometrical optics (WKB) stability theory which consists in studying the evolution of short-wavelength disturbances localized along the trajectories of the vortex. The first one corresponds to parametric resonances arising when a vortex is periodically compressed; the resulting instabilities are localized in the core of the vortex. On the contrary, in the second case, which corresponds to the generalization to compressible flows of the Rayleigh stability criterion for centrifugal instability, the growing perturbation surrounds the vortex at a given radius. In the latter case, the structure of the corresponding discrete eigenmodes may be described exactly, thus complementing and improving the WKB theory.

1 Introduction

Hydrodynamic instabilities in compressible flows remain relatively misunderstood for at least two reasons: first because only few equilibrium basic states solutions of the equations of motion are known, and secondly because the perturbation problem is more difficult than in the incompressible case as it involves an additional equation. Furthermore, as in the incompressible case, most of the hydrodynamic stability theories concern basic flows with relatively simple topology and some particular symmetries, allowing for simplification of the spectral problem [8,27].

Nevertheless, the theory of short-wavelength instabilities in inviscid flows provides a powerful tool for stability criteria of complex flows. It has known significant progress with the fundamental works of Eckhoff [9] and Lifschitz & Hameiri [24], who proposed a method based on the WKB approximation which allows to derive local stability criteria for any incompressible (see also [7,29]) or compressible time-dependent flow. Instead of characterizing the discrete eigenmodes with large wave numbers [21,2,27,18,28], or to construct localized solutions corresponding to the continuous spectrum [22,20], the *geometrical optics stability theory* consists in solving the initial value problem corresponding to a localized initial data. By expanding the solution on a WKB form [26], the linearized Euler equations are reduced to a system of ordinary differential equations evolving along the trajectories of the basic flow. As in geometrical optics, this system consists in the eikonal equation for the wave vector and the transport equations for the perturbation amplitude, which may often be solved analytically, otherwise by an elementary numerical integration. An unbounded solution of the transport equations is sufficient to prove instability.

The geometrical optics stability theory is now a standard tool, and in incompressible flows, the results are numerous (see [23,3] for reviews and references). Less is known on the stability of compressible flows, except for basic states with circular symmetry. With a normal mode approach, necessary conditions for instability have been derived [13,14,31]. With the short-wavelength approximation, Eckhoff & Storesletten [11,12] derived sufficient instability criteria, extending considerably Rayleigh's criterion for centrifugal instability and anticipating the incompressible result of Leibovich & Stewartson [21] for vortex breakdown (see discussion in [10]). Lebovitz & Lifschitz [20] studied localized instabilities in rotating fluid masses with both a description of the continuous spectrum and the geometrical optics method.

We present here some recent results on the short-wavelength instabilities in compressible planar inviscid vortices, that may be obtained with the WKB approximation (Sec. 2). Two mechanisms of instabilities are described: parametric resonances in time-dependent vortices (Sec. 3), and Rayleigh criterion for steady circular vortices (Sec. 4). Further details may be found in [16,17,19].

2 The geometrical optics stability theory

2.1 Compressible flows in an ideal gas

Let $[\boldsymbol{U}, R, P](\boldsymbol{x}, t)$ be the velocity, density and pressure fields of the subsonic compressible flow in an ideal (i.e. perfect and inviscid) gas filling a domain $\mathcal{D}_t$. The flow is governed by the Euler equations:

$$\frac{D\boldsymbol{U}}{Dt} + \frac{1}{R}\boldsymbol{\nabla} P = 0, \quad \frac{DR}{Dt} + R(\boldsymbol{\nabla}\cdot\boldsymbol{U}) = 0, \quad \frac{DP}{Dt} + \gamma P(\boldsymbol{\nabla}\cdot\boldsymbol{U}) = 0, \tag{1}$$

where $\gamma > 1$ is the constant ratio of specific heats, and $D/Dt = \partial/\partial t + \boldsymbol{U}\cdot\boldsymbol{\nabla}$ is the material derivative. The last equation is deduced from entropy conservation $DS/Dt = 0$ with $S = P/R^\gamma$. The flow is subject to slip boundary conditions. Taking the curl of the momentum equation, the compressible Helmholtz equation for the vorticity $\boldsymbol{W} = \boldsymbol{\nabla}\times\boldsymbol{U}$ reads [27]:

$$\frac{D\boldsymbol{W}}{Dt} = \boldsymbol{W}\cdot\boldsymbol{\nabla}\boldsymbol{U} - \boldsymbol{W}(\boldsymbol{\nabla}\cdot\boldsymbol{U}) + \frac{1}{R^2}\boldsymbol{\nabla} R\times\boldsymbol{\nabla} P. \tag{2}$$

2.2 Eckart's variables

Adding a small perturbation $[\boldsymbol{u}, \varrho, p](\boldsymbol{x}, t)$ to the basic flow $[\boldsymbol{U}, R, P](\boldsymbol{x}, t)$, injecting it into (1) and neglecting nonlinear terms yields a linear system of differential equations, that may be rewritten using Eckart's variables $[m, n](\boldsymbol{x}, t)$ defined by [11,24]:

$$\varrho = \frac{R}{C}(m+n), \; p = RCn, \tag{3}$$

where $C = (\gamma P/R)^{1/2}$ is the local sound celerity of the basic flow.

In terms of Eckart's variables, the linear system reads:

$$\frac{D\boldsymbol{u}}{Dt} + \boldsymbol{L}\boldsymbol{u} + C\left\{-\frac{\boldsymbol{\nabla} P}{\gamma P}m + \boldsymbol{\nabla} n + (\gamma-1)\left(\frac{\boldsymbol{\nabla} P}{\gamma P} + \frac{\mathrm{tr}\boldsymbol{L}}{2}\right)n\right\} = 0, \tag{4a}$$

$$\frac{Dm}{Dt} + (\gamma-1)\frac{\mathrm{tr}\boldsymbol{L}}{2}m + C\left(\frac{\boldsymbol{\nabla} R}{R} - \frac{\boldsymbol{\nabla} P}{\gamma P}\right)\cdot\boldsymbol{u} = 0, \tag{4b}$$

$$\frac{Dn}{Dt} + (\gamma-1)\frac{\mathrm{tr}\boldsymbol{L}}{2}n + C\left(\boldsymbol{\nabla}\cdot\boldsymbol{u} + \frac{\boldsymbol{\nabla} P}{\gamma P}\cdot\boldsymbol{u}\right) = 0. \tag{4c}$$

with $\boldsymbol{L} = \boldsymbol{\nabla}\boldsymbol{U}$ and $\mathrm{tr}\boldsymbol{L} = \boldsymbol{\nabla}\cdot\boldsymbol{U}$.

2.3 Short-wavelength perturbations

Following Eckhoff [9] and Lifschitz & Hameiri [24], we seek an asymptotic solution of (4a–c) on the WKB form:

$$[\boldsymbol{u}, m, n](\boldsymbol{x}, t) = \mathrm{e}^{\mathrm{i}\phi(\boldsymbol{x},t)/\varepsilon}[\tilde{\boldsymbol{u}}, \tilde{m}, \varepsilon\tilde{n}](\boldsymbol{x}, t), \tag{5}$$

where ϕ is a real-valued phase field and $\varepsilon \ll 1$ is a small real parameter at our disposal. Injecting (5) into (4a–c), and equating the various orders in ε yields, at leading order $O(1/\varepsilon)$:

$$\frac{D\phi}{Dt} = 0, \tag{6}$$

and $\boldsymbol{k}\cdot\tilde{\boldsymbol{u}} = 0$. Taking the gradient of (6) gives:

$$\frac{D\boldsymbol{k}}{Dt} = -\boldsymbol{L}^T\boldsymbol{k}, \tag{7}$$

where $\boldsymbol{k} = \boldsymbol{\nabla}\phi$ is the wave vector, and ^{T} stands for transpose.

At $O(1)$, after some manipulations, (4a–b) give respectively:

$$\frac{D\tilde{\boldsymbol{u}}}{Dt} = \left(\frac{2\boldsymbol{k}\boldsymbol{k}^T}{|\boldsymbol{k}|^2} - \boldsymbol{I}\right)\boldsymbol{L}\tilde{\boldsymbol{u}} + C\left(\boldsymbol{I} - \frac{\boldsymbol{k}\boldsymbol{k}^T}{|\boldsymbol{k}|^2}\right)\frac{\boldsymbol{\nabla} P}{\gamma P}\tilde{m}, \tag{8a}$$

$$\frac{D\tilde{m}}{Dt} = (1-\gamma)\frac{\mathrm{tr}\boldsymbol{L}}{2}\tilde{m} + C\left(\frac{\boldsymbol{\nabla} P}{\gamma P} - \frac{\boldsymbol{\nabla} R}{R}\right)\cdot\tilde{\boldsymbol{u}}, \tag{8b}$$

whereas Eckart's pressure follows:

$$\tilde{n} = \frac{\mathrm{i}\boldsymbol{k}\,\cdot}{|\boldsymbol{k}|^2}\left(\frac{2}{C}\boldsymbol{L}\tilde{\boldsymbol{u}} - \frac{\boldsymbol{\nabla} P}{\gamma P}\tilde{m}\right). \tag{9}$$

2.4 Lagrangian description

Equations (7) and (8a–b) involve only material derivative following the basic flow D/Dt. By using a Lagrangian representation $\boldsymbol{x} = \boldsymbol{x}(\boldsymbol{X}, t)$, where $\boldsymbol{X} = \boldsymbol{x}(\boldsymbol{X}, 0)$, and by introducing the trajectories of the basic flow, the partial differential system (7) and (8a–b) is transformed into a system of *ordinary* differential equations evolving along the trajectories [24]. Before this, we define $[\boldsymbol{a}, b](\boldsymbol{X}, t)$ as

$$\tilde{\boldsymbol{u}} = \frac{\boldsymbol{a}}{\sqrt{J}}, \quad \tilde{m} = \frac{b}{\sqrt{J}}, \tag{10}$$

where $J(\boldsymbol{X}, t) = \det(\partial \boldsymbol{x}/\partial \boldsymbol{X})$ is the Jacobian of the transformation which obeys:

$$\frac{\mathrm{d}J}{\mathrm{d}t} = J(\boldsymbol{\nabla} \cdot \boldsymbol{U}), \quad J(\boldsymbol{X}, 0) = 1. \tag{11}$$

The complete system of ordinary differential equations reads

$$\frac{\mathrm{d}\boldsymbol{x}}{\mathrm{d}t} = \boldsymbol{U}, \tag{12a}$$

$$\frac{\mathrm{d}\boldsymbol{k}}{\mathrm{d}t} = -\boldsymbol{L}^T \boldsymbol{k}, \tag{12b}$$

$$\frac{\mathrm{d}\boldsymbol{a}}{\mathrm{d}t} = \left(\frac{2\boldsymbol{k}\boldsymbol{k}^T}{|\boldsymbol{k}|^2} - \boldsymbol{I}\right) \boldsymbol{L}\boldsymbol{a} + \frac{\mathrm{tr}\boldsymbol{L}}{2}\boldsymbol{a} + C\left(\boldsymbol{I} - \frac{\boldsymbol{k}\boldsymbol{k}^T}{|\boldsymbol{k}|^2}\right) \frac{\boldsymbol{\nabla} P}{\gamma P} b, \tag{12c}$$

$$\frac{\mathrm{d}b}{\mathrm{d}t} = C\left(\frac{\boldsymbol{\nabla} P}{\gamma P} - \frac{\boldsymbol{\nabla} R}{R}\right) \cdot \boldsymbol{a} + (2 - \gamma)\frac{\mathrm{tr}\boldsymbol{L}}{2} b. \tag{12d}$$

In geometrical optics approximation, (12a) corresponds to the equation of *rays*, which are trajectories in the present case, (12b) is the *eikonal* (or Hamilton–Jacobi) equation, whereas (12c–d) are the *transport* (or Liouville) equations.

Choosing an initial position $\boldsymbol{x}(\boldsymbol{X}, 0) = \boldsymbol{X}$, the corresponding trajectory $\boldsymbol{x}(\boldsymbol{X}, t)$ may be integrated from (12a), then (12b) will give the wave vector $\boldsymbol{k}(\boldsymbol{X}, t)$, whereas the amplitudes $\boldsymbol{a}(\boldsymbol{X}, t)$ and $b(\boldsymbol{X}, t)$ are solutions of (12c–d). Initial conditions may be constructed such that $\boldsymbol{k}(\boldsymbol{X}, 0) \cdot \boldsymbol{a}(\boldsymbol{X}, 0) = 0$, ensuring that $\boldsymbol{k}(\boldsymbol{X}, t) \cdot \boldsymbol{a}(\boldsymbol{X}, t) = 0$ at any time, and thus $\boldsymbol{k} \cdot \tilde{\boldsymbol{u}} = 0$.

2.5 A sufficient condition for instability

Defining the amplitude vector $\boldsymbol{c} = (\boldsymbol{a}, b)^T$, it may be shown that *a sufficient condition for instability* is that $|\boldsymbol{c}(\boldsymbol{X}, t)|$ solution of (12c–d), together with (12a) and (12b), grows unboundedly along a given trajectory [9,24], *i.e.*

$$\sup_{\boldsymbol{X} \in \mathcal{D}_0} \lim_{t \to \infty} |\boldsymbol{c}(\boldsymbol{X}, t)| = \infty, \tag{13}$$

for some choice of initial conditions verifying $\boldsymbol{k}(\boldsymbol{X}, 0) \cdot \boldsymbol{a}(\boldsymbol{X}, 0) = 0$. $\mathcal{D}_0$ is the Lagrangian domain of reference.

The energy $E(t)$ of the perturbation reads, at leading order in ε:

$$E(t) = \int_{\mathcal{D}_t} \{|\boldsymbol{u}|^2 + m^2 + n^2\}\mathrm{d}\boldsymbol{x} \sim \int_{\mathcal{D}_t} \{|\tilde{\boldsymbol{u}}|^2 + \tilde{m}^2\}\mathrm{d}\boldsymbol{x} = \int_{\mathcal{D}_0} |\boldsymbol{c}(\boldsymbol{X},t)|^2\mathrm{d}\boldsymbol{X}, \tag{14}$$

when $\varepsilon \ll 1$. Evaluating the large time asymptotics of the kinetic energy is a delicate issue, depending of course on the behavior of $|\boldsymbol{c}(\boldsymbol{X},t)|$ on the different trajectories $\boldsymbol{X}$. In some particular cases, such as for the Rayleigh criterion, an explicit form may be given for $E(t)$ when $t \to \infty$.

3 Stability of periodically compressed vortices

3.1 Stagnation points

When the basic flow contains a stagnation point, *i.e.* a point $\boldsymbol{x} = \boldsymbol{X}_0$ on which $\boldsymbol{U}(\boldsymbol{x},t) = 0$, this point is a particular trajectory since $\boldsymbol{x}(\boldsymbol{X}_0,t) = \boldsymbol{X}_0$. In Lagrangian description $(\boldsymbol{X}_0,t)$, it is easy to see that the equilibrium conditions (1) give on the stagnation point [17]:

$$[\boldsymbol{\nabla} P](\boldsymbol{X}_0,t) = 0, \ \frac{\mathrm{d}}{\mathrm{d}t}\ln R(\boldsymbol{X}_0,t) = \frac{\mathrm{d}}{\mathrm{d}t}\ln P^{1/\gamma}(\boldsymbol{X}_0,t) = -\mathrm{tr}\boldsymbol{L}(\boldsymbol{X}_0,t). \tag{15}$$

For steady basic flows, the additional constraint $\mathrm{tr}\boldsymbol{L}(\boldsymbol{X}_0) = 0$ allows to determine completely the velocity gradient tensor $\boldsymbol{L}(\boldsymbol{X}_0)$ [24]. For time-dependent flows, an additional relation between $R(\boldsymbol{X}_0,t)$ and $\boldsymbol{L}(\boldsymbol{X}_0,t)$ may be given through Helmholtz equation (2) where the third term in the r.h.s. vanishes on a stagnation point.

Now, suppose that the basic flow is locally planar and two-dimensional in the vicinity of $\boldsymbol{X}_0$. In Cartesian coordinate, let $(\boldsymbol{e}_x, \boldsymbol{e}_y)$ be the plane of the local flow near $\boldsymbol{X}_0 = (X_0, Y_0)$, and $\boldsymbol{e}_z$ the spanwise direction, so that $\boldsymbol{U}(\boldsymbol{X}_0,t)\cdot\boldsymbol{e}_z = 0$. Let $\boldsymbol{L}(\boldsymbol{X}_0,t)$ be the 2×2 velocity gradient tensor. Locally, the basic vorticity $W(\boldsymbol{X}_0,t)\boldsymbol{e}_z$ is solution of the two-dimensional compressible Helmholtz equation which reads from (2) with (15):

$$\frac{\mathrm{d}}{\mathrm{d}t}\left(\frac{W}{R}\right)(\boldsymbol{X}_0,t) = 0. \tag{16}$$

On the stagnation point $\boldsymbol{X}_0$, it will be shown that the transport equations governing the short-wavelength perturbations may be reduced to a single ordinary differential equation of Hill–Schrödinger type, as in the incompressible case [3].

3.2 Reduction to a single Hill–Schrödinger equation

Let $\boldsymbol{\xi}(\boldsymbol{X}_0,t)$ and $\boldsymbol{v}(\boldsymbol{X}_0,t)$ denote respectively the projection of the wave vector $\boldsymbol{k}(\boldsymbol{X}_0,t)$ and of the amplitude $\boldsymbol{a}(\boldsymbol{X}_0,t)$ on the $(\boldsymbol{e}_x,\boldsymbol{e}_y)$-plane, and let $\mu(\boldsymbol{X}_0,t)$

and $w(\boldsymbol{X}_0, t)$ be respectively their projection along the $\boldsymbol{e}_z$-axis. Let $\boldsymbol{H}(\boldsymbol{X}_0, t)$ be the 2×2 tensor defined by [3]:

$$\boldsymbol{H} = \boldsymbol{L} \begin{pmatrix} 0 & 1 \\ -1 & 0 \end{pmatrix} . \tag{17}$$

Assuming that $\boldsymbol{\xi} \neq 0$ and $\mu \neq 0$, new variables $[p, q](\boldsymbol{X}_0, t)$ are defined as follows [17]:

$$p = \sqrt{R}\, \frac{|\boldsymbol{k}|}{|\boldsymbol{\xi}|}\, \boldsymbol{\xi} \cdot \boldsymbol{v}, \;\; q = \frac{1}{\sqrt{R}}\, \frac{|\boldsymbol{k}|}{|\boldsymbol{\xi}|}\, (\boldsymbol{k} \times \boldsymbol{a}) \cdot \boldsymbol{e}_z . \tag{18}$$

By manipulating equations (12b–d) and their projections on $(\boldsymbol{e}_x, \boldsymbol{e}_y)$ and $\boldsymbol{e}_z$, and by taking into account the previous relations, we get the following system of ordinary differential equations [17]:

$$\frac{\mathrm{d}p}{\mathrm{d}t} = \left(\frac{\mathrm{d}}{\mathrm{d}t} \ln \frac{|\boldsymbol{\xi}|}{|\boldsymbol{k}|} \right) p + \left(\frac{2R\mu^2}{|\boldsymbol{\xi}|^2 |\boldsymbol{k}|^2} \boldsymbol{\xi}^T \boldsymbol{H} \boldsymbol{\xi} \right) q, \tag{19a}$$

$$\frac{\mathrm{d}q}{\mathrm{d}t} = - \left(\frac{W}{R} \right) p - \left(\frac{\mathrm{d}}{\mathrm{d}t} \ln \frac{|\boldsymbol{\xi}|}{|\boldsymbol{k}|} \right) q. \tag{19b}$$

Time derivating (19b) and taking into account (16) and (19a), we get the following Hill–Schrödinger equation:

$$\frac{\mathrm{d}^2 q}{\mathrm{d}t^2} + Q(\boldsymbol{X}_0, t) q = 0, \tag{20}$$

where the potential $Q(\boldsymbol{X}_0, t)$ is analogous to its incompressible counterpart [3]:

$$Q(\boldsymbol{X}_0, t) = \frac{2W\mu^2}{|\boldsymbol{\xi}|^2 |\boldsymbol{k}|^2} \boldsymbol{\xi}^T \boldsymbol{H} \boldsymbol{\xi} + \frac{\mathrm{d}^2}{\mathrm{d}t^2} \ln \frac{|\boldsymbol{\xi}|}{|\boldsymbol{k}|} - \left(\frac{\mathrm{d}}{\mathrm{d}t} \ln \frac{|\boldsymbol{\xi}|}{|\boldsymbol{k}|} \right)^2 . \tag{21}$$

Note that for flows with linear velocity profiles, reduction to a single ordinary differential equation has already been achieved [30,25,6,15].

In general, and in the cases we are going to consider, $R(\boldsymbol{X}_0, t)$, $|\boldsymbol{\xi}(\boldsymbol{X}_0, t)|$ and $|\boldsymbol{k}(\boldsymbol{X}_0, t)|$ are bounded, so that the unbounded growth of $|q(\boldsymbol{X}_0, t)|$ is sufficient for instability [17].

3.3 Time-periodic compressions of a vortex core

In the vicinity of the vortex core $\boldsymbol{X}_0$, it may be checked that the local velocity gradient tensor [19]:

$$\boldsymbol{L}(\boldsymbol{X}_0, t) = \begin{pmatrix} -\dfrac{\mathrm{d}}{\mathrm{d}t} \ln R & -\dfrac{\Omega}{R} \\ \Omega R & 0 \end{pmatrix} , \tag{22}$$

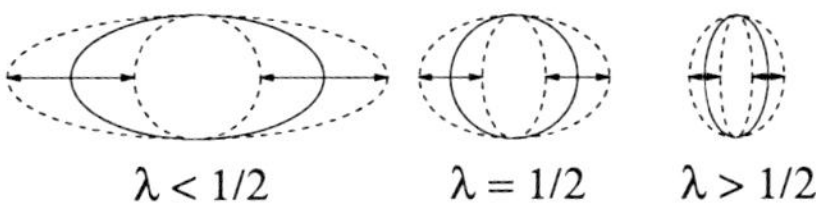

Fig. 1. Sketch of the local flows near the vortex core for various values of the parameter λ. Streamlines of the uncompressed vortices, $\nu = 0$ (solid lines), and envelopes of the trajectories in the compressed case, $\nu \neq 0$ (dashed lines). From [19].

verifies the equilibrium conditions (15) providing that (16) is satisfied. This implies $W(\boldsymbol{X}_0, t) = R(\boldsymbol{X}_0, t)$ and:

$$R(\boldsymbol{X}_0, t) = \left(\frac{\Omega}{1-\Omega}\right)^{1/2}, \quad \Omega(\boldsymbol{X}_0, t) = \frac{R^2}{R^2+1}. \tag{23}$$

We choose the periodic law:

$$\Omega(\boldsymbol{X}_0, t) = \lambda + \nu \cos(\omega t), \tag{24}$$

with $0 < \lambda < 1$ and $0 \leq \nu < \inf(\lambda, 1-\lambda)$, so that $0 < \Omega(\boldsymbol{X}_0, t) < 1$ and density $R(\boldsymbol{X}_0, t)$ is bounded at any time.

Near $\boldsymbol{X}_0$, the Eulerian basic velocity field reads

$$\boldsymbol{U}(\boldsymbol{x}, t) = [\boldsymbol{L}(\boldsymbol{X}_0, t)](\boldsymbol{x} - \boldsymbol{X}_0) + O(|\boldsymbol{x} - \boldsymbol{X}_0|^2), \tag{25}$$

so that instantaneous streamlines are either elliptical or circular, whereas the particle flow trajectories near $\boldsymbol{X}_0$ are given by:

$$\boldsymbol{x}(\boldsymbol{X}_0, t) - \boldsymbol{X}_0 = \left(\frac{1}{R}\cos\tau, \sin\tau, 0\right)^T, \quad \tau(\boldsymbol{X}_0, t) = \int_0^t \Omega \, \mathrm{d}t. \tag{26}$$

Thus when $\nu = 0$, the basic flow is elliptical (circular when $\lambda = 1/2$) and steady; λ is the rotation rate of fluid particles along these ellipses. When $\nu \neq 0$, the basic flow is periodically compressed along the $\boldsymbol{e}_x$-axis, and the trajectories are inscribed between two elliptical envelopes (Fig. 1). The parameter ν may be related to the amplitude of the compressions.

Now, on $\boldsymbol{X}_0$, explicit solutions of the eikonal equation (12b) may be expressed for flows with local velocity gradient tensor $\boldsymbol{L}(\boldsymbol{X}_0, t)$ given previously. When the flow is compressed axially, the solution of (12b) together with (22) reads [19]:

$$\boldsymbol{k}(\boldsymbol{X}_0, t) = \left(\sqrt{1-\mu^2} R \cos\tau, \sqrt{1-\mu^2}\sin\tau, \mu\right)^T, \quad \tau(\boldsymbol{X}_0, t) = \int_0^t \Omega \, \mathrm{d}t, \tag{27}$$

where $0 \leq \mu \leq 1$ without loss of generality. When $\mu = 0$, the perturbation is planar and two-dimensional since $\boldsymbol{k} \cdot \boldsymbol{a} = 0$. Otherwise it is three-dimensional. We recall that when $\mu = 0$ or 1, the variables in (18) are no more defined and particular analyses are required [17].

3.4 Parametric resonances in circular vortex cores

We first consider the case of a locally circular vortex ($\lambda = 1/2$) subjected to weak periodic compressions ($\nu \ll 1$). Recall that in the steady case ($\nu = 0$), a circular vortex core is stable, even in a compressible flow [24].

Introducing $\delta = 2\nu$, on the stagnation point $\boldsymbol{X}_0$, we have from (23) and (24):

$$R(\boldsymbol{X}_0, t) = 1 + \delta \cos(\omega t) + O(\delta^2). \tag{28}$$

The potential (21) in the Hill–Schrödinger equation (20) reads, from (28), (22) and (27):

$$\begin{aligned} Q(\boldsymbol{X}_0, t) = \mu^2 + \delta\mu^2 \ & \left\{ \left(1 + \mu^2 - \frac{\omega^2}{2}\right) \cos(\omega t) \right. \\ & + \left(\frac{\mu^2}{2} - \frac{\omega^2+5}{4} - \omega\right) \cos\{(1+\omega)t\} \\ & \left. + \left(\frac{\mu^2}{2} - \frac{\omega^2+5}{4} + \omega\right) \cos\{(1-\omega)t\} \right\} + O(\delta^2). \end{aligned} \tag{29}$$

The potential is either periodic (when ω is a rational number), or almost-periodic (when ω is irrational). Thus, the resulting equation (20) with (29) is not a Mathieu equation, but the principal resonances may be studied with a multiple scale analysis.

For small amplitudes, Craik & Allen [6] obtained a similar equation in time-periodic incompressible linear flows, and argued that each forcing term in the potential (29) is responsible for a region of instability centered on half its own frequency. In short, it means that three different resonances arise from:

$$|\mu| = \frac{|\omega|}{2}, \ |\mu| = \frac{|1+\omega|}{2}, \ |\mu| = \frac{|1-\omega|}{2}. \tag{30}$$

Since $0 < \mu < 1$ and $\omega > 0$, this means that the leading order parametric instabilities cannot happen unless:

$$0 < \omega < 3. \tag{31}$$

However, this qualitative argument gives no indication on the growth rates. So we use multiple scale asymptotics to characterize precisely the three various parametric instabilities [4]. Furthermore, various resonances may interact for some particular values of the parameters. Details of the calculations are given in [17] and the results may be summarized as follows (to order δ).

The resonance $\omega = 2\mu$ happens when $0 < \omega < 2$, and the perturbation grows exponentially with growth rate:

$$\sigma = \frac{\delta}{8}\left(1 - \frac{\omega^2}{4}\right)\omega. \tag{32}$$

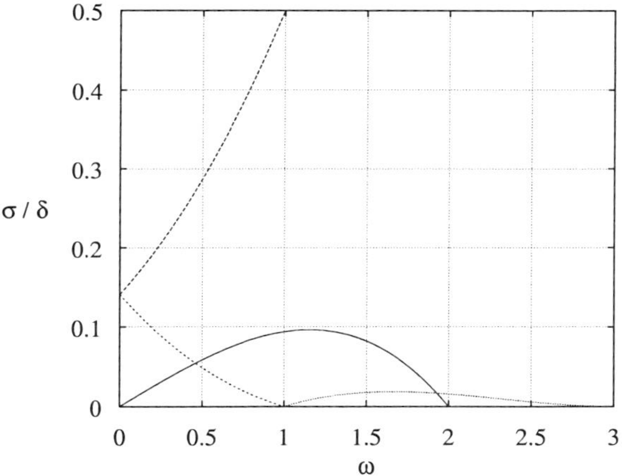

Fig. 2. Growth rates of the various principal parametric instabilities in the case of weak compressions, given by expressions (32), (33) and (34). From [17].

The resonances $1 \pm \omega = 2\mu$ happen both when $0 < \omega < 1$ with respective growth rates:

$$\sigma = \frac{\delta}{64}(1 \pm \omega)(3 \pm \omega)^2. \tag{33}$$

The resonance $\omega - 1 = 2\mu$ happens for $1 < \omega < 3$, and the corresponding growth rate is:

$$\sigma = \frac{\delta}{64}(\omega - 1)(\omega - 3)^2. \tag{34}$$

These resonances are plotted in Fig. 2. Some particular cases may arise when two of these bandwidth interact [17].

Finally, two-dimensional perturbations ($\mu = 0$) are stable, whereas the most unstable configuration is reached for spanwise wave vectors ($\mu = 1$) with

$$\omega = 1, \ \sigma = \frac{\delta}{2}. \tag{35}$$

3.5 Compressions of arbitrary amplitudes

In the general case, system (19a–b) involves either periodic (when ω/λ is rational) or almost-periodic (when ω/λ is not rational) coefficients, and requires a numerical time-integration. Floquet theory and Prufer technique are used respectively in the periodic and almost-periodic case [3]. The problem has been solved numerically with a fourth order Runge–Kutta temporal scheme [19], and the results are represented for fixed values of λ and ω as maps of growth rate in function of $0 \leq \mu \leq 1$ and $0 \leq \delta < 1$, the normalized amplitude of compression: $\delta = \nu/\lambda$ when $0 < \lambda < 1/2$, and $\delta = \nu/(1 - \lambda)$ when $1/2 \leq \lambda < 1$.

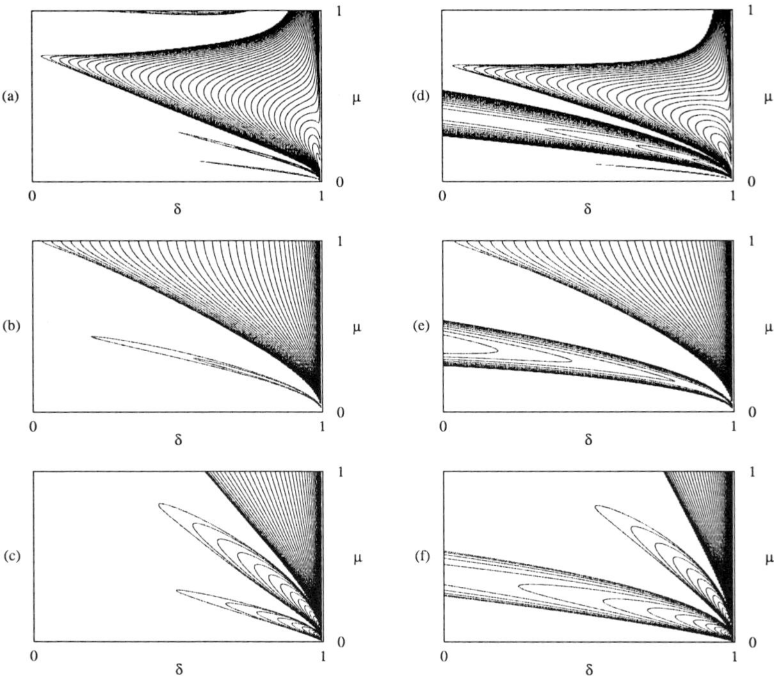

Fig. 3. Isolevels of growth rates σ for various values of the ratio ω/λ. Left (a–c): circular case ($\lambda = 1/2$). Right (d–e): elliptical case ($\lambda = 3/4$). (a) and (d): $\omega/\lambda = 1$; (b) and (e): $\omega/\lambda = 2$; (c) and (f): $\omega/\lambda = 4$. From [19].

In the circular case (stable when $\delta = 0$), results are plotted on Figs. 3(a–c). In the resonant case $\omega/\lambda = 2$ [Fig. 3(b)], modes with $\mu = 1$ are the most dangerous and exist for all amplitudes. A second unstable band with weaker growth rates also arises. When $\omega/\lambda < 2$ [Fig. 3(a)], the large unstable band still exists for all $0 < \delta < 1$, but for oblique wave-vectors. Other thinner resonances arise, and complexity of the instability maps increases when ω/λ decreases (Fig. 5). When $\omega/\lambda > 2$ [Fig. 3(c)], the large resonant band seems to leave the map, so that modes with $\mu = 1$ become unstable for finite amplitudes of compression. This threshold increases for increasing ω/λ, and the other unstable band becomes thinner, so that large frequency of compression weakens this instability mechanism. Finally, for low amplitudes $\delta \ll 1$, results have been found in perfect agreement with the asymptotic results given previously, and the ratio $\omega/\lambda = 2$ has been found to be the most favorable case for instability, see (35).

In the elliptical case [Fig. 3(d–f)], an additional instability bandwidth appears in the maps, which correspond to the modification of the three-dimensional elliptical instability by the periodic compression. When $\delta = 0$, the incompressible case is recovered [5,1,30], and when δ increases, the unstable wave-vectors tend to lie down the plane of the basic flow. The other unstable bands correspond qualitatively to the ones of the circular case, and when $\omega/\lambda = 2$ the behavior predicted by the asymptotic analysis is recovered for small amplitudes. In each case considered, the unstable bands extend to all wave-vectors when $\delta \to 1$, for which the largest growth rates are reached. This limit is however singular when $\lambda \geq 1/2$ because it corresponds to infinite compression. Two-dimensional unstable modes ($\mu = 0$) have never been put forward by our calculations when $0 \leq \delta < 1$. The elliptical case with $\lambda < 1/2$ presents qualitatively similar behaviors, except that the limit $\delta \to 1$ now corresponds to compressions between the circle and the infinitely elongated ellipse.

4 Rayleigh stability criterion for compressible flows

4.1 Vortex breakdown in compressible flows

We consider now the stability of a steady swirling basic flow with circular symmetry: $[\boldsymbol{U}, R, P](r)$ with $\boldsymbol{U}(r) = V(r)\boldsymbol{e}_\theta + W(r)\boldsymbol{e}_z$. Trajectories are concentric helices whose projections on the $(\boldsymbol{e}_r, \boldsymbol{e}_\theta)$-plane are circles that may be parametrized by r; (12a) is thus solved. The eikonal equation (12b) is easily solved explicitly, as well as the transport equations (12c–d). In particular, when

$$(-V' + V/r)\,\boldsymbol{k}\cdot\boldsymbol{e}_\theta - W'\boldsymbol{k}\cdot\boldsymbol{e}_z = 0, \tag{36}$$

where $' = \mathrm{d}/\mathrm{d}r$, the system governing the amplitude vector $\boldsymbol{c}(r,t) = (\boldsymbol{a}, b)^T$ can be reduced along the basic flow trajectories into an ordinary autonomous linear system $\mathrm{d}\boldsymbol{c}/\mathrm{d}t = \boldsymbol{M}(r)\boldsymbol{c}$. The system is unstable if $\boldsymbol{M}(r)$ admits a positive real eigenvalue $\sigma(r)$, i.e. if

$$-\sigma^2(r) = \frac{2V}{r}\left(V' - \frac{V}{r}\right)\frac{W'^2 + V'^2 - \left(\frac{V}{r}\right)^2}{W'^2 + \left(V' - \frac{V}{r}\right)^2} + \frac{V^2}{r}\left(\frac{R'}{R} - \frac{V^2}{rC^2}\right) < 0, \tag{37}$$

somewhere in the flow domain [11]. As shown by Eckhoff [10], this criterion is a generalization of the incompressible condition for vortex breakdown [21].

4.2 The case of planar circular vortices

The rest of the paper examines the case $W = 0$, for which (37) corresponds to the compressible Rayleigh criterion for centrifugal instability:

$$-\sigma^2(r) = \frac{2V}{r}\left(V' + \frac{V}{r}\right) + \frac{V^2}{r}\left(\frac{R'}{R} - \frac{V^2}{rC^2}\right) < 0, \tag{38}$$

somewhere.

In that case, choosing the wave vector $\boldsymbol{k}(r,t) = k\boldsymbol{e}_z$ ensures that (36) is verified everywhere. We assume that the WKB growth rate verifies $\sigma(r) > 0$ in the set $[r_1, r_2]$ with a quadratic maximum $\sigma_0 = \sigma(r_0)$ at $r_1 < r_0 < r_2$, so that $\sigma_0' = 0$ and $\sigma_0'' < 0$. It may be shown using Laplace method that the large time asymptotics of the energy of the perturbation (14) reads, when $\varepsilon \ll 1$ and $t \to \infty$ [16]:

$$E(t) \sim X_0^2 \left(\frac{-\pi}{\sigma_0'' t} \right)^{1/2} \mathrm{e}^{2\sigma_0 t}, \tag{39}$$

with $X_0 = X(r_0)$, $X(r)$ being the norm of the eigenvector associated with $\sigma(r)$, the positive eigenvalue of $\boldsymbol{M}(r)$.

Thus, the existence of discrete unstable eigenmodes is expected. Indeed, it will be shown below that the maximum WKB growth rate σ_0 is the exact growth rate of the corresponding spectral problem for large wave numbers. This extends in some sense Bayly's incompressible result [2].

4.3 Description of the normal modes at large wave numbers

Particularizing system (4a–c) for axisymmetric normal modes

$$[\boldsymbol{u}, m, n](\boldsymbol{x}, t) = \mathrm{e}^{st} \mathrm{e}^{-\mathrm{i}(\omega t + kz)} [\bar{\boldsymbol{u}}, \bar{m}, \bar{n}](r), \tag{40}$$

with s the growth rate, ω the pulsation, and k the axial wave number, one can derive the following system for $\bar{u}$ (the radial velocity) and $\bar{n}$ (the Eckart pressure):

$$(\lambda^2 + \sigma^2)\bar{u} + \mathrm{i}\lambda \left\{ \left(\frac{V^2}{rC} \left(\frac{\gamma}{2} - 1 \right) + \frac{CR'}{2R} \right) \bar{n} + C\frac{\mathrm{d}\bar{n}}{\mathrm{d}r} \right\} = 0, \tag{41a}$$

$$\lambda \left\{ \left(\frac{C}{r} + \frac{V^2}{rC} \right) \bar{u} + C\frac{\mathrm{d}\bar{u}}{\mathrm{d}r} \right\} + \mathrm{i} \left\{ C^2 k^2 - \lambda^2 \right\} \bar{n} = 0, \tag{41b}$$

where $\lambda = \mathrm{i}s + \omega$.

We now seek for the solution of the above system at large wave number $k \gg 1$ [16]. In order to satisfy the principle of dominant balance, ω is necessarily zero, and the variables of (25–b) are expanded as:

$$s = \sigma_0 - \frac{\mathcal{S}}{k},\ \bar{u} = \mathcal{U}(\eta),\ \bar{n} = \frac{1}{k^{3/2}} \mathcal{N}(\eta),\ \eta = (r - r_0) k^{1/2}. \tag{42}$$

In the vicinity of r_0, when the maximum for the WKB growth rate $\sigma(r)$ is reached, we have, by Taylor expansion:

$$\sigma^2(r) = \sigma_0^2 + \frac{\sigma_0 \sigma_0'' \eta^2}{k} + \cdots . \tag{43}$$

At leading order, system (25–b) reduces to:

$$C_0 \frac{\mathrm{d}\mathcal{N}}{\mathrm{d}\eta} - (2\mathcal{S} + \sigma_0''\eta^2)\mathcal{U} = 0, \tag{44a}$$

$$\sigma_0 \frac{\mathrm{d}\mathcal{U}}{\mathrm{d}\eta} + C_0\mathcal{N} = 0, \tag{44b}$$

with $C_0 = C(r_0)$. This further results into an equation for $\mathcal{U}$:

$$\frac{\mathrm{d}^2\mathcal{U}}{\mathrm{d}\eta^2} + \left(2\frac{\mathcal{S}}{\sigma_0} + \frac{\sigma_0''}{\sigma_0}\eta^2\right)\mathcal{U} = 0, \tag{45}$$

which is the quantum harmonic oscillator equation [4]. With He_j the Hermite polynomials and $\mathcal{A} = (-4\sigma_0''/\sigma_0)^{1/4}$, the jth most amplified mode presents the following eigenvalue and eigenfunction:

$$\mathcal{S}_j = (-\sigma_0\,\sigma_0'')^{1/2}(j + 1/2),\ \mathcal{U}_j(\eta) = \exp\left\{-(\mathcal{A}\eta/2)^2\right\}\mathrm{He}_j(\mathcal{A}\eta). \tag{46}$$

The most amplified mode is reached for $j = 0$, for which the corresponding eigenfunction is the Gaussian.

In terms of primitive variables, the corresponding radial perturbation velocity profile behaves as, for the most amplified mode $j = 0$:

$$u(\boldsymbol{x}, t) \sim \mathrm{e}^{(\sigma_0 - \mathcal{S}_0/k)t}\mathrm{e}^{-ikz}\exp\left\{-k\mathcal{A}^2(r - r_0)^2/4\right\}, \tag{47}$$

when $k \gg 1$. Similar expressions for the other variables may be easily obtained from the original system [16]. Clearly, at leading order, the energy of these modes reads $E(t) \sim \exp(2\sigma_0 t)$ as expected from (39).

The asymptotic structure of the *discrete* eigenmodes (47) is characteristic of short-wavelength instabilities, and gives an assessment to the geometrical optics stability theory: the growth rate tends asymptotically to the WKB maximum σ_0, given by (38), with a correction in $1/k$, whereas the corresponding eigenfunction tends asymptotically to the Dirac-delta function $\bar{u}(r) \sim \delta(r - r_0)$ (up to a multiplicative normalization factor), showing that the instability is centered in the vicinity of the trajectory r_0 corresponding to the most amplified growth rate, exactly as suggested by the WKB theory. Similar behaviors were observed in incompressible flows [2,18,28].

4.4 Numerical computation of the eigenmodes

The analytical results on the growth rates and modes structure are checked against a numerical study, with a spectral collocation method. The eigenfunctions are normalized so that the maximum of the radial velocity perturbation $\bar{u}(r)$ is 1. Basic flows satisfying (38) with a quadratic maximum at r_0 inside the domain may be easily constructed [16].

As the axial wave number k increases, the actual growth rate s of the eigenmodes asymptotically reaches the maximum WKB growth rate σ_0 given by (38).

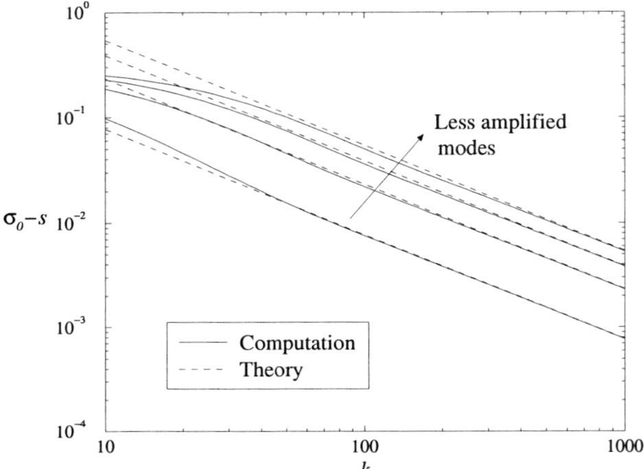

Fig. 4. Computations of $\sigma_0 - s$, the correction to the maximum WKB growth rate σ_0. Theory corresponds to $\mathcal{S}_j/k$ for $j = 0, \ldots, 3$ given by (46). From [16].

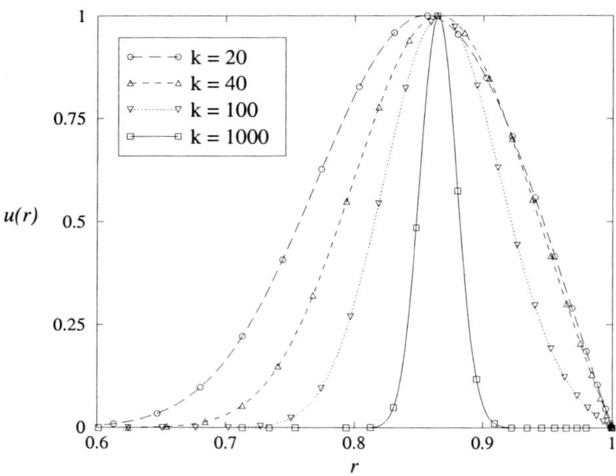

Fig. 5. Perturbation radial velocity profile $\bar{u}(r)$ of the most amplified eigenmode for different axial wave numbers k. From [16].

Furthermore, the growth rate correction $\sigma_0 - s$ is evaluated numerically and compared against the high wave number analytical correction $\mathcal{S}/k$ given by (46). Both are in excellent agreement for $k \geq 100$, as shown in Fig. 4.

The structure of the radial perturbation velocity is shown in Fig. 5 and perfectly matches the eigenmodes (47). For increasing k, this illustrates the decreased lateral spreading about the point where the WKB growth rate is maximal.

5 Discussion

Generic mechanisms of instability have been described in compressible vortices of an ideal gas flow. These mechanisms are local in the sense that they are only dependent of the local properties of the considered basic flows. In the case of the parametric resonances in periodically compressed vortex cores, the present results have been applied to the interaction between a circular vortex and a couple of acoustic waves propagating in opposite direction, leading to a parametric destabilization of the vortex core [17].

Furthermore, it is not difficult to construct approximate solutions in the low-Mach number approximation, such as compressed elliptical vortices [19] or compressed Taylor–Green cells [17]. Thus the present results may be useful in industrial applications, such as in internal combustion engines, since the stability properties of a periodically compressed flow may have important implications, since an efficient air/fuel mixing is expected for a maximum destabilisation [25].

The case of the compressible Rayleigh criterion for steady circular vortices gives an additional assessment to the geometrical optics stability theory, since discrete eigenmodes have been shown to be unstable with a growth rate which tends asymptotically to the one predicted by the WKB theory. This is not obvious at all, the methods being completely different. It is known that the WKB theory describes parts of the continuous spectrum of the Euler equations [22,20] and gives the asymptotic growth rate of the corresponding Green's function [7,29]. But the link between the short-wavelength approximation and the discrete spectrum of the linearized Euler operator remains an open question.

References

1. B.J. Bayly: Phys. Rev. Lett. **57**, 2160 (1986)
2. B.J. Bayly: Phys. Fluids **31**, 55 (1988)
3. B.J. Bayly, D.D. Holm, A. Lifschitz: Phil. Trans. R. Soc. Lond. A **354**, 895 (1996)
4. C.M. Bender, S.A. Orszag: *Advanced Mathematical Methods for Scientists and Engineers* (McGraw-Hill, New York 1978)
5. C. Cambon, C. Teissèdre, D. Jeandel: J. Méc. Théo. Appl. **4**, 629 (1985)
6. A.D.D. Craik, H.R. Allen: J. Fluid Mech. **234**, 613 (1992)
7. S.Yu. Dobrokhotov, A.I. Shafarevich: Mat. Zametki **51**, 72 (1992)
8. P.G. Drazin, W.H. Reid: *Hydrodynamic Stability* (Cambridge University Press, Cambridge 1981).
9. K.S. Eckhoff: J. Diff. Eqns. **40**, 94 (1981)
10. K.S. Eckhoff: J. Fluid Mech. **145**, 417 (1984)
11. K.S. Eckhoff, S. Storesletten: J. Fluid Mech. **89**, 401 (1978)
12. K.S. Eckhoff, S. Storesletten: J. Fluid Mech. **99**, 433 (1980)
13. R.F. Gans: J. Fluid Mech. **68**, 403 (1975)
14. D.P. Lalas: J. Fluid Mech. **69**, 65 (1975)
15. S. Le Dizès, M. Rossi, H.K. Moffatt: Phys. Fluids **8**, 2084 (1996)
16. A. Le Duc, S. Leblanc: Phys. Fluids **11**, 3563 (1999)
17. S. Leblanc: J. Fluid Mech. (submitted)
18. S. Leblanc, C. Cambon: Phys. Fluids **9**, 1307 (1997)

19. S. Leblanc, L. Le Penven: Phys. Fluids **11**, 955 (1999)
20. N. Lebovitz, A. Lifschitz: Proc. R. Soc. Lond. A **438**, 265 (1992)
21. S. Leibovich, K. Stewartson: J. Fluid Mech. **126**, 335 (1983)
22. A. Lifschitz: Phys. Lett. A **152**, 199 (1991)
23. A. Lifschitz: Adv. Appl. Math. **15**, 404 (1994)
24. A. Lifschitz, E. Hameiri: Phys. Fluids A **3**, 2644 (1991)
25. N.N. Mansour, T.S. Lundgren: Phys. Fluids A **2**, 2089 (1990)
26. V.P. Maslov: *Théorie des Perturbations et Méthodes Asymptotiques* (Dunod, Paris 1972)
27. P.G. Saffman: *Vortex Dynamics* (Cambridge University Press, Cambridge 1992)
28. D. Sipp, E. Lauga, L. Jacquin: Phys. Fluids **11**, 3716 (1999)
29. M.M. Vishik, S. Friedlander: J. Math. Pures Appl. **72**, 145 (1993)
30. F. Waleffe: Phys. Fluids A **2**, 76 (1990)
31. F.W. Warren: J. Fluid Mech. **68**, 413 (1975)

Three-Dimensional Instabilities of a Counterrotating Vortex Pair

Thomas Leweke[1] and Charles H. K. Williamson[2]

[1] IRPHE, CNRS/Universités Aix-Marseille, Marseille, France
[2] Mechanical & Aerospace Engineering, Cornell University, Ithaca, NY, USA

1 Introduction

The dynamics of a pair of parallel counter-rotating vortices has been the object of a large number of studies in the last three decades. The continued interest in this flow is, to a great extent, due to its relevance to the problem of aircraft trailing wakes. In addition, the counterrotating vortex pair represents one of the simplest flow configurations for the study of elementary vortex interactions, which can yield useful information for the understanding of the dynamics of more complex transitional and turbulent flows.

The first three-dimensional stability analysis of a vortex pair was performed by Crow [1], who showed that such a pair is unstable to perturbations which are symmetric with respect to the plane separating the vortices, and which have an axial wavelength of around eight vortex spacings. This result is in good agreement with the vortex structures observed in the contrails of real aircraft.

When the amplitude of these long-wavelength perturbation grows sufficiently large, the vortex cores touch at the location of maximum inward displacement, leading to a break-up of the pair into a series of vortex rings. This scenario is also mainly known from observations of full-scale aircraft wakes [15], and it was also observed in recent numerical simulations [10]. Very little is known, however, about the subsequent evolution of the flow, i.e. about the dynamics and persistence of the rings in the late stages of the Crow instability.

The transition between two wavy vortices and a series of vortex rings happens via a periodic cross-linking, or reconnection, of vorticity. This interesting phenomenon involves a change in the topology of the vortex lines and represents a fundamental interaction between concentrated vortices. So far, experimental results concerning this process are mostly qualitative [9,5]. More detailed knowledge about vortex reconnection comes from numerical simulations (see Kida & Takaoka [2] for a recent review). Melander & Hussain [7] and Shelley *et al.* [12] chose as initial conditions configurations which are very similar to a vortex pair perturbed by the Crow instability. The interacting vortices of the present study are therefore also an ideal candidate for the investigation of the reconnection phenomenon.

Crow [1] identified two more bending-wave modes of instability, symmetric and antisymmetric, with a much shorter wavelength of around one vortex spacing. However, Widnall, Bliss & Tsai [16] pointed out that these modes are spurious, since the expression for the self-induced rotation of these perturbations, used

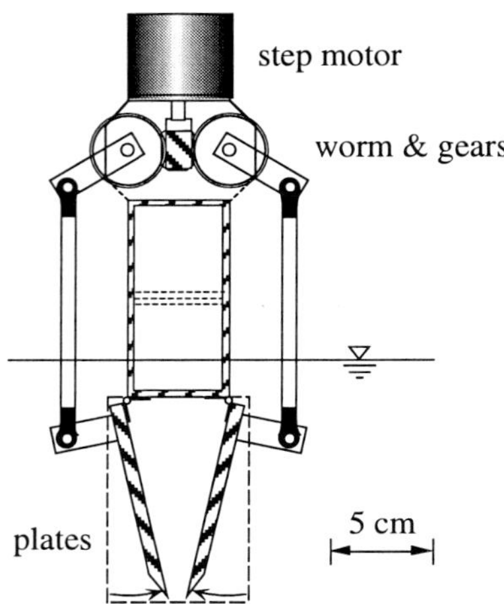

Fig. 1. Schematic of the vortex pair generator used in the experiments. End plates (dashed lines) were fitted to both lateral ends of the plates

in Crow's calculations, is asymptotically valid for long waves only. Widnall *et al.* [16] recalled, however, that other more complicated disturbance modes exist, whose low self-induced rotation would allow their growth in an external strain field, leading to instability. This result from a qualitative argument was later confirmed by more rigorous stability analyses of a vortex filament in an externally imposed strain field [8,14,11]. On the experimental side, short–wavelength instabilities have been observed qualitatively on vortex rings [3,6,17], and vortex pairs [13], but no detailed analysis of their structure and their consequences on the vortex pair dynamics existed prior to the present study.

2 Experimental details

The vortex pair is generated in a water tank at the sharpened parallel edges of two flat plates, hinged to a common base and moved in a prescribed symmetric way by a computer-controlled step motor (see Fig. 1). The vortices are typically separated by a distance of 2.5 cm, and their length is about 170 cm. This high aspect ratio of the vortex pair is necessary to limit the influence of end effects that spread rapidly into the central part of the flow. Visualization is achieved using fluorescent dye (Fluorescein), illuminated by the light of a 5 W Argon laser.

A vortex pair is characterized by the circulation Γ of each vortex, the separation b between the vortex centres, and a characteristic core size a, which is the radius of the tube around the vortex centre containing most of the vorticity. These characteristics were determined from flow field measurements, using Digital Particle Image Velocimetry (DPIV), which showed that the initial velocity

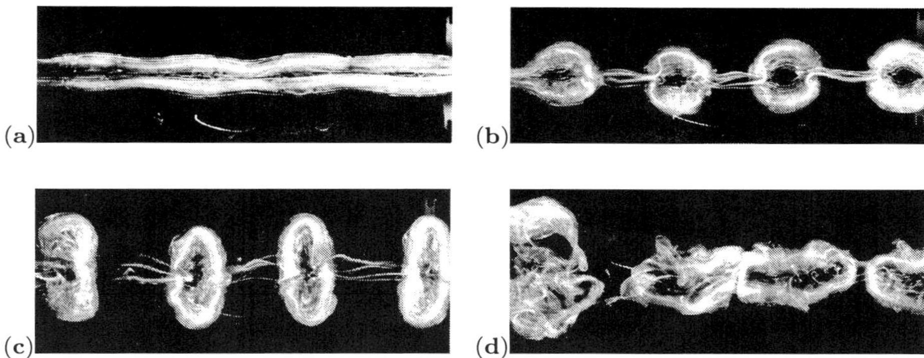

Fig. 2. Visualization of the long-wavelength instability of a vortex pair for $Re = 1450$ and $a/b = 0.23$. The pair is moving towards the observer. (**a**) $t^* = 3.2$, (**b**) $t^* = 6.5$, (**c**) $t^* = 9.3$, (**d**) $t^* = 14.4$

profiles of the vortices are very well represented by the one of a Lamb-Oseen vortex with a Gaussian vorticity distribution. The Reynolds number based on the initial circulation ($Re = \Gamma_o/\nu$) was in the range between 1500 and 2500 in this study. In the following, time t is non-dimensionalized by the time it takes the initial vortex pair to travel one vortex spacing b, i.e. $t^* = t\ (\Gamma/2\pi b^2)$. More details about the experimental set-up and techniques are given in [4].

3 Long-wavelength instability and reconnection

Figure 2 shows the overall features of the long-wavelength vortex pair instability. The initially straight and parallel vortices develop a waviness, which is symmetric with respect to the plane separating the vortices, and whose axial wavelength is about 6 times the initial vortex separation. Photographs taken simultaneously from two perpendicular directions show that the plane of the waves is inclined by about 45° with respect to the plane containing the pair (see also Fig. 3a, in agreement with Crow's [1] theoretical prediction and previous observations.

This waviness is amplified in time until the vortex cores touch, break up, and reconnect to form periodic vortex rings, which initially look almost circular in the front view of Fig. 2b. Side-view visualizations show, however, that these rings are not contained in a plane; they are bent upwards in the transverse direction. The visualizations also show that the large-scale rings are still linked by thin strands of dye. Subsequently, the rings stretch out into oval vortices exhibiting a well-known [5] oscillatory behaviour (Figs. 2c–d), whereby the principal axes of the oval are periodic

Figure 3 shows the flow structure in the reconnection plane, i.e. the cross-section of the pair at the location of minimum separation, during the cross-linking process. Initially the pair is symmetric, and the cores are well separated. In Fig. 3a, the amplitude of the perturbation has grown quite large. The plane of the waves can be seen in the background, illuminated by scattered light from the

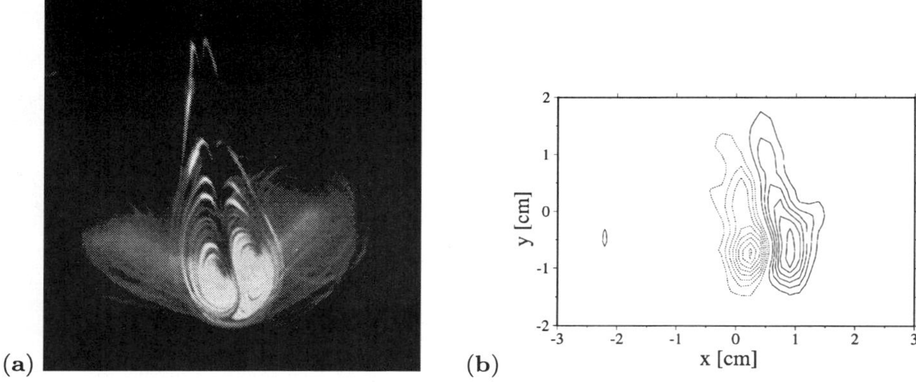

Fig. 3. Flow in the reconnection plane at $t^* \approx 5$. (**a**) Dye visualization using a laser sheet, (**b**) contours of measured axial vorticity

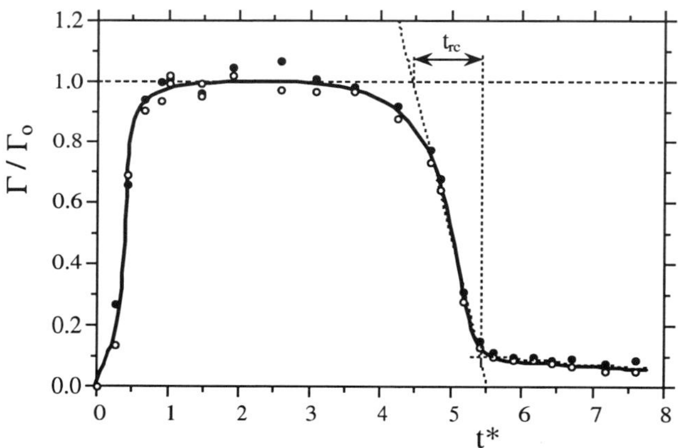

Fig. 4. Evolution of the circulation in the reconnection plane for $Re = 2340$, and definition of the reconnection time t_c

reconnection plane. The vortex cores have come much closer and have started to elongate vertically. At the same time, a "tail" of dye is developing behind the descending pair.

This visualization is complemented by measurements of the time-dependent vorticity distribution in the same plane with the help of DPIV (Fig. 3b. These measurements confirmed the qualitative observations in Fig. 3a, in particular the formation of a tail of vorticity during the cross-linking. At later times, a small, slowly decaying pair of counterrotating vortices remains, corresponding to the dye threads linking the vortex rings in Fig. 2b–c. These results are in good qualitative agreement with the numerical results in [7].

From the vorticity measurements, the circulation in each half of the reconnection plane can be calculated (Fig. 4). After the initial roll-up, the circulation

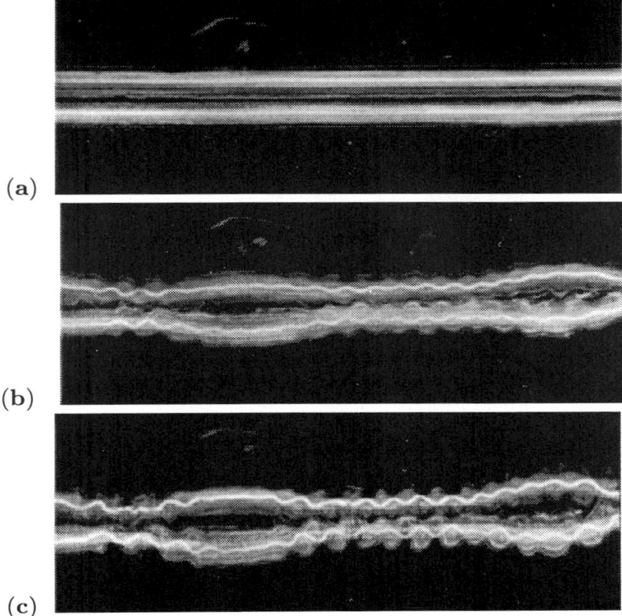

Fig. 5. Visualization of vortex pair evolution under the combined action of long-wavelength (Crow) and short-wavelength instabilities. $Re = 2750$. The pair is moving towards the observer. **(a)** $t^* = 1.7$, **(b)** $t^* = 5.6$, **(c)** $t^* = 6.8$

remains constant for some time. Once the cores touch, it decreases rapidly due to the cancellation and reorientation of vorticity. At later times, it reaches a new, almost constant value of about 10% of the initial circulation Γ_o, which corresponds to the circulation of the threads between the large-scale vortex rings.

Following [7], one can deduce from Fig. 4 a characteristic reconnection time t_c, using extrapolations of the different almost linear parts of the circulation history. From our experimental measurements we find $t_c \approx 0.9$. If the result found in the DNS study of Melander & Hussain [7] is transformed to the non-dimensional units used in this study, one obtains a reconnection time $t_c \approx 0.7$ for their case, which again is in reasonably good agreement with the present result.

4 Short-wavelength instability

In this chapter the overall qualitative features of the short-wavelength instability of a vortex pair, appearing at higher Reynolds numbers, are presented, mostly using results from flow visualization.

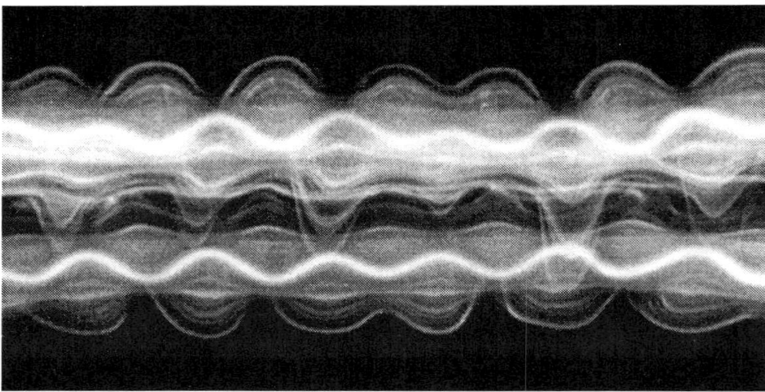

Fig. 6. Close-up view of the short-wavelength perturbation, showing the characteristic internal vortex deformations, as well as the distinct phase relationship between the perturbations of the two vortices. $Re = 2750$, $t^* = 6.2$

4.1 Initial stages and physical origin

We show, in Fig. 5, visualizations of different stages of the vortex pair evolution, at a Reynolds number $Re = 2750$. The photograph in Fig. 5a was taken immediately after the end of the plate motion. The vortices are straight and uniform along their axes. Due to the impulsive start of the plate motion, the centreline of each vortex is marked by a particularly bright filament of dye. In Fig. 5b two distinct instabilities have started to deform the pair, as follows. The symmetric large-scale deformation, which has an axial wavelength of about $6b$, is due to the Crow instability discussed in the previous section. Simultaneously, another perturbation with a much smaller wavelength $\lambda < b$ is growing. It must be emphasized that this second disturbance is not simply a wavy displacement of the vortex without change in the cross-sectional structure, as for the Crow instability. The close-up view in Fig. 6 shows that the short-wave instability modifies the internal structure of the vortex cores. In particular, the inner and outer layers of each vortex are displaced in opposite radial directions, and between these one observes the existence of an "invariant" streamtube on each vortex.

The axial wavelength λ of the short waves was found to be close to 80% of the initial vortex spacing b. For comparison, Thomas & Auerbach [13], who studied vortex pairs with similar geometric features, measured wavelengths in the range $0.6 < \lambda/b < 0.8$.

An interesting and unexpected observation relates to the phase relationship between the short-wave perturbations on the two vortices. In the close-up front view of Fig. 6, the vortex centres are, at each axial position, displaced in the same direction. This means that the initial reflectional symmetry of the flow, with respect to the plane separating the vortices, is lost. In simultaneous side views, however, the displacements of the two vortex centres are found to be *out of phase*. This behaviour was observed systematically; it is not a random coincidence, and we should reiterate that the flow has not been forced in any

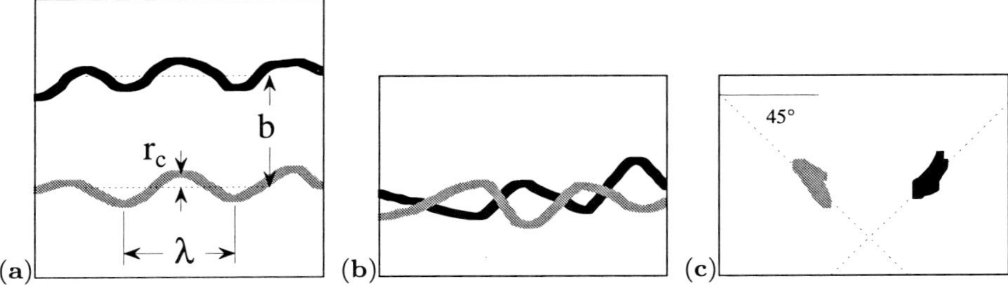

Fig. 7. Example of a three-dimensional reconstruction of the deformed vortex center lines, from simultaneous perpendicular views. (**a**) front view, (**b**) side view, (**c**) end view

manner. The instabilities on the two vortices are therefore found to evolve in a coupled, or cooperative, fashion.

In order to complete the picture of the three-dimensional orientation of the disturbance, a third perpendicular view along the primary vortex axes is needed. Since this was not accessible simultaneously in the experiments, due to the arrangement of the laser optics, it was constructed digitally from the two other views. From simultaneous front and side view visualizations, the two vortex centres were seized as two-dimensional lines (using a digitizing tablet), as shown in Fig. 7. By identifying two corresponding points in the two views, the lines can be correlated to give three-dimensional curves representing the vortex centres. Figure 7c shows the projection of these curves on a plane perpendicular to the primary vortex axes, which corresponds to the missing end view of the pair. The centrelines are displaced in planes inclined by approximately 45° with respect to the horizontal. The orientation of these planes coincides with the directions of maximum stretching associated with the strain fields that the vortices induce on each other. This is an indication that the short-wavelength instability is a direct consequence of this strain, in a way similar to the mechanism anticipated by Widnall *et al.* [16]. In this respect, it should also be emphasized that the perturbation does not rotate around the vortex axis; its structure is stationary while its amplitude keeps growing.

The structure of the perturbed vortex pair at a particular axial position is shown in the light-sheet visualization in Fig. 8. The displacement of the left vortex centre to the lower right, and the one of the right centre to the upper right clearly demonstrates the antisymmetric nature of this coupled, or "cooperative", vortex instability.

An analysis and comparison of different theories treating vortex pair instabilities and the dynamics of strained vortices, shown in detail in [4], have led us to identify this phenomenon as a manifestation of an elliptic instability of the vortex cores. The streamlines in the core of each vortex have an elliptical shape due to the interaction of the vorticity with the mutually-induced strain, and this kind of flow is known to be receptive to three-dimensional instability. Comparisons between our experimental observations and predictions derived from the

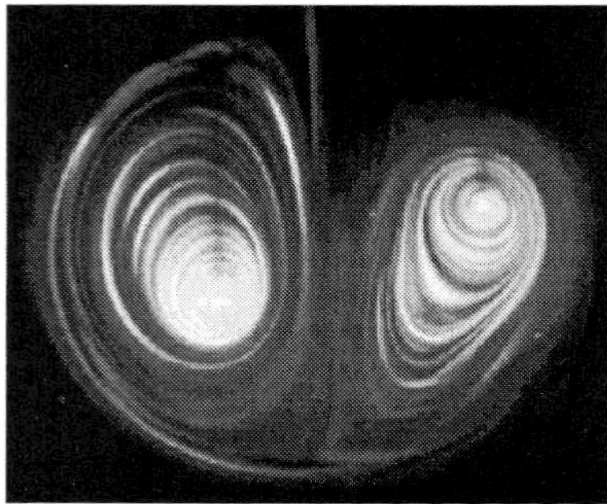

Fig. 8. Dye visualization in a cross-cut plane of the perturbation due to the short-wavelength instability at $Re = 2400$

theory on elliptic instability show good qualitative and quantitative agreement, concerning the perturbation mode shape, its aspect ratio and wavelength, and the instability growth rate. The elliptic instability is here observed clearly for the first time in a real open flow.

4.2 Long-term evolution of the pair

The late-time development of the vortex pair in the presence of the short-wavelength instability is quite distinct from the previous case with only the long-wavelength Crow instability. The fact that a significant change is occurring in the flow is illustrated by the evolution of the overall descending motion of the pair. Up to a non-dimensional time $t^* \approx 6$, which corresponds approximately to the time of Fig. 5b, the descent speed is constant. The pair then slows down considerably over a short period of time, and finally continues to move down at a speed which is only about one third of the initial speed. In the case of the Crow instability, the vortex system keeps on descending with the same speed for much longer times.

The origin of this change in the flow lies in the breakdown of the complicated, but nevertheless organized disturbance structure described in the previous section, which begins approximately at the time when the pair starts to slow down. The visualization in Fig. 9a shows the early stage of this breakdown. We see that, near the front edge of the pair, the parts of the fluid initially orbiting each vortex individually, begin to mix. In a periodic, interlocking way, tongues of fluid from one side are drawn around the respective other vortex.

This cross-over is a direct consequence of the internal vortex deformations, and can be illustrated with the help of Fig. 8. On the one hand, the centre of rotation of the left vortex is displaced toward the front (lower) stagnation point of the pair, whereas the one on the right is pushed away. On the other hand, we know that the outer layers of each vortex are displaced in the opposite directions. This means that the outer parts of the right vortex are actually brought closer to the stagnation point, and to the approaching left vorticity maximum; the latter will eventually capture some of this fluid and pull it to the left. The vortex

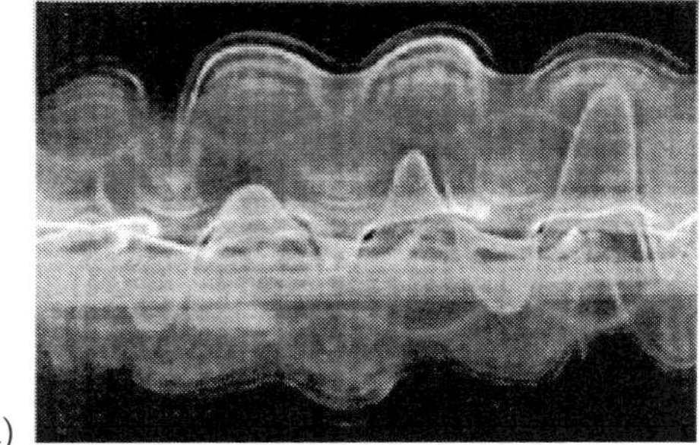
(a)

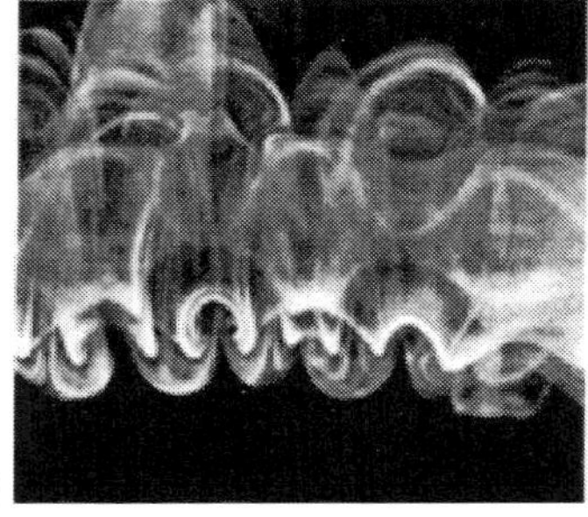
(b)

Fig. 9. Breakdown of the short-wavelength perturbation. (**a**) front view, (**b**) side view

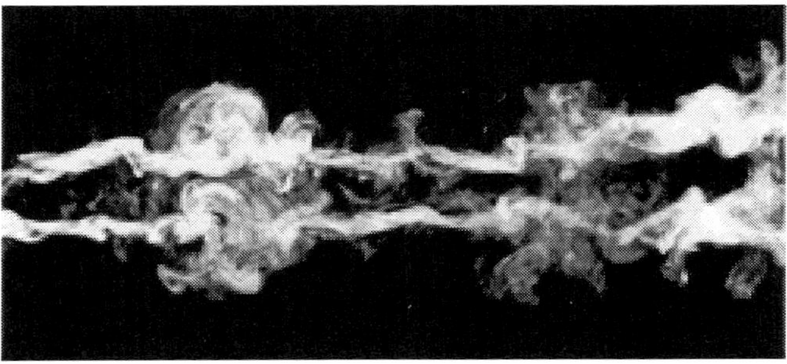

Fig. 10. Late-time evolution of a vortex pair with both short- and long-wavelength instabilities. $Re = 2750$, $t^* = 12.1$. The pair is moving towards the observer

centrelines are also faintly visible in Fig. 9a, and the phase of their displacement is consistent with this scenario. Due to the orientation of the deformation (see Fig. 7c), which is linked to the stretching axes of the strain, it is clear that this phenomenon can only occur at the leading edge of the pair.

Once the tongues have crossed over, they are quickly pulled apart by the stagnation point flow and wrapped around the primary vortices. This results in the formation of an array of transverse counter-rotating secondary vortex pairs. Axial vorticity is pulled to the left with the fluid, and is subsequently tilted and stretched by the stagnation point strain field. By this process the circulation of the primary vortices is reduced, which could explain the decrease in the self-induced translation velocity of the pair.

In Fig. 9b, the same phenomenon is seen from the side. The wavelength of the initial elliptic instability can still be identified from the inner core deformation, which, at this point, is still relatively ordered. This view shows that, for each wavelength, *two* secondary vortex pairs are generated at the lower edge.

The formation of the secondary vortex pairs has a strong effect on the further development of the overall flow. The visualizations in (Fig. 5) show that the elliptic instability develops faster where the Crow instability brings the vortices together. At these periodic locations along the axes, the organized structure of the primary vortices breaks down first, due to the presence of the perpendicular vortices in the outer layers. From further flow visualizations, it appears that the primary vortices are stripped of these outer layers, which seem to move away from where the Crow instability pulls the pair together, to leave only a skeleton of the initial vortex pair (Fig. 10). One might interpret this axial flow as a consequence of a pressure gradient set up between the less coherent primary vortex in the regions of break-up, to the more coherent parts of the vortex not yet broken up. This process may be seen as the result of an 'attempted' vortex reconnection, a phenomenon which, in the absence of short waves, was shown to lead to periodic vortex rings. As a result of this axial motion, the dyed fluid concentrates in the large clusters visible in Fig. 10, whose internal motion is apparently relatively unorganized, and whose spacing is dictated by the initial large-scale Crow instability.

5 Summary

We have investigated experimentally the evolution of a pair of counterrotating vortices, and have observed two distinct three-dimensional instabilities in this flow. In the case of the long-wavelength Crow instability, with an axial wavelength of several times the inter-vortex spacing), the vortices undergo a wavy deformation before they cut and reconnect into a series of large-scale vortex rings. At higher Reynolds numbers, a second, short-wave instability appears, with a wavelength scaling on the vortex core size. It leads to internal deformations of the vortices and can be explained by an elliptic instability of the vortex cores. In the presence of short waves, the long-term evolution of the pair is distinctly altered: the average propagation speed of the vortex system decreases drastically, and the large-scale vortical structures are destroyed much more effectively than in the case of a pure Crow instability.

References

1. S.C. Crow: AIAA J. **8**, 2172 (1970)
2. S. Kida, M. Takaoka: Annu. Rev. Fluid Mech. **26**,169 (1994)
3. C.-H. Krutzsch: Ann. Phys. (5) **35**, 497 (1939)
4. T. Leweke, C.H.K. Williamson: J. Fluid Mech. **360**, 85 (1998)
5. T.T. Lim, T.B. Nickels: 'Vortex Rings'. In: *Fluid Vortices*, ed. by S.I. Green (Kluwer, Dordrecht 1996), pp. 95–153
6. T. Maxworthy: J. Fluid Mech. **51**, 15 (1972)
7. M.V. Melander, F. Hussain: Phys. Fluids *A* **1**, 633 (1989)
8. D.W. Moore, P.J. Saffman: Proc. R. Soc. Lond. *A* **346**, 413 (1975)
9. Y. Oshima, S. Asaka: J. Phys. Soc. Japan **42**,708 (1977)
10. R.E. Robins, D.P. Delisi: AIAA J. **35**, 1552 (1997)
11. A.C. Robinson, P.J. Saffman: J. Fluid Mech. **142**, 451 (1984)
12. M.J. Shelley, D.I. Meiron, S.A. Orszag: J. Fluid Mech. **246**, 613 (1993)
13. P.J. Thomas, D. Auerbach: J. Fluid Mech. **265**, 289 (1994)
14. C.-Y. Tsai, S.E. Widnall: J. Fluid Mech. **73**, 721 (1976)
15. M. Van Dyke: *An Album of Fluid Motion* (Parabolic, Stanford 1982)
16. S.E. Widnall, D.B. Bliss and C.-Y. Tsai: J. Fluid Mech. **66**, 33 (1974)
17. S.E. Widnall, J.P. Sullivan: Proc. R. Soc. Lond. *A* **332**, 335 (1993)

Acoustic Characterization of a Stretched Vortex in an Infinite Medium

Sébastien Manneville[1], Agnès Maurel[1], Frédéric Bottausci[2] and Philippe Petitjeans[2]

[1] Laboratoire Ondes et Acoustique, UMR CNRS 7587 ESPCI, 10 rue Vauquelin, 75005 Paris, France
[2] Laboratoire Physique et Mécanique des Milieux Hétérogènes, UMR CNRS 7636, ESPCI, 10 rue Vauquelin, 75005 Paris, France

Abstract. A new experimental device is presented, that allows to isolate and control a stretched vortex in an "infinite" medium. Acoustic measurements based on the ultrasound-flow interaction yield the main vortex characteristics (position, circulation, core size). This global and non-invasive method also allows a dynamical tracking of the vortex. Experimental results on the mean vortex characteristics as a function of the control parameters are presented, together with some examples of transitory regimes and of precession motion.

1 Introduction

Experimental measurements of the vorticity and of the dynamics of vortical structures in fluid flows are essential to the understanding of hydrodynamic instabilities and, more generally, of turbulence. Indeed, in turbulent flows, experimental observations [2] as well as numerical simulations [2,3] have revealed the presence of very intense rotational structures that concentrate most of the vorticity of the flow. Such structures of small diameter but very long along their rotation axis often display a strong "stretching" (large longitudinal velocity gradients).

In general, it is almost impossible to isolate those "vorticity filaments" from the surrounding turbulent flow and their role in small scale intermittency in turbulence is not yet clearly understood [4,5,7]. That's why many recent experiments try and create isolated stretched vortices in various geometries [8–11], mostly inspired from Von Kármán flows generated between two rotating discs in a confined medium. To break free from restrictive boundary conditions, we set up a new experimental device that allows to control a stretched vortex in an "infinite" medium. This device is described in the next section.

To avoid the use of a probe or the seeding of the flow like in classical measurement techniques (hot wire anemometry, Laser Doppler Velocimetry, dye or particles flow visualization, Ultrasound Doppler Velocimetry), we tested a new acoustic technique to measure the vortex characteristics [12,3,14,15]. This technique based on the sound-flow interaction is briefly recalled in the third section.

The last section of this paper shows the experimental results. In particular, the acoustic measurement tool is shown to provide a thorough characterization of a very strongly stretched vortex in both its mean behaviour and its dynamics.

2 Experimental setup

A stretched vortex is generated in water between two identical rotating discs by applying a pumping at the center of each disc. In this experiment, rotation and stretching are thus applied independently and at the same points by rotating tubing. The rotation frequency $\Omega/2\pi$ is fixed by a feedback motor (Yaskawa DR2 Servopack) and tunable between 0 and 25 Hz. In the experiments presented below, the disc radius is 10 cm and the discs are corotating along the same axis.

Suction is applied at the center of each disc through a small hole of diameter 5 mm by a pump (Eheim IP67). The suction flowrate Q is set by a gate controlled by a step-by-step motor and ranges from 0 to 6.4 L/min. In our experiments, the pumping is symmetrical so that the flowrate for each disc ranges from 0 to 3.2 L/min, which corresponds to longitudinal velocities up to 5 m/s at the suction holes. To avoid stirring of the surrounding fluid, the rotating tubing is inserted into fixed PVC tubing.

The initial vorticity generated by the discs is amplified by the pumping [11] and this setup gives birth to a very intense stretched vortex ("vorticity filament"). Its length, which is also the distance D between the two discs, can be tuned up to about 30 cm. The large volume of the water tank (300 L) allows us to neglect the effects of the boundaries on the vortex: in the experiments described below, the nearest wall is 20 cm away from the vortex and the free surface is at 30 cm. Unlike experiments in cylindrical geometries [10][12][16] where the disc diameter is about the size of the water tank, boundary conditions -except those on the discs- are of no influence in the present experiment.

Figure 1 shows two photographs of the vortex visualized by injecting a dye directly in the vortex core . Most of the time, the vortex appears as a straight filament linking the centers of the two discs (fig. 1(a)). Depending on the control parameters, the vortex also follows a small precession motion around the rotation axis of the discs. Observations also show large fluctuations of the "stagnation point" (point on the vortex axis where the longitudinal velocity v_z is zero) and even a layered radial structure of v_z and of the stretching $\partial v_z/\partial z$.

During transitory regimes (formation of the vortex, vortex breakdown after a perturbation or depending on the experimental conditions), the vortex can undergo large deviations far from the rotation axis of the discs (fig. 1(b)). In such cases, transitory precession motions of large amplitude are observed, followed by a relaxation towards a more stable state.

3 Acoustic measurement of the vortex characteristics

Even if Doppler methods (Laser or Ultrasound Doppler Velocimetry) avoid placing a probe inside the vortex core, those methods require that the flow be seeded by scatterers. The dynamics of such scatterers in a vorticity filament was shown to be very complex and to give rise to demixtion phenomena where the scatterers migrate towards low pressure areas [18]. Moreover, the size of the measurement volume in Doppler techniques is sometimes too large as compared to the size of the vortex filament, which restricts the precision in the vortex core.

(a)

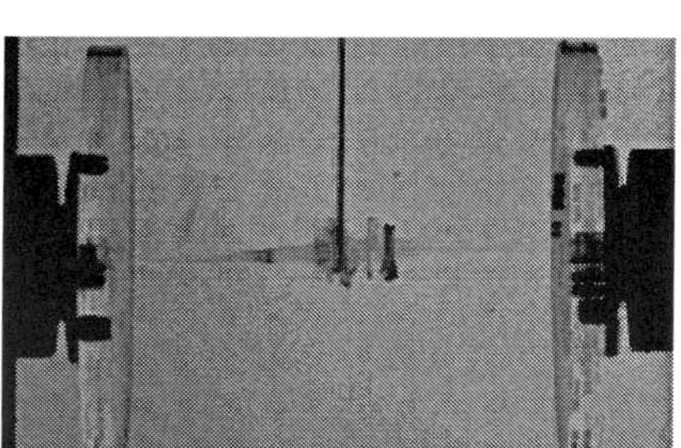

(b)

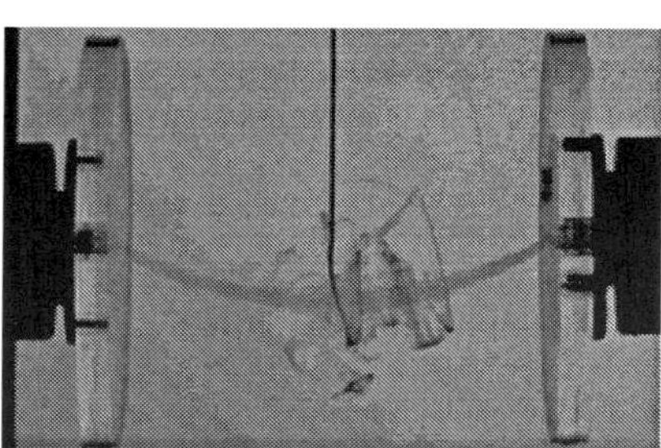

Fig. 1. Visualization of the vortex generated by the "double rotating suction system". (a) Stable regime. (b) Large departure from the rotation axis of the discs during the transitory regime of formation of the vortex. The dark line in the middle of the picture corresponds to the dye injector.

Our measurement method relies on the transmission of an ultrasonic wave in a moving medium without seeding and derives from classical acoustic tomography techniques used in oceanography for instance [19,20] .

3.1 The sound-flow interaction

An acoustic wave that crosses a moving medium is advected and distorted by the flow velocity field. If geometrical acoustics are valid, that is when the flow is slowly varying in space compared to the acoustic wavelength λ, the distortion simply results from a velocity composition: the wave is either sped up or slowed down depending on the area of the flow it crosses (see fig. 2). In that case, the ray theory for acoustic propagation can be used and many authors have investigated the trajectories of acoustic rays in a vortex [21–23]. Experiments on "large" vortices -with a characteristic size much larger than λ- [3,14] have shown the importance of geometric distortions of an acoustic wavefront due to a vortex and confirmed that fully non-invasive measurements of its characteristics were possible.

In the opposite limit where the fluid velocity field has strong velocity gradients localized on sizes smaller than λ, the incident wave is scattered by the flow. The theoretical framework provided by the Born approximation has led to numerous studies on sound scattering by a vortex [24–28] to predict the pressure field radiated far from an isolated vortex or from a turbulent flow modelled by a random gas of vortices. Experimentally, those studies triggered the first vorticity measurements by scattering of sound in hydrodynamics [29–31].

In our experiment, the order of magnitude of the control parameters lead to a vortex whose core is only a few λ so that the experimental situation is in between the two limits presented above and mixes both geometric effects and sound scattering.

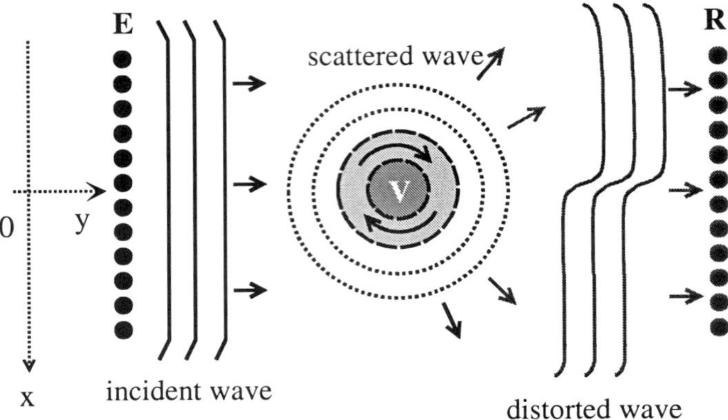

Fig. 2. Sound-flow interaction: the incident wave is emitted by the transducer array (E) and distorted by the flow velocity field and, depending on the vortex, scattered by the vortex core (V). The receiving array (R) records the pressure field transmitted through the vortex.

3.2 Measurement technique

A plane ultrasonic wave is emitted from a linear array of 64 piezzo-electric transducers with a pitch of 0.83 mm and with a central frequency of f =3.5 MHz (see fig. 2). The sound speed in water at rest is taken to be c =1480 m/s, which leads to a wavelength of λ =0.42 mm. A second array of 64 transducers, parallel to the emitters and about 30 cm apart, records the pressure field after one crossing of the vortex. The Mach number $M = u/c$ where u is a characteristic velocity of the vortex (for instance, the maximum orthoradial velocity) is of order $10^{-3}$and for the times-of-flight and for the propagation distances involved here, it can be considered that (i) the flow is frozen during the time of acoustic propagation, (ii) acoustic rays remain straight from a transducer array to another, and (iii) no frequency shift occurs for the incident wave.

A first experiment in the fluid at rest is performed to get some reference pressure signals. Those signals are then compared to the signals received after crossing the vortex: by taking the ratio of the Fourier transforms of the signals obtained in the presence of a vortex and in the fluid at rest, a phase and amplitude information is recovered on the wavefront distortion due to the crossing of the vorticity filament [12,3] . This information is a function of the position x of the receiver. Thus, with a single plane emission, the use of transducer arrays yields a spatial information on the flow along the x-axis. Let us also note that when the vortex is axisymmetric, the measurement depends only on the orthoradial component $v_\theta(r,z)$ of the velocity field (where r is the distance to the vortex center and z the position along the vortex axis) and carries no information on the longitudinal and radial components $v_z(r,z)$ and $v_r(r,z)$.

3.3 Acoustic data analysis

Figure 3 shows the typical phase and amplitude distortions averaged on 500 samples in the case of a stable vortex (for which the precession motion can be neglected). In spite of the low Mach number, the effect of the vortex on the incident wave is noticeable and precisely mesurable. Error bars correspond to the standard variation of the measurements and are due to (i) the weak instationarity of the flow, (ii) the vibrations of the experimental setup, and (iii) the electronic noise surrounding the experiment.

The main feature of the phase distortion Φ(fig. 3(a)) is the phase jump $\Delta\Phi = \Phi(x \geq 15\text{ mm}) - \Phi(x \leq 15\text{ mm})$ when crossing the vortex. This corresponds to the geometric advection of the acoustic wave by the vortex velocity field: depending on the side of the vortex it crosses, the wave is slowed down ($\Phi \leq 0$) or sped up ($\Phi \geq 0$) compared to the propagation in the fluid at rest. For an irrotational velocity field $v_\theta(r,z) = \Gamma/2\pi r$ far from the vortex core, the geometrical acoustics computation yields

$$\Delta\Phi = \frac{2\pi f}{c^2}\Gamma$$

where Γ is the vortex circulation . From the phase distortion Φ, we can thus deduce the position x_0 of the vortex center along the x-axis ($\Phi(x_0) = 0$) as well as the vortex circulation. In the case of figure 3, the phase jump yields $\Gamma = 145\text{ cm}^2/\text{s}$.

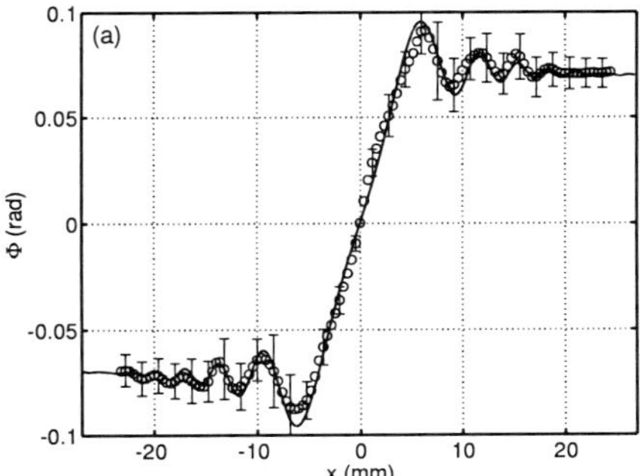

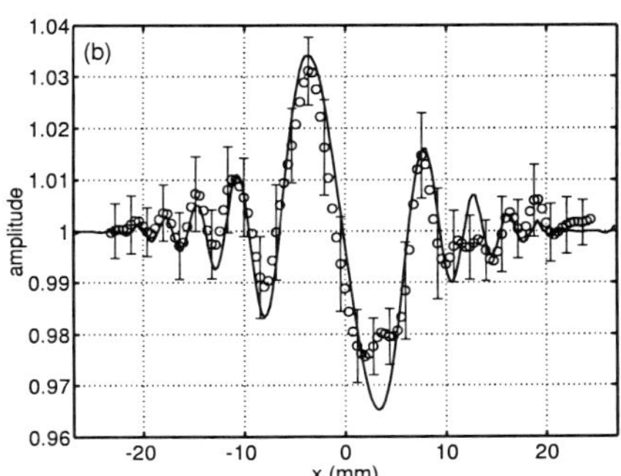

Fig. 3. Analysis of the mean distortion of a plane ultrasonic wave after crossing of a stretched vortex ($\Omega/2\pi$=1.7 Hz, D=80 mm, Q=5.6 L/min). (a) Phase distortion. (b) Amplitude distortion. The abscissa x is the transducer position along the receiver array. The origin $x = 0$ was arbitrarily taken as the position of the vortex center. Solid curves are the best theoretical fits of the data based on analytical calculations of sound scattering by a vortex for which Γ=145 cm^2/s and r_0=1.3 mm.

Another feature of the wavefront distortion is the presence of oscillations both on the phase (fig. 3(a)) and on the amplitude (fig. 3(b)) signals. Those oscillations result from the interference between the wave scattered by the vortex core and the incident wave advected by the flow as soon as the core radius r_0 is smaller than a few λ (typically 5λ). In that case, analytical calculations [12] for a perfect plane wave generated at infinity and incident on a velocity field given by

$$v_\theta(r,z) = \frac{\Gamma}{2\pi r} \text{ for } r \geq r_0,$$

$$v_\theta(r,z) = r\omega \text{ for } r \leq r_0 \text{ with } \omega = \frac{\Gamma}{2\pi r_0^2}$$

allows to account very well for those oscillations by using the value of Γ measured above and by adjusting only the value of r_0. The fits shown in fig. 3 yield a circulation $\Gamma = 145 \pm 5$ cm^2/s and a core radius $r_0 = 1.3 \pm 0.5$ mm$\approx 3\lambda$. Even if r_0 is not known very precisely, it is important to note that this method allows some measurements inside a very small vortex core, which is not accessible to usual techniques.

4 Experimental results

We checked that the results presented below are weakly dependent on the position z of the transducer arrays along the vortex axis. Thus, the orthoradial velocity $v_\theta(r,z)$ is taken as independent of z [6].

4.1 Mean characteristics as a function of the control parameters

Figure 4 shows the vortex circulation Γ as a function of the control parameters: the rotation speed of the discs Ω, the distance between the discs D, and the total suction flowrate Q.

The scaling law obtained for $\Gamma(\Omega)$ (fig. 4(a)) with an exponent close to 3/4 differs from the $\sqrt{\Omega}$ scaling predicted and observed in confined geometries [8][16]. This scaling law together with pressure measurements allows to deduce scaling for $u_{\theta,max}$ and r_0 as a function of Ω [17]. The scaling law $\Gamma \propto D^{-1/2}$ (fig. 4(b)) for small values of D is qualitatively explained by the following energetic argument: the power dissipated in the volume of the vortex outside the boundary layers depends only on Ω and Q, thus $\int(\partial v/\partial r)^2 dV \sim \Gamma^2 D/r_0^2 = f(\Omega, Q)$, which leads to $\Gamma \propto D^{-1/2}$(assuming that r_0 is independent of Γ and D). Those first two results indicate that the geometry of our experiment gives rise to a vorticity filament which is noticeably different from a confined vortex where forcing by the Eckman layers is essential.

Finally, the weak linear dependence of Γ upon Q is similar to the one observed in confined vortices, which may indicate that the suction essentially plays a role in the boundary layers close to the discs. Complementary measurements, for instance of the stretching , could lead to a consistent model for a stretched vortex in an infinite medium.

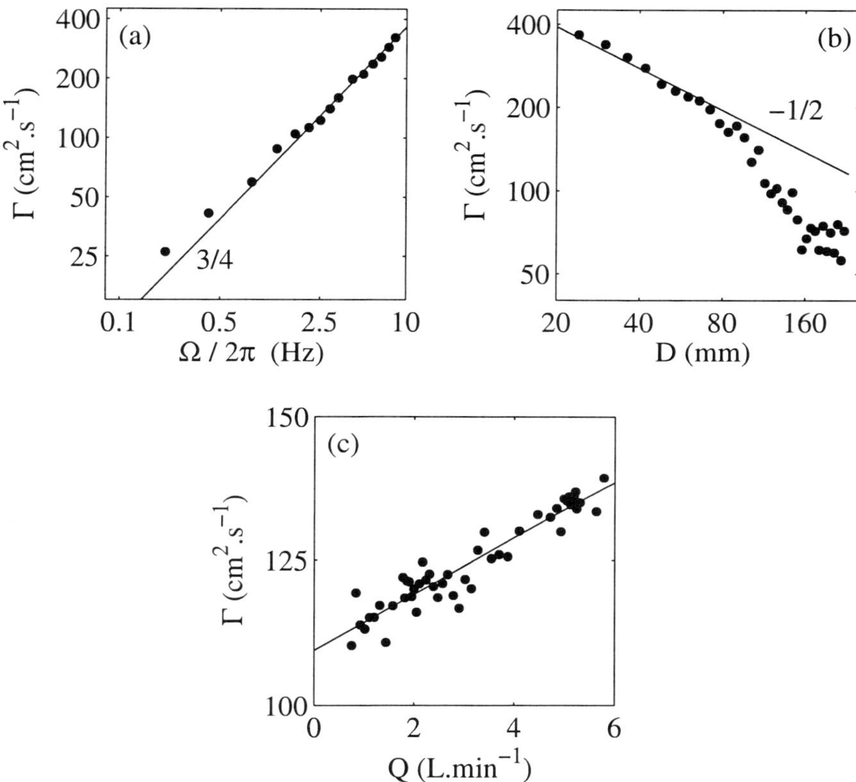

Fig. 4. Vortex circulation as a function of (a) the rotation speed of the discs Ω (for $D = 120$ mm and $Q = 3.0$ L/min), (b) the distance between the discs D (for $\Omega/2\pi =$ 1.7 Hz and $Q = 5.6$ L/min), and (c) the suction flowrate Q (for $\Omega/2\pi = 1.7$ Hz and $D = 80$ mm).

4.2 Dynamical tracking of the vortex

The measurements presented above can also be performed on an unstable or non-stationary vortex, with a sampling frequency of about 30 Hz. This yields a dynamical tracking of the vortex position $x_0(t)$ and of its circulation $\Gamma(t)$. The measurement of the vortex size as a function of time is also possible but it implies to adjust r_0 in the analytical calculation for each sample. Our results are qualitatively similar to previous observations on a confined vorticity filament [10,9,18].

- Example of transitory regime: figure 5 shows dynamical measurements performed on a stable vortex for which the sign of the disc rotation is abruptly changed at time t_0. After the perturbation, the vortex is formed far from the rotation axis of the discs ($x_0 = 0$) and oscillates and stabilizes around

$x_0 = 0$ in about 10 s. The circulation goes from Γ_0 to $-\Gamma_0$ in the same time interval and its behaviour is well modelled by an exponential relaxation with a characteristic time of about 5 s.

- Precession motion and 2D trajectory of the vortex: a slight modification of the measurement setup leads to the trajectory of the vortex in the incidence plane of the acoustic wave. Here, we use a circular array of 128 transducers and of diameter 20 cm that completely surrounds the vortex. Three spherical waves are simultaneously emitted from three transducers at 120° on the circular array. The coordinates $(x_0(t), y_0(t))$ are then computed by triangulation from the three angles corresponding to the zero phase-shifts of the incident spherical waves. An example of such measurements is shown in figure 6. A small precession frequency f_p is clearly measurable from the time series $x_0(t)$ and $y_0(t)$ (fig. 6(a) and (b)) and the trajectory of the vortex is close to a circle of radius r_p travelled in the same direction as that of the disc rotation. The radius of the precession increases with the distance between the discs and the motion is truly periodic only above a certain value of D of about 14 cm (fig. 6(c) and (d)).

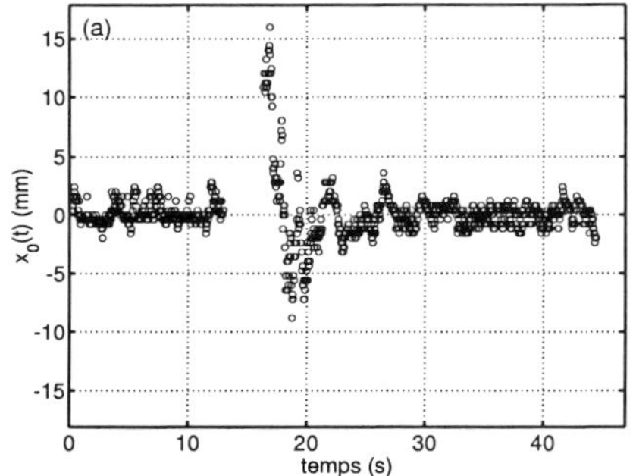

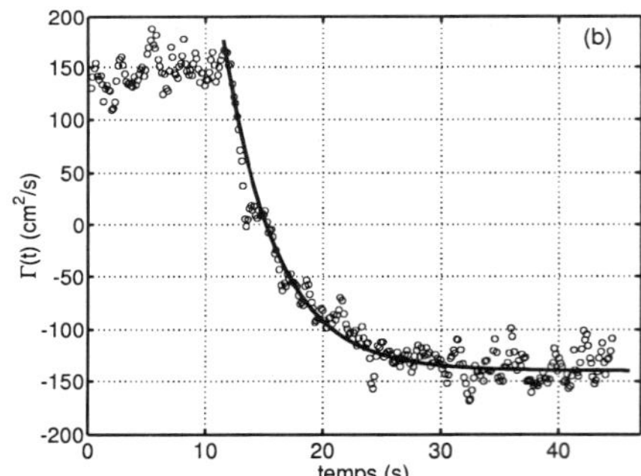

Fig. 5. Transitory regime after changing the sign of the disc rotation at time $t_0 = 12$ s ($\Omega/2\pi = 1.7$ Hz, $D = 80$ mm, $Q = 5.6$ L/min). (a) Vortex abscissa $x_0(t)$. (b) Vortex circulation $\Gamma(t)$; the solid line is an exponential fit $2\Gamma_0(\exp((t_0 - t)/\tau) - 1/2)$ for $t \geq t_0$, with $\tau = 4.5$ s.

5 Conclusions and perspectives

A new expriment was set up, that allows to investigate systematically the effects of the control parameters on both the mean characteristics and on the dynamics of a stretched vortex in a non-confined medium. The behaviour of such a vortex turns out to be slightly different from that observed in confined geometries and may be more representative of the characteristics of the vorticity filaments in turbulent flows.

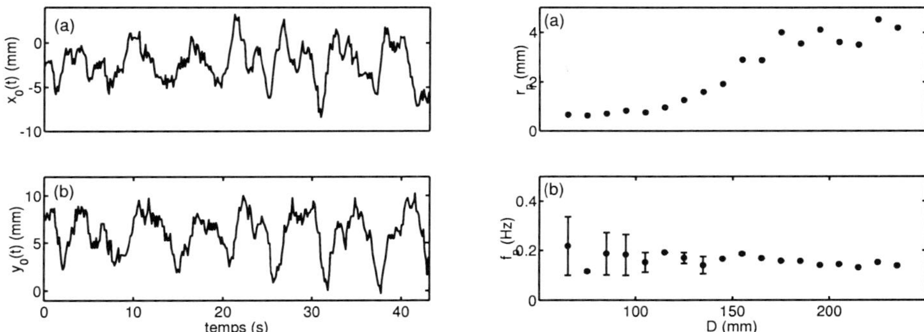

Fig. 6. Precession motion of the vortex ($\Omega/2\pi = 2.5$ Hz, $D = 165$ mm, $Q = 5.6$ L/min). Vortex coordinates (a) $x_0(t)$ and (b) $y_0(t)$ obtained by triangulation. (c) Radius r_p and (d) frequency f_p of the precession as a function of the distance between the discs D.

Our acoustic method based on the use of transducer arrays is non-invasive and yields directly a spatial information even in the core of a vortex which is only a few acoustic wavelengths. This method allows fast dynamical measurements and the sampling frequency could be increased from 30 Hz to about 1 kHz in the next future.

Finally, the systematic use of spherical waves emitted from a circular transducer array surrounding the vortex is under study, in order to improve the precision on the vortex core measurement and to allow a 2D tomography of the flow.

References

1. Cadot O., Douady S., Couder Y., *Phys. Fluids*, **7**, 630–646 (1995).
2. Brachet M.-E., *Fluid Dyn. Res.*, **8**, 1–8 (1991).
3. Jimenez J., *Phys. Fluids*, **4**, 652–654 (1991).
4. Belin F., Maurer J., Tabeling P., Willaime H., *J. Phys. II France*, **6**, 573–583 (1996).
5. Chainais P., Abry P., Pinton J.-F., submitted to *Phys. Fluids* (1999).
6. Bottausci F., Petitjeans P., Wesfreid J.E., Maurel, A. Manneville, S., Structure and Dynamics of vortices, Springer-Verlag, (2000)
7. Roux S., Muzy J.-F., Arneodo A., to appear in Eur. Phys. J. B (1999).
8. Mory M., Yurchenko N., *Eur. J. Mech. B*, **6**, 729–747 (1993).
9. Andreotti B., Maurer J., Couder Y., Douady S., *Eur. J. Mech. B*, **17**, 451–470 (1998).
10. Pinton J.-F., Chillà F., Mordant N., *Eur. J. Mech. B*, **17**, 535–547 (1998).
11. Petitjeans Ph., Robres J.-H., Wesfreid J.-E., Kevlahan N., *Eur. J. Mech. B*, **17**, 549–560 (1998).
12. Roux Ph., de Rosny J., Tanter M., Fink M., *Phys. Rev. Lett.*, **79**, 3170–3173 (1997).

13. Labbé R., Pinton J.-F., *Phys. Rev. Lett.*, **81**, 1413–1416 (1998).
14. Manneville S., Maurel A., Roux Ph., Fink M., *Eur. Phys. J. B*, **9**, 545–549 (1999).
15. Manneville S., Robres J.-H., Maurel A., Petitjeans Ph., Fink M., *Phys. Fluids*, **11**, 3380–3389 (1999).
16. Andreotti B., Ph. D thesis, University Paris VII (1999)
17. Moisy F, Petitjeans P, Structure and Dynamics of vortices, Springer-Verlag (2000)
18. Wunenburger R., Andreotti B., Petitjeans Ph., soumis à *Exp. Fluids* (1999).
19. Winters K. B., Rouseff, *IEEE Ultrason. Ferroelec. Freq. Control*, **40**, 26–33 (1993).
20. Johnson S. A., Greenleaf J. F., Tanaka M., Flandro G., *ISA Trans.*, **16**, 3–15 (1997).
21. Landau L. D., Lifshitz E. M., *Fluid Mechanics*, 2nd edition, chap. 8, (MIR, Moscou, 1989).
22. Salant R. F., *J. Acoust. Soc. Am.*, **46**, 1153–1157 (1969).
23. Georges T. M., *J. Acoust. Soc. Am.*, **51**, 206–209 (1971).
24. Lighthill M. J., *Proc. R. Soc. London A*, **211**, 564–587 (1952).
25. Kraichnan R. H., *J. Acoust. Soc. Am.*, **25**, 1096–1104 (1953).
26. Fetter F. L., *Phys. Rev. A*, **136**, 1488–1493 (1964).
27. Fabrikant A. L., *Sov. Phys. Acoust.*, **29**, 152–154 (1983).
28. Sakov P. V., *Acoust. Phys.*, **39**, 280–282 (1993).
29. Lund F., Rojas C., *Physica D*, **37**, 508–514 (1989).
30. Baudet C., Ciliberto S., Pinton J.-F., *Phys. Rev. Lett.*, **67**, 193–195 (1991).
31. Pinton J.-F., Laroche C., Fauve S., Baudet C., *J. Phys. II France*, **3**, 767–773 (1993).

Merging and Three-dimensional Instability in a Corotating Vortex Pair

Patrice Meunier and Thomas Leweke

Institut de Recherche sur les Phénomènes Hors Équilibre
UMR 6594 CNRS/Universités Aix-Marseille I & II
12 Avenue Général Leclerc, F-13003 Marseille, France.

Abstract. We investigate experimentally the interaction between two parallel laminar vortices of equal circulation. They are created by the impulsive displacement of two flat plates in a fluid initially at rest. The temporal evolution of the pair is analyzed with dye visualizations and quantitative measurements using Particle Image Velocimetry. For low Reynolds numbers, the vortices remain two-dimensional and merge into a single one. The size of the cores increases due to viscosity up to a certain limit, approximately a fourth of the separation distance, at which merging occurs. At higher Reynolds numbers, a three-dimensional instability is discovered, showing the characteristic features of an elliptic instability of the vortex cores. The two vortices still merge, but the final vortex is now turbulent.

1 Introduction

The dynamics of a corotating vortex pair is a basic example of interactions between laminar vortices. Most previous studies on this configuration (see, *e.g.*, [1,2]) modeled the vortices as two surfaces of uniform vorticity (patches) in a two-dimensional inviscid flow. They have established that merging into a single vortex occurs if the size of the cores, scaled on their separation distance, exceeds a certain limit; otherwise, the two initial vortices continue to rotate around each other indefinitely. Experimental observations made by Griffiths *et al.* [3] seem to be in agreement with this criterion. However, the effect of viscosity, leading to the increase of core size in time, as well as the possibility of a three-dimensional instability of the vortex pair, may significantly change this picture and so far have received little or no attention. In recent experiments [4], two different instabilities have been observed in a counterrotating vortex pair: a long wavelength instability, treated theoretically by Crow [5], and a short-wavelength instability, identified as the so-called elliptic instability [6–8] of the vortex cores. A corotating vortex pair was shown to be stable with respect to long-wavelength perturbations [9]. However, no theoretical prediction for a possible short-wavelength instability of this flow has been established so far.

In this study, we investigate experimentally the two- and three-dimensional dynamics of a pair of laminar parallel corotating vortices, with special emphasis on the viscous and three-dimensional effects mentioned above.

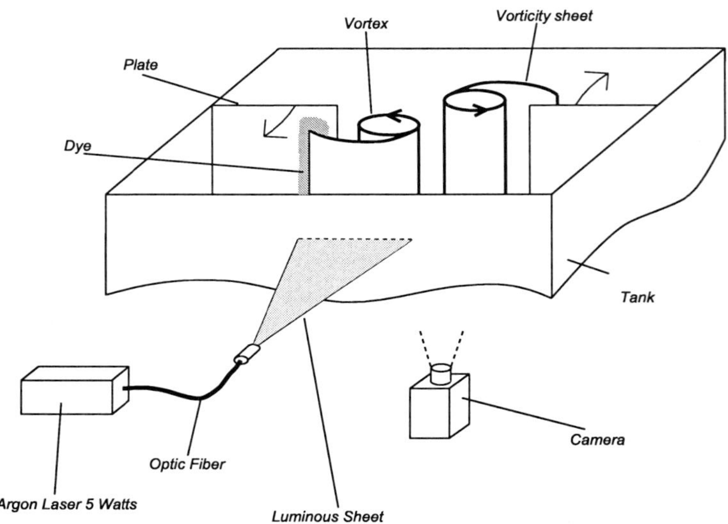

Fig. 1. Experimental device used for the generation of the corotating vortices

2 Experimental device

The vortices are generated in a water tank using two flat plates with sharpened edges, impulsively started from rest, using computer controlled step motors. The rotation of each plate around its vertical axis induces a flow around the plate. The vorticity created in the boundary layer rolls up into a vertical vortex as shown in Fig. 1. The plates have a high aspect ratio, to localize end effects inducing axial velocity in a small area at the bottom of the tank. The flow is visualized close to the surface of the water, since axial velocities are small there for long times. The vortices are thus laminar, two-dimensional and without axial velocity at the beginning.

The temporal evolution of the flow is visualized using two different dyes: Fluoresceine and Rhodamine B. The plates are painted with the dyes outside the water before the experiment. They are then slowly lowered into the tank. When the plates rotate, the dye rolls up into the cores of the vortices, marking their positions. An Argon ion laser illuminates the dyes either in volume for a sideview, or in a horizontal sheet for the visualization of a cross-cut section. For quantitative velocity measurements, Digital Particle Image Velocimetry has been used, by introducing particles of diameter 100 μm. It is possible to use such relatively big particles, since they are still two orders of magnitude smaller than the smallest lengthscale of the flow, *i.e.* the size of the vortex cores, which is of the order of 1 cm. A high-resolution camera Kodak Megaplus E.S. 1.0 of 1008×1018 pixels was used to capture pairs of images in horizontal sections. The pairs of images are then treated with an algorithm of intercorrelation, improved for high vorticity fields. The algorithm calculates a rapid approximation of the flow, for a first rough evaluation of the velocities and velocity gradients. The interrogation windows are then displaced and deformed, in order to keep a high correlation peak in areas of high velocity gradients. Such a slow but accurate algorithm is needed because velocity gradients are important in the vortex cores, exactly where the velocities need to be known with precision.

The vortex pair is characterized by the circulation Γ of each vortex, the separation b between the two vortex centers and a characteristic vortex core size a (see section 3.3).

The dynamics depend on two non-dimensional parameters. The Reynolds number, based on the circulation Γ, varies between 700 and 4000, and the initial ratio a_0/b_0 of core size and separation distance varies between 0.1 and 0.2. The two vortices rotate around each other with a period $t_c = 2\pi^2 b^2/\Gamma$, which is used to non-dimensionalize time, $t^* = t/t_c$, starting at the beginning of the plate motion.

3 Two-dimensional merging

For Reynolds numbers lower than 2000, the vortices remain two-dimensional and laminar. They deform in an elliptic way, get closer, and then merge into a single vortex in a rapid transition.

3.1 Dye visualizations

Figure 2 presents cross-cut visualizations of the flow at different stages during the merging process. Visualizations help to qualitatively follow the evolution of the vortices, even if the interpretation of the results must be done carefully. The comparison between visualization and velocity measurement will be done in the next section.

At the beginning (Fig. 2a), the vortices are far enough to remain practically axisymmetric. As their core sizes increase due to viscosity, they deform in an elliptic way (Fig. 2b–c), and create a tip on their inner side. Each tip is attracted by the opposite vortex, but the separation between the two centers remain constant. In Fig. 2d, two other tips are created at the outer side of the vortices. These tips are ejected in a radial direction, forming two arms of dye. Meanwhile, the two vortices get closer and finally merge into a single pattern that resembles the Chinese Yin-Yang symbol (Fig. 2e). The two arms roll up around the central pattern, forming a spiral of dye at later stages (Fig. 2f). In the final vortex, the two different colors are still distinguishable in the core, since the core is almost in solid body rotation. This distinction does not represent the structure of the vorticity field, because the two colors correspond to vorticity of the same sign.

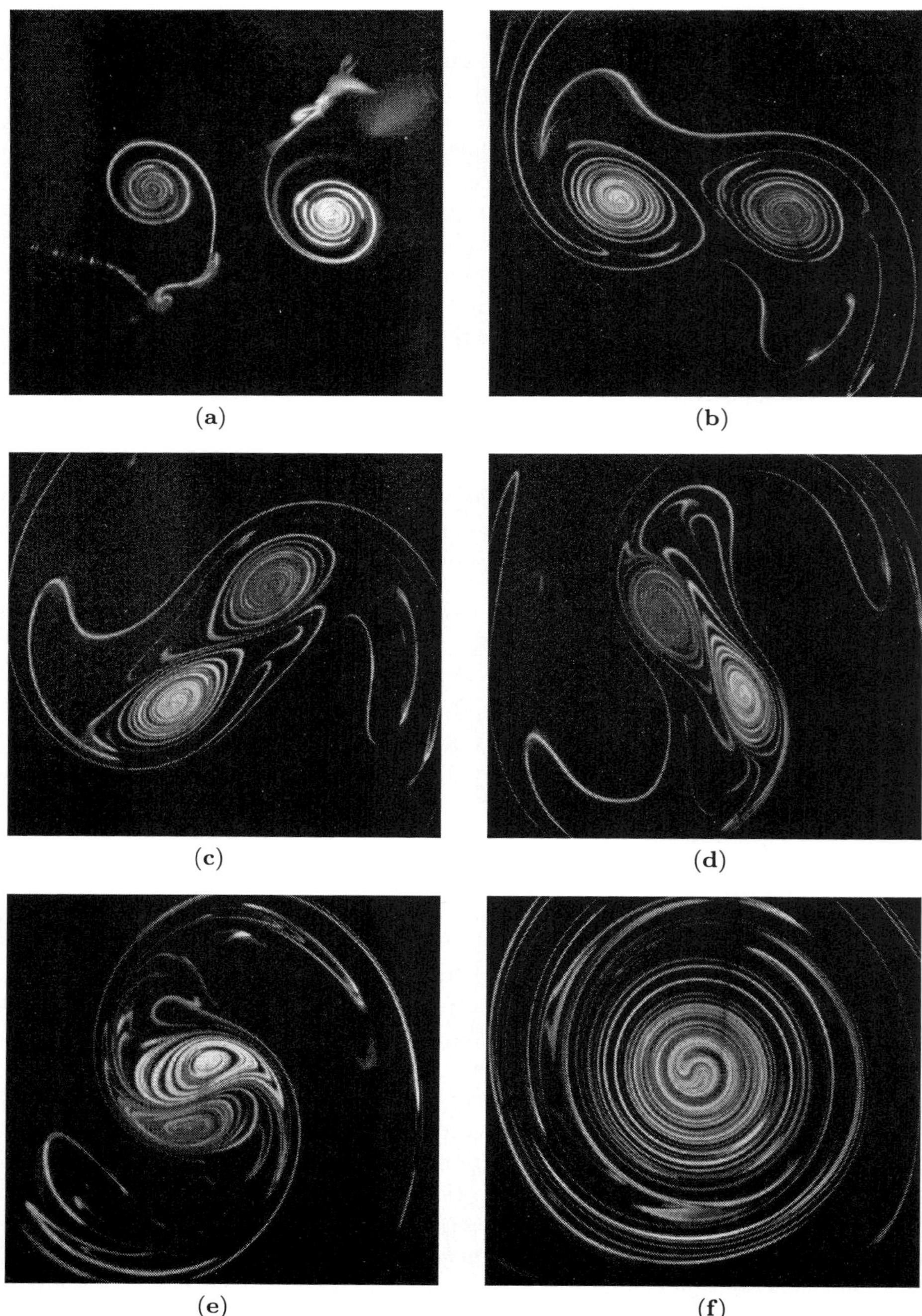

Fig. 2. Visualization of the 2D merging of two corotating vortices. (**a**) $t^* = 0$, (**b**) $t^* = 0.5$, (**c**) $t^* = 0.6$, (**d**) $t^* = 0.7$, (**e**) $t^* = 0.9$, (**f**) $t^* = 2$

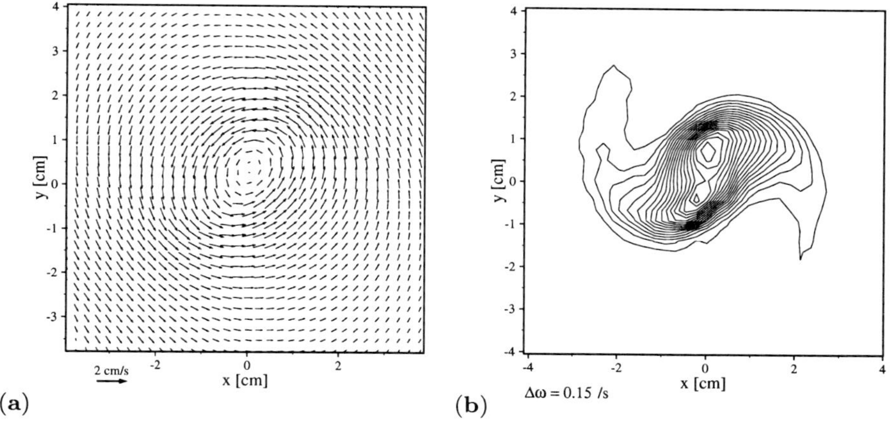

Fig. 3. Velocity field (**a**) and vorticity field (**b**) obtained with PIV measurements. $Re = 700$, $t^* = 0.94$

3.2 Dye pattern versus vorticity distribution

Although dye visualizations allow a good qualitative understanding of the merging phenomenon, it is necessary to carry out velocity measurements in the same cross-cut planes for two reasons. First, it is not possible to measure the size a of the vortex cores with dye visualizations. Second, the evolution of the dye and of the vorticity can be different. It is thus necessary to make careful comparisons. The velocity field of a cross-cut plane is presented in Fig. 3a. The axial vorticity is then calculated with a 9-point finite-difference algorithm, without additional smoothing (Fig. 3b). Although the velocity field seems to be axisymmetric, the vorticity reveals an organized structure far from axisymmetric. Two vorticity peaks are still distinguishable in a single vortex core, surrounded by two vorticity arms that start to roll up into a spiral. The global pattern of the vorticity strongly looks like the Fig. 2e, confirming the analogy between dye and vorticity.

In two dimensions, the evolution equations for vorticity and a passive scalar (dye) are the same, except for the values of the diffusivities (viscosity ν and scalar diffusivity κ for vorticity and dye, respectively). In addition, the initial conditions are similar here, since dye and vorticity are created at the same place in the beginning, *i.e.* on one side of the plate. Differences in the two evolutions arise from the high ratio between the diffusivities of vorticity and dye in the water, *i.e.* the Schmidt number : $Sc = \nu/\kappa \approx 1000$. Two main differences between Fig. 2e and Fig. 3b can be noted. First, the dye vortices remain distinct whereas the two vorticity peaks merge into a single vorticity distribution. Second, the vorticity arms cannot grow very long because they immediately diffuse in a low-vorticity background. The dye arms, on the contrary, create a very long spiral because their diffusion is negligible on the timescales of the experiment. This is why the vortex of Fig. 2f has very long spirals with a discontinuous radial structure, whereas the vorticity at the same stage has a smooth distribution very close to Gaussian.

In conclusion, the evolution of the vorticity and of a passive scalar such as dye are close, but present a few differences at late stages because of the high value of the Schmidt number for dye in water.

3.3 Temporal decomposition of the merging process

With PIV measurements, it was possible to calculate the core size a of the vortices and the separation distance b between the two vorticity peaks (Fig. 4). a was calculated by minimization of the quadratic difference between the experimental vorticity field and a Gaussian vorticity distribution defined by (1).

$$\omega = \frac{\Gamma}{\pi a^2} e^{-r^2/a^2} \tag{1}$$

The fitted field was composed of two vortices before merging and of one vortex after merging. There is no data for the merging stage, because the velocity field cannot be fitted by any of these two configurations. Viscosity introduces two differences between the theory of vortex patches and the experiment. First, the vorticity distribution is smooth. Second, there can be no stationary (or more precisely: periodic) state due to viscous diffusion, and merging will always occur eventually.

The evolution of a and b helps defining three stages in the merging process. In the first stage, the vortices rotate around each other almost like two point vortices. The separation distance b remains constant (see Fig. 4a) and the period of rotation is almost equal to t_c. The influence of one vortex on the other is negligible. This corresponds to Fig. 2a–c where they are only slightly elliptically deformed. Each vortex core grows with a viscous law (2), represented by the straight lines of Fig. 4b (for different t_0).

$$a^2 = 4\nu(t - t_0) \tag{2}$$

The duration of this stage scales on a diffusive time, *i.e.* on the inverse of the Reynolds number. The second stage begins when the vortices reach a certain size, scaled on the separation between the two vortices. The critical ratio above which merging begins is found experimentally to be:

$$\frac{a_c}{b_0} = 0.25, \tag{3}$$

which is of the order of the value (4) given in [1,2] for vortex patches.

$$\frac{a_c}{b_0} = 0.316. \tag{4}$$

At this critical ratio a_c/b_0, the separation distance b starts to decrease (Fig. 4a). This configuration corresponds to Fig. 2d where two vorticity arms appear and lead to the rapid merging of the vortices. From a number of different

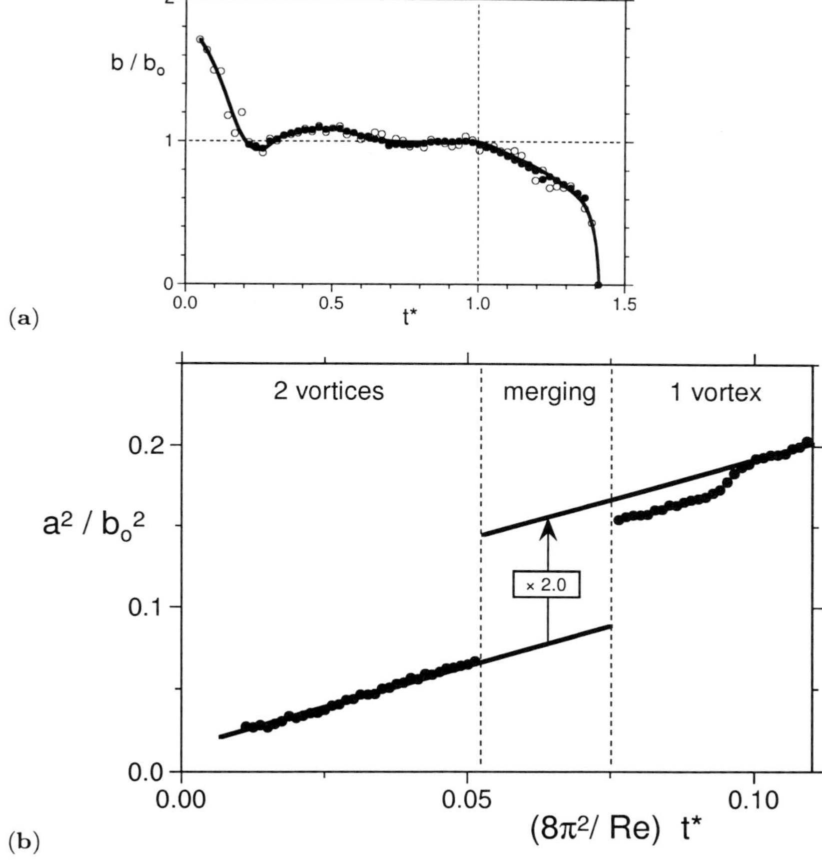

Fig. 4. (**a**) Evolution of the distance b between the vortices before merging. (**b**) Evolution of the core size a of the vortices before and after merging

measurements, it appears that the duration of the second stage seems to scale on a convective time, although the Reynolds number can have a small effect. After the merging, a third stage begins, where the final vortex undergoes an axisymmetrization process through the rolling up of the vorticity arms. Diffusion smoothes the vorticity filaments and makes the vortex core more and more Gaussian, evolving eventually again as in (2).

4 Three-dimensional instability

When increasing the Reynolds number and/or reducing the dimensionless core size a/b, the initial viscous phase before merging lasts longer, and there is time for a three-dimensional instability to develop while the two vortices are still separated. This phenomenon is here observed for the first time in the corotating

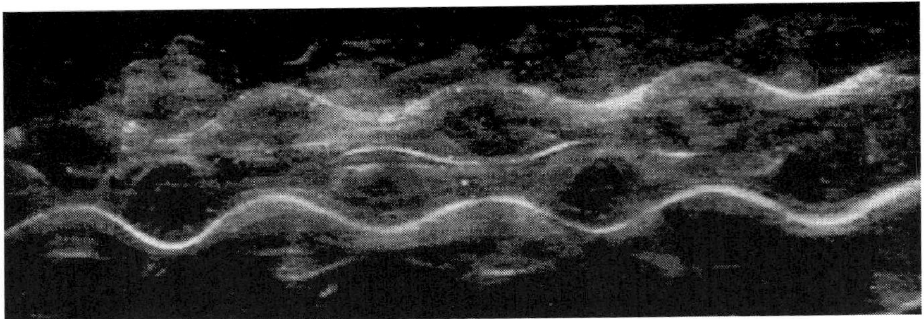

Fig. 5. Side view of the elliptic instability on a corotating vortex pair. $Re = 4140$

vortex pair. The following results were obtained at $Re = 4140$ and $a_o/b_o \approx 0.15$.

4.1 Structure of the instability

The sideview visualization in Fig. 5 shows a clear observation of the three-dimensional instability. At this particular instant, the vortices, which spin around each other, are in a plane perpendicular to the view direction. The two bright dye filaments correspond to the two vortex centerlines. These are deformed in a sinusoidal way with a wavelength λ close to the vortex separation b, and the perturbation on both vortices are in phase. Figure 5 also shows that the outer layers of dye orbiting each vortex are displaced in opposite radial directions than the centerline. The two regions are separated by an invariant tube of diameter d_{inv}. This characteristic internal deformation of the vortices is very similar to what was recently observed in counterrotating vortex pairs (see the paper by Leweke & Williamson in the present collection), where it was found that it was the consequence of a cooperative elliptic instability of the vortex cores. Although the vortex parameters and their separation are similar to those in the study of counterrotating vortices, it is still surprising that this instability develops so clearly in the rotating vortex system. Further visualizations and measurements (see below) have confirmed that the instability mode is stationary in the rotating frame of reference of the vortex pair, and that the planes of the wavy centerline perturbations are well aligned with the stretching direction of the mutually induced strain (at 45° with respect to the line joining the vortices, see Fig. 6a). The theory of the elliptic instability of a constant-vorticity vortex in an infinite strain [7] predicts a ratio λ/d_{inv} of 1.99, whereas this ratio is $\lambda/d_{inv} = 2.0 \pm 0.2$ in the experiment.

All this indicates that the three-dimensional instability observed in Fig. 5 is indeed an elliptic instability of the strained vortical flow in the cores.

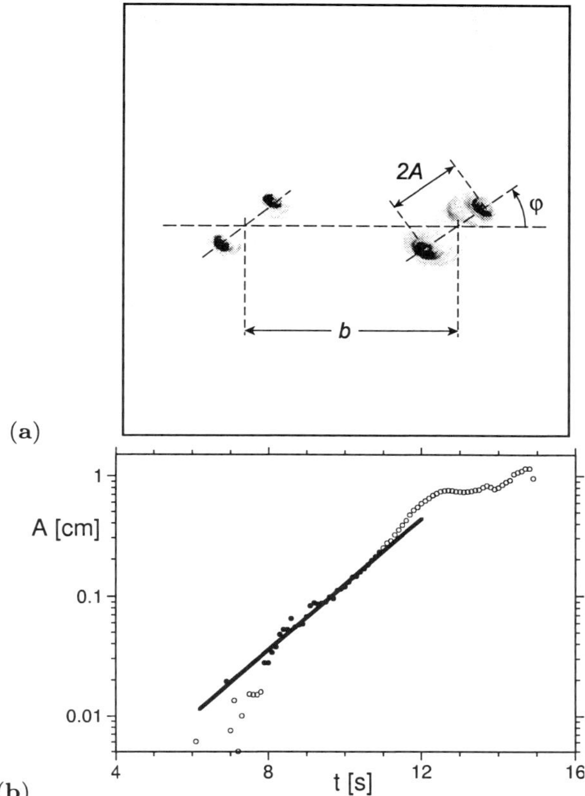

Fig. 6. (**a**) Visualization of the two vortex centers on two different horizontal sections. (**b**) Amplitude of the centerline oscillation.

4.2 Growth rate

Measurements were carried out to determine the growth rate of the instability. For this purpose, the phase of the unstable mode was fixed by sticking small pieces of thin tape on the plates generating the vortices, at intervals corresponding to the naturally developing wavelength. By simultaneously illuminating two cross-cut planes, separated by half a wavelength and located at the maximum and minimum of vortex displacement, one obtains a visualized image like the one in Fig. 6a (shown here as a negative). From a series of such images, one can deduce, among other quantities, the average vortex separation b and the displacement amplitude A as function of time. The latter can be related to the mode amplitude; for small values, both are proportional. The result is shown in

Fig. 6b. The growth is indeed exponential over a certain period, and an appropriate least-squares fit allows a determination of the growth rate σ. The result is:

$$\sigma = 0.623 \pm 0.014 s^{-1} \tag{5}$$

A theoretical prediction can be obtained, using the results in [8] for the elliptic instability of a Gaussian vortex in an infinite uniform non-rotating strain, and those in [10]. For the present case, using the value of the mutually induced strain at the center of each vortex, and the experimentally observed wavelength, the prediction for the dimensional growth rate is $0.55 s^{-1}$ which is reasonably close to the measured value in 5.

When the perturbation gets larger, nonlinearities introduce a change in the growth rate and in the structure of the instability. The perturbations lose their organised radial structure, and the external layers leave their initial vortex to roll up around the opposite vortex. The exchange of dye reveals an exchange of vorticity, and the creation of arms of vorticity which again accelerate the merging of the vortices. This merging appears to happen earlier than for low Reynolds numbers, *i.e.* at smaller ratios a/b, and the resulting vortex is characterized by turbulent small-scale motion in its core.

5 Conclusion

We have presented experimental results on the interaction between two identical, laminar, parallel vortices. For low Reynolds numbers, the two vortices merge into a single laminar vortex. This process can be decomposed into three stages. In a first stage, the core sizes grow by viscous diffusion. When the ratio of core radius and separation distance a/b_0 reaches a critical value of approximately 0.25, a second stage begins, scaling on a convective time, where the appearance of two vorticity arms causes the rapid merging of the vortices. In a third stage, the final vortex undergoes axisymmetrization by rolling up its arms of vorticity around the central pattern into a spiral. Diffusion helps the final vortex to tend to a Gaussian vortex again. For high Reynolds numbers, an instability appears, which has the same complex radial structure found in the first mode of the elliptic instability. The measured growth rate is in good agreement with theoretical predictions. The two vortices finally merge, earlier than without instability, but lead to a turbulent final vortex.

Further investigations on the precise influence of the three-dimensional elliptic instability on the merging characteristics are under way.

References

1. P. Saffman, R. Szeto: Phys. Fluids **23**, 12 (1980)
2. E. A. Overman, N. J. Zabusky: Phys. Fluids **25**, 8 (1982)
3. R. W. Griffiths, E.J. Hopfinger: J. Fluid Mech. **178**,73 (1987)

4. T. Leweke, C. H. K. Williamson: J. Fluid Mech. **360**, 85 (1998)
5. S. C. Crow: AIAA J. **8**, 2172 (1970)
6. C.-Y. Tsai, S. E. Widnall: J. Fluid Mech. **73**, 721 (1976)
7. F. Waleffe: Phys. Fluids A **2**, 76 (1990)
8. C. Eloy, S. Le Dizès: J. Fluid Mech. **378**, 145 (1999)
9. J. Jimenez: Phys. Fluids **18**, 1580 (1975)
10. M. J. Landman, P. G. Saffman: Phys. Fluids **30**, 2339 (1987)

Pressure Measurements in a Stretched Vortex

Frédéric Moisy[1] and Philippe Petitjeans[2]

[1] Laboratoire de Physique Statistique, ENS, 24 rue Lhomond, 75231 Paris Cedex 05 (France)
[2] Laboratoire de Physique et Mécanique des Milieux Hétérogènes, ESPCI, 10 rue Vauquelin, 75231 Paris (France)

Abstract. We report new results obtained with a strong laboratory vortex, generated between rotating disks with an axial stretching, providing a model of strong vortex structures present in turbulent flows. We first characterize the strength of the vortex in this geometry by means of a global pressure measurement, and give direct evidence of a mechanism of stretching saturation due to strong rotation. Pressure measurements are then performed with a miniature probe sensor close to the vortex, in order to study the relevance of bulk measurements with small but invasive sensors. We analyze the influence of some geometric characteristics of the probe (size, distance from the vortex) on the measured pressure, together with the influence of the presence of a probe on the vortex strength. We briefly discuss consequences of these observations on pressure measurements in the bulk of turbulent flows.

1 Introduction

Vorticity filaments are known to play a central role in turbulent flows [1], and particularly in the intermittency of energy dissipation. Experimentally, it is tempting to investigate these filaments through their core depression [7]. This pressure is related to the characteristics of the vortex such as vorticity, diameter, and stretching. However, because of the invasive character of conventional pressure probes, very little is known about the pressure fluctuations in the bulk of turbulent flows. Most of the time, measurements are performed on the wall [12,11,5], and are not easily related to the bulk properties. Let us mention here recent progress in non invasive quantitative measurements, with acoustic [6–8] and cavitation [1,10] techniques.

In this paper, we first describe a non intrusive method to measure the mean pressure of an isolated stretched vortex generated between rotating disks. Then, classical local measurements with a pressure probe are analyzed and compared with the other technique described here.

2 Experimental set-up

The experiment is performed in a large water tank, and the flow is driven between two smooth disks, 10 cm in diameter, rotating in the same sense (see figure 1). The frequency can be adjusted from 0 to 25 Hz. In the center of each disk, a 0.5 cm diameter hole is made which is used to create an axial pumping. The

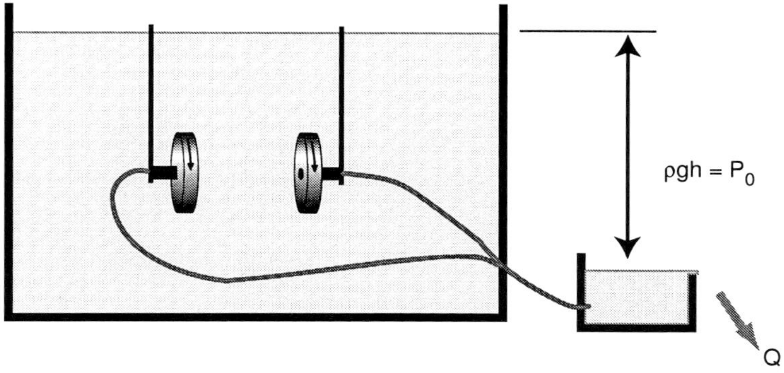

Fig. 1. Experimental set-up.

suction is generated by gravity, from a constant level auxiliary tank, under the main tank level. The imposed level difference h, between 0 and 90 cm, allows to impose a depression $P_0 = \rho g h$ between 0 and 90 mbar.

The axial velocity from the center to a hole due to the suction ranges from 0 to 5 m/s, leading to a mean axial velocity gradient $\langle \gamma_z \rangle \simeq \langle \partial U_z / \partial z \rangle$ from 0 to 50 s^{-1}. The combined effects of rotation and stretching generate a strong vortex (see figure 2), slowly precessing in the same direction as its rotation. The stability of this vortex has been extensively studied by Manneville *et al.* [7], by means of an ultrasound scattering technique [8].

The water leaving the auxiliary tank allows an accurate measurement of the flow rate Q (within 1 %). This flow rate is related to the mean axial stretching,

$$\langle \gamma_z \rangle = \frac{Q}{4Sd},$$

where S is the hole area and d the distance between the disks. The measured flow rate lies between 0 and 3 L/min, and depends both on the imposed suction pressure and on the frequency of the disks. In the absence of rotation of the disks, the imposed depression P_0 leads to a flow rate Q_0. Varying the height of the auxiliary tank we perform a calibration $Q_0 = F(P_0)$. Inverting this calibration allows to deduce, from the measurement of Q, the actual pressure P in the presence of the vortex.

The control parameters are (Ω, P_0, d), where $\Omega = 2\pi f$ is the disk frequency. In the experiments described in this paper the distance d has been kept fixed to 10 cm, and only (Ω, P_0) are varied.

3 Global measurement and vortex characterization

In the absence of disk rotation, a suction pressure P_0 is imposed and a constant flow rate Q_0 is measured. When the disks begin to rotate, after a few seconds, the flow rate decreases from Q_0 to a smaller value Q. From this change we can

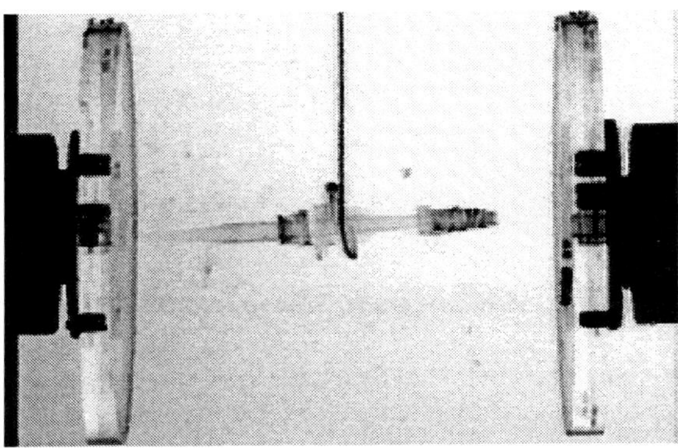

Fig. 2. Visualization of the stretched vortex. Dye is injected in the periphery of the vortex, through the vertical tube in the center.

deduce the pressure change at the axis level, inverting the calibration $Q = F(P)$ described above. The new pressure at the axis is then given by

$$P'_{\text{axis}} = P_{\text{axis}} + P_0 - F^{-1}(Q),$$

We then define $\Delta P = |P'_{\text{axis}} - P_{\text{axis}}|$ as the magnitude of this depression, which can be seen as the depression caused by the presence of the stretched vortex. We will hereafter note ΔP the vortex depression; we have ΔP=0 in the absence of disk rotation $[Q = Q_0 = F(P_0)]$. Since this procedure is based on a calibration covering the whole range of pressure and flow rate, it is free of assumption and gives confidence in the determination of the vortex depression.

Figure 3 shows measurements of the vortex depression ΔP for increasing disks velocity, at two different suction depressions P_0. We can see, for each case, two different regimes: the first one, called the "linear regime", for low disks velocity, corresponds to vortex depressions roughly behaving as $\Delta P \propto f$. After some frequency limit, depending on the suction depression, a new behavior arises, called the "saturated regime", where the vortex depression remains constant as the frequency is increased. This vortex depression ΔP is found to be slightly lower than the suction depression P_0.

The vortex depressions, normalized by the suction depressions, are found to collapse well, when plotted versus the non dimensional ratio

$$\beta = \left(\frac{\rho \Omega^2 R^2}{P_0} \right)^{1/2}, \qquad (1)$$

where R is the disk radius. $\rho \Omega^2 R^2$ can be interpreted as the depression caused by a solid-body rotation due to the disk rotation, in the absence of axial pumping. However, the R dependence of β has to be checked in more detail, performing new experiments with different disk diameters. Figure 4 shows measurements of the normalized vortex depressions versus β, for 8 different suction depressions

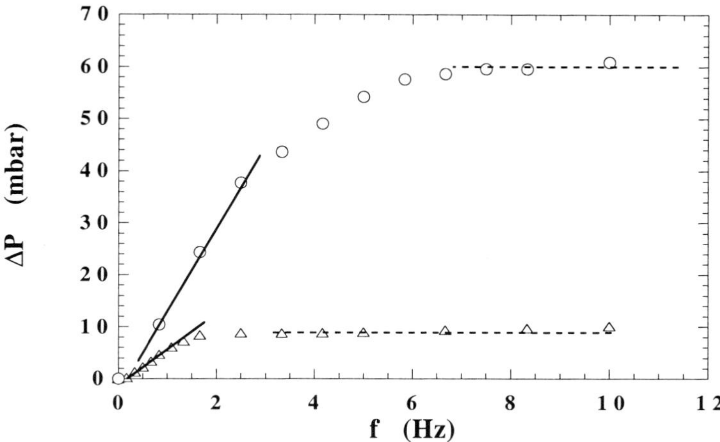

Fig. 3. Vortex depression ΔP versus disk frequency f, for two different suction pressures. $\triangle$: $P_0 = 10$ mbar and $\circ$: $P_0 = 70$ mbar. Solid lines indicate the linear regimes and dashed lines the saturated regimes.

P_0, ranging from 20 to 90 mbar. These curves are well fitted by the empirical formula

$$\frac{\Delta P}{P_0} = \sigma \tanh\left(\frac{\beta - \beta_s}{\beta_c}\right), \tag{2}$$

where β_s represents the lower threshold, below which there is no measurable vortex depression, and β_c represents the cross-over between the linear and the saturated regimes. For large values of β, the normalized depression saturates to the limiting value σ. We obtain $\beta_s \simeq 0.02$ and $\beta_c \simeq 0.38$, with no noticeable dependence on P_0. Values of σ are found to lie around 0.85 ± 0.05. We observe a very slow increase of σ for increasing P_0, but this effect seems to be very weak (less than 10 % for the range of P_0 we have spanned) and may allow to consider σ as a constant. The fact that $\sigma < 1$ means that the vortex never becomes strong enough to invert the flow rate, *i.e.* to pump the auxiliary tank water through the holes.

The saturation of $\Delta P/P_0$ for $\beta > \beta_c$ can be understood in terms of rotation–stretching interaction. For low frequencies ($\beta < \beta_c$), the axial suction is very efficient and highly enhances the low vorticity level injected by the disks. In this regime the vortex becomes relatively strong. For higher values of the frequency, the vorticity becomes important and tends to bidimensionalize the flow, reducing the axial strain. The stretching becomes less efficient, and unable to enhance vorticity any more. The vortex pressure tends to saturate at some ratio of the suction depression, given by $\sigma \simeq 0.85$. This saturating value may result from the equilibrium between the enhancement of rotation by stretching and the decrease of stretching by rotation.

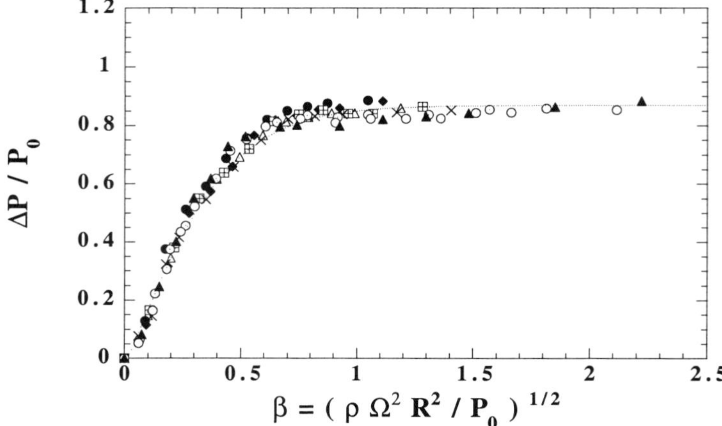

Fig. 4. Normalized vortex depressions $\Delta P/P_0$ versus the rotation–stretching ratio β. The different symbols represent 8 different suction depressions P_0 from 20 to 90 mbar. The full line is a best fit by the empirical formula (2), with $\beta_s = 0.02$, $\beta_c = 0.38$ and $\sigma = 0.85$.

The $\Delta P \propto \Omega$ behavior of the linear regime allows to deduce some scaling properties of this vortex. Manneville *et al* [7] have shown that in this regime the circulation Γ behaves as $\Omega^{3/4}$. Writing the depression as $\Delta P \sim \rho u_{\theta\max}^2$, where $u_{\theta\max}$ is the maximum azimuthal velocity, and the circulation as $\Gamma \sim u_{\theta\max} r_c$ (r_c is the vortex core radius), we can deduce

$$u_{\theta\max} \sim \Omega^{1/2}, \qquad r_c \sim \Omega^{1/4}.$$

Nothing is known about the vortex properties in the saturated regime. New experiments, using both acoustic and pressure measurements, are needed to provide more insight into these behaviors, and to characterize in more details the transition between the two regimes.

When the frequency is increased to higher values ($\beta \simeq 5$), we observe a new behavior (not shown here): The normalized vortex pressure begins to decrease from the saturated value σ down to much lower values (around $\sigma/2$). This may be due to vortex instabilities occurring at this higher Reynolds numbers: The vortex becomes unstable, breaks up and another one is generated. The measured pressures then result from a time-averaged flow rate, and can be interpreted as a mean value of the pressure due to the break up and reforming vortex. Indeed, in order to confirm this scenario, dye visualizations are needed, but are hard to perform because of the high speeds involved. Another explanation may come from a change in the forcing efficiency, due to the thinning of the boundary layer at higher frequencies.

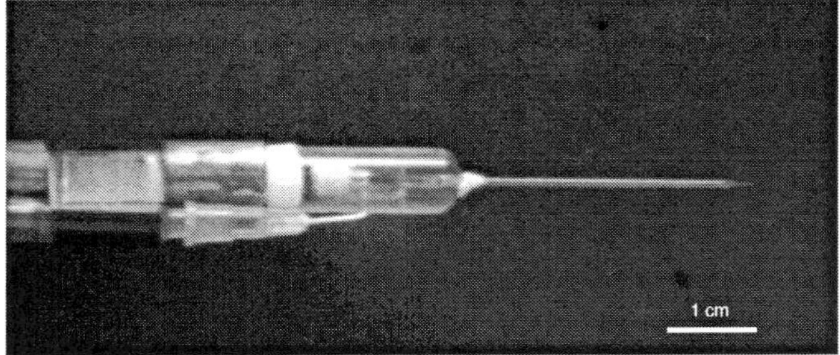

Fig. 5. The pressure probe consists of a piezoelectric transducer, placed at the end of a 200 µm diameter capillary tube. The needle, connected to a syringe, allows to expel residual air bubbles in the cavity, causing pressure fluctuation damping.

4 Pressure sensor and local measurements

We are now interested in more classical local measurement ot the pressure near the vortex core. Here, the aim is to characterize the measurement itself by an invasive detector, and additionally to estimate the vortex perturbation by the presence of the probe in this configuration.

The pressure measurements are performed using a piezoelectric transducer (see figure 5), 5 mm in diameter, placed in a water filled cavity, connected to a small glass tube. In order to minimize perturbations due to the presence of this probe in the flow, we make use of a fine capillary tube, whose initial diameter is 1 mm, which can be stretched down to 50 µm. We denote ϕ the tube external diameter (the internal diameter is around 0.7 ϕ). The viscous damping frequency is around 10^5 Hz, and the vortex shedding frequency around 1 kHz for the typical velocities involved here. The most limitating constraint is given by the resonnance frequency of the cavity, which is above 100 Hz for the smallest diameter. The pressure probe position is controlled by a 3D displacement system, with 1 µm accuracy. The distance from the probe to the disk axis is denoted r_s, and for the present experiments the probe is kept at mid-distance between the disks.

Because of the slow precession, the vortex frequently sweeps in the close vicinity of the pressure probe, causing a sharp pressure drop. Figure 6 shows a typical pressure time series, when the pressure sensor is at a distance r_s=1 mm from the axis. In this example, the probe diameter at the tube end is ϕ = 260 µm. The pressure drops can reach −80 mbar, a value rather low compared to the solid-body rotation pressure $\frac{1}{2}\rho U_{\text{disk}}^2 \simeq 13$ mbar and to the mean dynamic pressure caused by the axial stretching, $\frac{1}{2}\rho U_z^2 \simeq 30$ mbar. The mean waiting time between successive events is 1.9 s, leading to a mean precession frequency $f_p \simeq 0.5$ Hz $\simeq f_{\text{disk}}/10$, a value in good agreement with ultrasound scattering measurements [7].

Two distributions of the pressure fluctuations are plotted on Figure 7, for two different probe distances, r_s = 1 and 4 mm. Pressure drops caused by the vortex

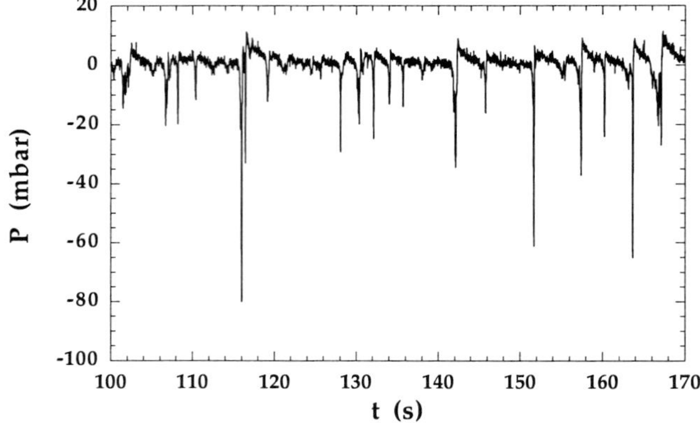

Fig. 6. Typical time series of pressure. The slight overshoot following the sharp drops are due to residual air microbubbles in the sensor cavity. f_{disk}=5.1 Hz, $Q = 2{\times}3$ L/min, and $\phi = 260$ µm.

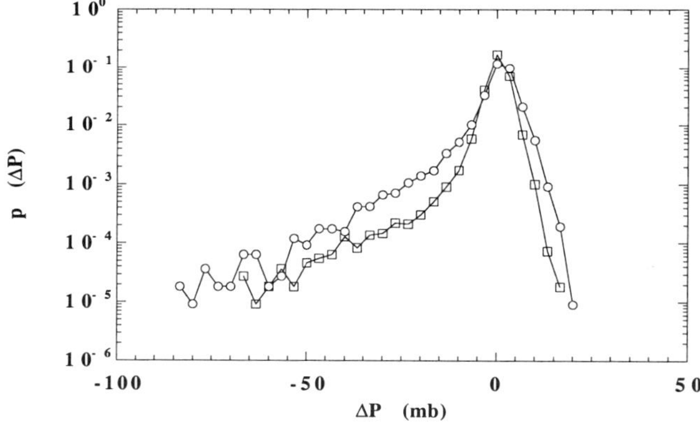

Fig. 7. Distribution of pressure for two distances, circle: $r_s = 1$ mm and square: $r_s =$ 4 mm. f_{disk}=5.1 Hz, $Q = 2 \times 3$ L/min, and $\phi = 260$ µm.

sweeping account for the large negative tails of these distributions, which can be roughly fitted by exponentials for $\Delta p < -15$ mbar. Their width, around 3 mbar, give a rough measure of the noise level. The slight positive tail is mainly due to an overshoot following the pressure drops, possibly caused by the presence of residual air microbubbles in the sensor cavity.

We can see that the further the probe is from the axis, the rarer the pressure drops. However, the slope of the negative tails remains unchanged for different values of r_s. This suggests that, for large r_s, the pressure drops typically reach the same values, but the vortex sweeps less frequently in the probe vicinity. As r_s is increased, the negative tail decreases and vanishes for $r_s \geq 8$ mm; this value is found to be independent of the the disk frequency, and gives a rough estimate of the precession radius.

From the mean width Δt of the pressure drops, we can estimate the core size r_c of the vortex. For the measurements performed with the $\phi = 260$ μm sensor, we obtain $\Delta t = 160 \pm 60$ ms, leading to

$$r_c = \frac{2\pi R_p f_p}{\Delta t} \simeq 1 \text{ mm}.$$

where $2\pi R_p f_p$ is the mean precession velocity, typically 5 mm/s, obtained from a mean precession radius $R_p \simeq 6$ mm and the precession frequency $f_p \simeq 0.5$ Hz. Although the uncertainty is high, this estimation of $r_c \simeq 1$ mm is in qualitative agreement with measurements of Manneville *et al.*[7]. It does not seem to depend on the probe diameter, at least for $\phi < 1$ mm.

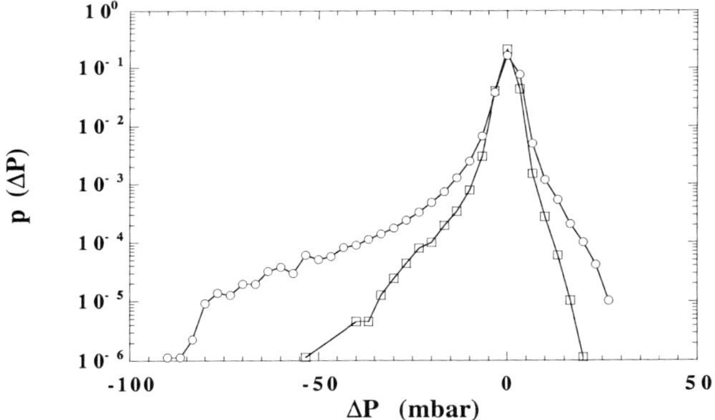

Fig. 8. Distribution of pressure for two probe sizes, at $r_s = 3$ mm. Circle: $\phi = 260$ μm, square: $\phi = 1$ mm. The best fit by an exponential $\exp[\Delta P/\alpha]$ for $\Delta p < -15$ mbar leads to $\alpha = 17$ mbar for $\phi = 260$ μm, and $\alpha = 6.7$ mbar for $\phi = 1$ mm.

We will now address the delicate issue of the influence of the probe on the vortex characteristics. Figure 8 shows two pressure distributions, at the same distance from the axis $r_s = 3$ mm, for two different probe sizes $\phi = 0.26$ mm and 1 mm. As in figure 7, the negative tails can be well fitted by exponentials, for $\Delta p < -15$ mbar. But, in contrast with figure 7, where the exponential slope is not affected by the distance r_s, now we can see that it is strongly decreased in the case of the bigger probe. This clearly indicates the strong influence of the probe size on the local pressure measurement.

It is important to determine whether this influence originates from the probe characteristics, or from a change in the vortex itself due to the presence of the probe; both are relevant in the context of pressure measurements in the bulk of turbulent flows. A detailed comparison between global measurements, described

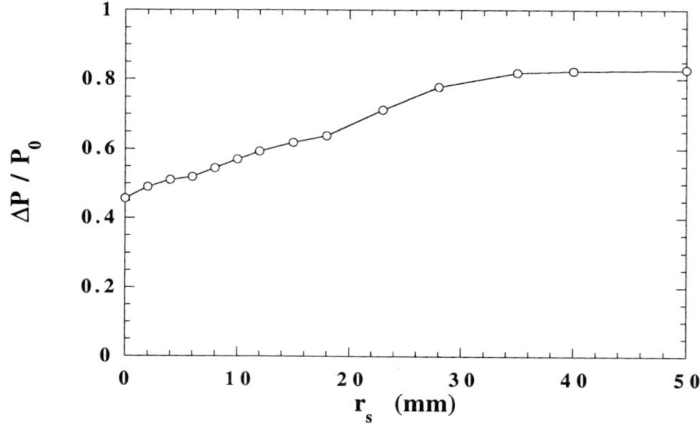

Fig. 9. Vortex depression ΔP normalized by the imposed suction depression P_0 in the saturated regime ($\beta = 1.05$), versus the probe distance from the axis r_s (diameter $\phi = 0.7$ mm).

in the previous section, and local measurements is in progress. Preliminary results are given on figure 9, which represents the change of the global vortex depression with the probe distance from the axis r_s. For $r_s > 40$ mm, we obtain $\Delta P/P_0 = \sigma \simeq 0.83$, in good agreement with the figure 4 in the saturated regime ($\beta = 1.05 \sim 3\beta_c$). When r_s is decreased down to 0, the rescaled depression falls down to $\Delta P/P_0 = 0.45$, meaning that the vortex (time-averaged) depression is roughly twice as small as withous the probe. Even when the probe is farther than the mean precession radius ($R_p \simeq 8$ mm), the vortex depression is strongly decreased.

Visualization by means of dye injected in the periphery of the vortex (as in figure 2) shows that no instability occurs when r_s is larger than a few probe diameter ϕ. These observations suggest that, even when the vortex remains stable, the presence of the probe at a distance $r_s < 10r_c$ significantly decreases its strength. This clearly illustrates the long-range sensitivity of such vortex to localized perturbation. Synchronized visualizations together with local pressure measurements are needed to get more insight into this behavior.

5 Discussion and Conclusion

In this paper we have reported first results obtained with a new experiment allowing to generate a strong and controlled stretched vortex. Two kinds of pressure measurements have been performed, global and local.

The global time-averaged pressure measurement allows to characterize the vortex strength for different rotation rates Ω and suction depressions P_0. We give evidence of a transitional behavior between two regimes, labeled with the non-dimensional ratio $\beta = (\rho\Omega^2 R^2/P_0)^{1/2}$, measuring the relative magnitude of the rotation effect compared to the stretching effect. For $\beta < \beta_c \simeq 0.38$, stretching highly enhances the small level of vorticity injected by the disks, and the vortex depression is found to scale as β. For $\beta > \beta_c$, the strong rotation tends to saturate the stretching efficiency, which is unable to strengthen the vortex any more. In this second regime, the vortex depression ΔP saturates at the final value σP_0, where P_0 is the imposed suction depression and σ is found to be 0.85 ± 0.05. These observations are in good agreement with those of Nore *et al.* [11], who investigated both numerically and experimentally the rotation–stretching interaction mechanisms. The originality of our experimental set-up is to keep the flow rate Q (thus the stretching) free to adjust with the actual vorticity, giving clear evidence of the stretching saturation effect of the strongly rotating flow. We plan to use this global pressure measurement together with the ultrasound scattering technique [8,7] to investigate the circulation, vorticity and vortex radius evolution in these two regimes, and in particular to understand the observed values of the threshold β_c and the saturation ratio σ.

Local measurements, by means of conventional invasive probes, have been performed too, in order to investigate the relevance of bulk measurements in turbulent flows. In the particular configuration of a strong vortex in a quiet background flow, we show that such a method provides a reasonable measurement of the pressure drops, whose magnitude is in qualitative agreement with values obtained via the global measurements described above. However, we give evidence of a strong influence on both the probe size and the vortex–probe distance. These two effects are presumably due to the probe characteristics itself, and to the perturbation of the flow caused by the presence of the probe. This second point is illustrated by the strong decrease (by a factor 2) of the vortex depression in the presence of the probe, even when it remains farther than the mean precession radius.

These observations are of great importance in the context of pressure measurements in the bulk of turbulent flows. It appears that depressions caused by vortices whose radii are smaller than the probe size are likely to be missed, not only because of a spatial averaging effect, but essentially by perturbation of the flow caused by the probe itself. We mention here that the use velocity measurements by means of hot wire anemometry to detect small scale filaments [12], the mean advection velocity (typically one order of magnitude higher than the fluctuating velocity) makes the advection time lower than the typical timescale of perturbation. In this case, the measurement is believed not to be affected by the finite probe size. More work has to be done in order to clarify what is due to a pure probe size effect, and what is due to the perturbation of the vortex in this geometry.

The authors acknowledge G. Baque and F. Botausci for experimental helps, and S. Manneville, A. Maurel, V. Emsellem, P. Tabeling and H. Willaime for fruitful discussions.

References

1. U. Frisch: Turbulence, the legacy of A.N. Kolmogorov. Cambridge University Press, Cambridge (1995).

2. S. Douady, Y. Couder, and M.E. Brachet: Direct observation of the intermittency of intense vorticity filaments in turbulence *Phys. Rev. Lett.* **67** (8), 983–986 (1991).
3. O. Cadot, S. Douady and Y. Couder: Characterization of the low pressure filaments in a 3D turbulent shear flow *Phys. Fluids* **7**, 630 (1995).
4. P. Abry, S. Fauve, P. Flandrin, and C. Laroche : Analysis of pressure fluctuations in swirling turbulent flows *J. Phys. II France* **4** 725–733 (1994).
5. P. Chaisnais, P. Abry, and J.-F. Pinton: Intermittency and coherent structures in a swirling flow: a wavelet analysis of joint pressure and velocity measurements, to appear in *Phys. Fluids* (1999).
6. B. Dernoncourt, J.-F. Pinton and S. Fauve: Experimental study of vorticity filaments in a turbulent swirling flow, PHYSICA D (1998).
7. S. Manneville, A. Maurel, F. Bottausci, and Ph. Petitjeans: The "double rotating suction system": Acoustic characterization of a vortex in infinite medium, in this Volume. See also S. Manneville, A. Maurel, C. Prada, M. Tanter and M. Fink: Ultrasound Propagation Through a Rotational Flow: Numerical Methods Compared to Experiments, in proceeding of the 9th International Conference on Theoretical and Computational Acoustics, submitted to The Journal of Computational Acoustics (1999).
8. S. Manneville, J. H. Robres, A. Maurel, P. Petitjeans, and M. Fink: Vortex dynamics investigation using an acoustic technique, Physics of Fluids **11** (11), 3380-3389 (1999).
9. A. La Porta, G.A. Voth, F. Moisy and E. Bodenschatz: Using cavitation to measure statistics of low-pressure events in large-Reynolds-number turbulence, submitted to *Phys. Fluid* (1999).
10. F. Moisy, A. La Porta, G.A. Voth and E. Bodenschatz: Using cavitation as a local pressure probe in turbulence, in this Volume.
11. C. Nore, B. Andreotti, S. Douady and M. Abid: On Burgers' vortex: the missing mechanism, in this Volume.
12. F. Belin, J. Maurer, P. Tabeling and H. Willaime: Observation of intense filaments in fully developed turbulence, *J. Phys II* **6**, 573–584 (1996).

Using Cavitation as a Probe of Low-Pressure Filaments in Turbulence

Frédéric Moisy[1], Arthur La Porta[2], Greg Voth[2] and Eberhard Bodenschatz[2]

[1] Laboratoire de Physique Statistique, ENS, 24 rue Lhomond, F-75231 Paris Cedex 05

[2] Laboratory of Atomic and Solid State Physics, Cornell University, Ithaca, New York 14853-2501

Abstract. We report new observations of low-pressure filaments in a turbulent flow between counter-rotating disks [1]. Cavitation from microscopic gas bubbles seeding the water is used to probe the structure of the pressure field. The spatial structure of the low-pressure events, mainly vertical filaments standing along the disks axis, are visualized using a high speed video system. The negative tail of the probability density functions of pressure is determined from light scattering measurements performed with a fast photo detector, and is found to be exponential. These observations highlight the importance of the large scales on the pressure fluctuations.

1 Introduction

The importance of vortex structures with concentrated vorticity in turbulent flows has been recognized for a long time. Direct numerical simulations of homogeneous and isotropic turbulence clearly show that the most intense of these vortex filaments, the so-called worms [2,3], are responsible for the non-gaussian statistics of velocity derivatives [4]. Moreover, worms are also found to be responsible for the skewed pressure distributions [5], although this can not really be viewed as a signature of structures in the flow.

The basic idea of the experiment described here arises from the link between high vorticity regions and local pressure minima *via* the Poisson equation

$$\frac{2}{\rho}\nabla^2 p = \omega^2 - \sigma^2$$

for an incompressible fluid (of density ρ), where ω^2 and σ^2 are the squared vorticity and rate of strain tensors. This link, highlighted by Brachet (1990) [6], has motivated several experiments [7,8], where migration of gas bubbles against the pressure gradient allows visualization of low-pressure regions in turbulent water flows. However, quantitative pressure measurements usually suffer from the invasive character of conventional sensors [9,10]. For this reason, such measurements are usually performed by probes mounted on the walls [11,12], and little is known about the pressure field in the bulk of turbulent flows.

In order to provide information about pressure fluctuations in the case of liquids, it is tempting to use cavitation. Cavitation in low pressure vortices has been extensively studied in the context of fluid machinery [13–16]. Arndt [13] shows how the pressure inside the vortex core can be estimated from cavitation inception. Ran and Katz [17,18] show that it is possible to obtain a non intrusive pressure measurement from (non cavitating) bubble size measurements.

In the experiment we present here, we make use of cavitation to investigate both the structural organization and the statistics of pressure fluctuations in the turbulent flow between counter-rotating disks. Cavitation is triggered where small air bubbles seeding the flow enter in regions where the local pressure becomes below a fixed pressure threshold. This pressure threshold is given by the vapor pressure and a surface tension correction due to the air bubble size seeding the flow. The purpose of this paper is to focus on the large scale properties on the flow in this geometry, while the results reported in La Porta *et al* [1] detail the statistical aspects.

2 Experimental set-up

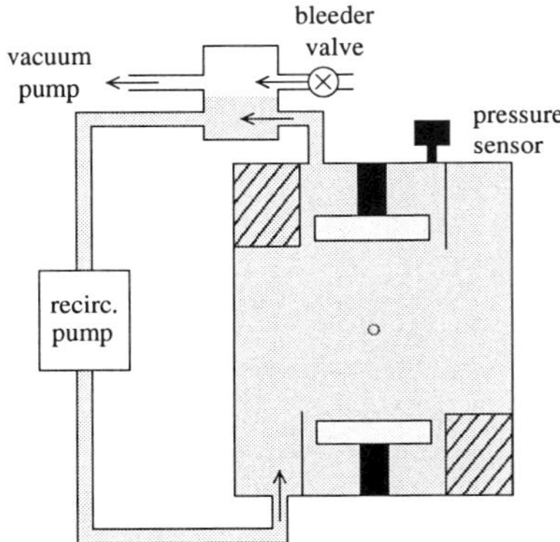

Fig. 1. Left: Picture of the experimental apparatus. Right: Pressure control system. The dashed circle in the center represents the 3 cm^3 imaged region used in section 5.

The turbulent flow we study takes place in a 110 L water tank, driven by two counter-rotating disks (see figure 1). A large plane window is mounted on the side to allow undistorted visualization, and a smaller circular one at 90° for illumination. The disks are open-ended cylinders, 20 cm in diameter and 4.3 cm heigh, with twelve radial vanes to provide efficient stirring. They are spaced 33 cm apart, driven by two electronically controlled DC motors, spanning a frequency rate ranging from 0 to 9 Hz within ± 1%. More experimental details can be found in Voth *et al* [19] and La Porta *et al* [1].

In order to use cavitation to probe the pressure field, it is crucial to control precisely the mean hydrostatic pressure; this is achieved using the arrangement shown in figure 1. It consists of an auxiliary chamber, mounted above the main fluid volume, both two being included in a recirculating loop. Pumping on the water/air interface in the auxiliary chamber, and adjusting the air flow rate through a servo controlled bleeder valve, allows us to finely regulate the hydrostatic pressure in the main tank. Pressure fluctuations have to be large enough to generate significant cavitation, imposing a minimum frequency rate of 7 Hz. Table 1 sums up the flow characteristics for frequency rates of 7, 8 and 9 Hz. The mean hydrostatic pressure, noted p_0, is taken at the center of the flow, and includes the shift from the top to the center, $\rho g h = 30$ mbar.

f (Hz)	$Re = \Omega R^2/\nu$	τ_η (ms)	η (μm)	λ (mm)	$\frac{1}{2}\rho U_d^2$ (mbar)	R_λ
7	$4.40 \cdot 10^5$	0.67	26	2.12	96	1658
8	$5.03 \cdot 10^5$	0.54	23	1.99	126	1772
9	$5.65 \cdot 10^5$	0.45	21	1.89	160	1880

Table 1. Range of turbulence parameters. The disk velocity U_d is given by $2\pi R f$, where R =10 cm is the disk radius. Values of the Taylor scale (λ), Kolmogorov scales (η, τ_η) and microscale Reynolds number (R_λ) are determined from measurements in the center of the flow [19].

The mean flow between counter rotating disks can be seen as the superposition of two components, a pumping mode and a shearing mode. Fluid near the top and the bottom rotates with the disks, producing a shear layer in the mid plane. Additionally, the axial pumping and the radial ejection, due to the disk rotation, creates two toroidal recirculation rolls. The mean velocity of this pumping mode has been estimated by a position sensitive detector, and is found to be around 20% of the maximum azimuthal velocity.

The flow is illuminated using light sources at 90°, and low-pressure events are detected as light scattered from cavitating bubbles. Two imaging systems have been used:

- Qualitative characterization of the low-pressure events is achieved by recording scattered light with a 400 Hz video camera. The exposure time is given by the sampling rate when continuous illumination is used (tungsten lamp), but it can be decreased down to 20 μs using a stroboscopic illumination, synchronized with the video system. These observations are described in section 4.
- In order to extract more quantitative information about the probability density function of the pressure fluctuations, a 30 kHz PIN photodiode collects light scattered from a 3 cm^3 central circular area. Assuming cavitation is triggered when local pressure reaches a given threshold, we show, in section 5, how pressure statistics can be deduced from the fraction of the time that cavitation occurs.

Both pure and salted water have been used to perform the visualization. Salted water (0.27 mol/L NaCl) makes the air microbubbles more stable, and allows to decrease the hydrostatic pressure further in order to visualize weaker filaments. The quantitative measurements presented in section 5 have been performed only in pure water, but consistent results have been obtained in salted water.

3 Using cavitation bubbles as pressure sensors

In the following we only discuss the case of cavitation in pure water, from which quantitative information on the pressure fields has been obtained. The salted water case seems to be much more complex — the bubbles seem to be more stable, and less likely to cavitate or coalesce.

Pure cavitation is expected to occur when the local pressure in a liquid goes below the vapor pressure: a vapor bubble will grow by conversion of liquid to vapor at the bubble interface [20]. However, the surface tension energy needed to nucleate an arbitrary small vapor bubble from pure liquid water considerably decreases this pressure threshold. In most of the cases, cavitation occurs from impurities, the roughness of the wall, or pre-existing microbubbles in the water, allowing to reach the surface tension barrier.

In our set-up, water is finely purified and cavitation in the center of the flow arises from air microbubbles seeding the water. In the absence of disks rotation, examination with a long working distance microscope shows that the radius of these stable air bubbles is 35 ± 10 µm. Note that this value is of the order of the Kolmogorov scale, ranging from 21 to 26 µm (see table 1). Assuming that the air filling the seeding bubbles behaves as an isothermal ideal gas [17], we can deduce the cavitation pressure threshold:

$$p_c \simeq 20 \text{ mbar}. \tag{1}$$

We note that this value is close to the the water vapor pressure ($p_v = 29$ mbar at 25°C). Migration of gas bubbles into low-pressure regions precedes cavitation of vapor bubbles, a mechanism already noted by Ran and Katz [18].

In turbulent flows, the pressure distribution is believed to be negatively skewed [5]; the most negative pressure drops locally reach the well defined pressure threshold (1) and trig cavitation. In our experiment, for small cavitation rate (*i.e.* for high hydrostatic pressure), cavitation can be considered as a passive trigger of low pressure regions. Vapor bubbles are indeed unstable and quickly collapse after their nucleation from the seeding bubbles. In this case, cavitation consists essentially of brief flashes of light scattering. A reasonable timescale of the growth and collapse process can be estimated from the resonant frequency of the air bubbles, around 100 µs for the 35 µm radius bubble (see the Ref. [17] for a detailed study of the time response of a gas bubble). This time scale is then found to be lower than the Kolmogorov timescale (around 500 µs, see table 1), allowing us to consider that cavitation is triggered instantaneously by pressure fluctuations. Of course, the same does not go for higher cavitation rates (lower hydrostatic pressure), where big vapor cavities are found to fill the vortex cores [13]. Indeed, vapor bubbles' response time can be much higher, and it is only for low cavitation rates that cavitation can be seen as a passive trigger.

4 Visualization of coherent structures

The large scale structure of the flow is observed using the high speed video system, by setting the hydrostatic pressure slightly above the pressure threshold ($p_0 \simeq 70$ to 120 mbar). At the lowest hydrostatic pressure, cavitation bubbles are created at a high rate throughout the fluid volume, leading to a high density of stable gas bubbles seeding the flow. In this case, visualization is due to cavitation itself, as well as migration of gas bubbles towards local minima of the pressure. For higher hydrostatic pressure, the small seeding gas bubbles are trapped into the vortex core, increasing the cavitation likelyhood. However, in this case, gas bubbles are not detected by our imaging system and visualization is only due to cavitation of vapor bubbles from the preexisting nuclei.

Fig. 2. Few cavitating bubbles aligned along a vertical filament, in stroboscopic lighting (exposure time 20 μs), vertical field of view is 6.8 cm. From one picture to the other one, vapor bubbles are different but clearly cavitate from the same region.

Cavitation highlights regions of low pressure in three regions :

- In the center, mostly long vertical filaments are shown. Their length can reach 10 cm, and their mean life time is about 30 ms, up to 100 ms, of the order of the turnover time $1/f \simeq 110$ to 140 ms. Their diameter can be estimated around 2 mm. This value is found to be close to the Taylor scale (see table 1), however the scaling with Reynolds number can not be tested. At high hydrostatic pressures, when only the deepest depression trigger cavitation, these events appear less often and only consist in a few bubbles, aligned along a vertical "necklace" (see figure 2).
- Near the central shear layer, more disorganized flame-like cavitation events appear from periphery to the center. Sometimes they roll up, leading to a well defined horizontal filaments, advected to the center. The mean waiting time between such events can also been estimated to $1/f \sim 100$ ms.
- At last, we can mention that cavitation also occurs near the blades of the disks, seeding the flow with disorganized sprays of air microbubbles. Careful observations reveal the existence of strong localized vortices between each blades, moving together with the disks (see figure 3). However, because of the high velocity at the periphery of the disks, we believe that there is no link between these blades vortices and the long vertical filaments in the center of the flow.

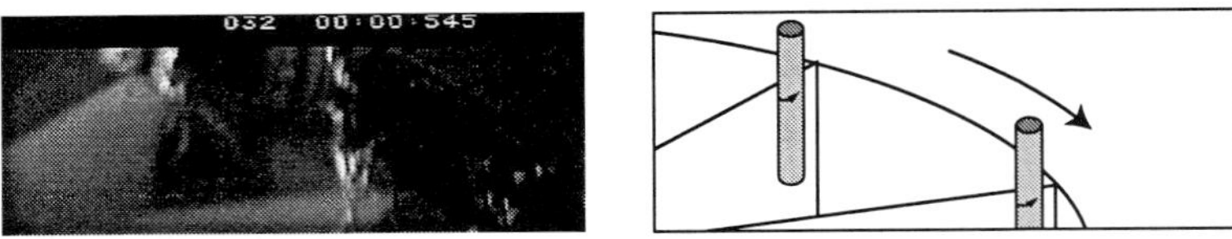

Fig. 3. Picture and sketch of stable vortices localized between each disk blade.

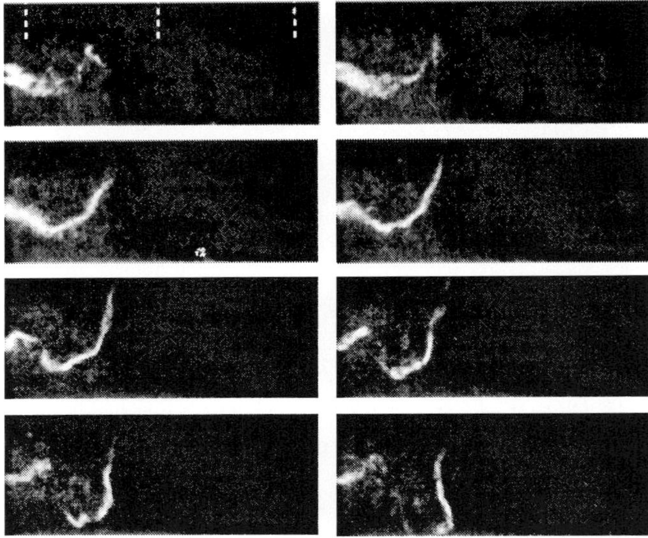

Fig. 4. Standing up of a filament, created on the horizontal shear layer (on the left), and advected upward. Dashed lines on the first picture indicates the center and the borders of the disks. Continuous lighting in salted water, the horizontal field of view is 23 cm.

The first and second observations lead us to suggest a mechanism of formation and amplification of the filaments: The shear layer rolls up from a (radial) Kelvin-Helmholtz instability , leading to horizontal filaments. This filaments stand up to the center of the flow, advected by the mean recirculation flow and stretched by the mean radial and axial strain, leading to the strong vertical filaments observed in the center (see figure 4, sketched on figure 5). Both the mean waiting time and the mean life time ($\simeq 1/f$) suggest that large scale features are responsible of the formation and amplification of these objects.

The maximum hydrostatic pressure $p_0(max)$ for which first cavitation events appear gives an estimation of the typical depression associated to the strongest

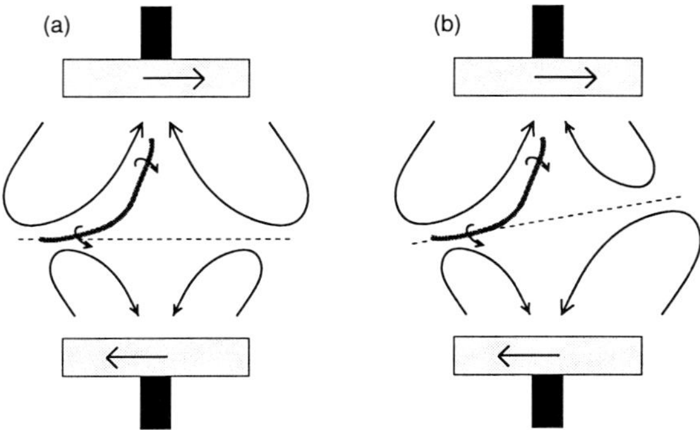

Fig. 5. Scenario of standing up of a filament. The unstable shear layer shifts (a) or tilts (b) (dashed line), on timescale of one or a few turnover times.

filaments. The so-called cavitation inception index,

$$\sigma_i = \frac{{p_0}^{(max)} - p_c}{\frac{1}{2}\rho {U_d}^2},$$

is a rough measure of the strongest depression giving rise to cavitation, compared to the typical large-scale pressure $\frac{1}{2}\rho U_d^2$ (see the values in table 1). This index is found to lie roughly between 1.2 and 1.5 (with a slight trend to increase with the Reynolds number), meaning that the strongest depressions are controlled by the largest scales. From this observation it is clear that the azimuthal velocity of the strongest filaments is of the order of the disk velocity, *i.e.* the highest velocity injected in the flow. This is found to be qualitatively consistent with other investigations in different flow configurations [13]. It is also in good agreement with the wall measurements of Fauve (1993) [21] and Cadot (1995) [12].

Lower hydrostatic pressures allow to investigate weaker filaments, whose depression are less deep. Observations in this regime reveal a population of smaller filaments, apparently more isotropically distributed. We can suppose that these weaker filaments are less affected by the large scale features of the flow.

Destabilization of filaments can be investigated using salted water in this low pressure regime. Making bubbles more stable and less likely to coalesce, it allows to keep track of the structures even if their depression increase. Sometimes destabilization appears to be due of helical undulations, self-advected along the vortex axis, whose typical wavelength seems to decrease until the vortex looses coherence (see figure 6).

Vertical filaments at the center of the flow are sometimes observed to cluster in braid-like structures. The figure 7 shows two filaments, mutually advected and orbiting around each other. There is no evidence of link between these braid-like structures and destabilization nor smaller filaments.

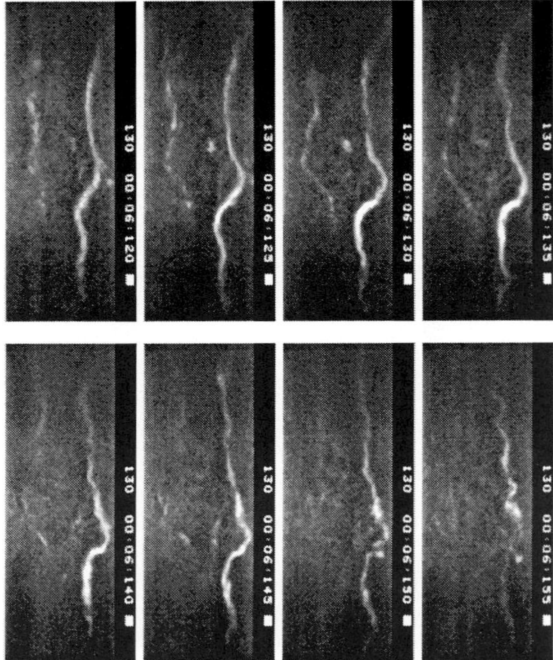

Fig. 6. Large vertical filament undergoing undulations, leading to its destabilization. The vertical field of view is 16 cm, and the time interval between frame is 5 ms.

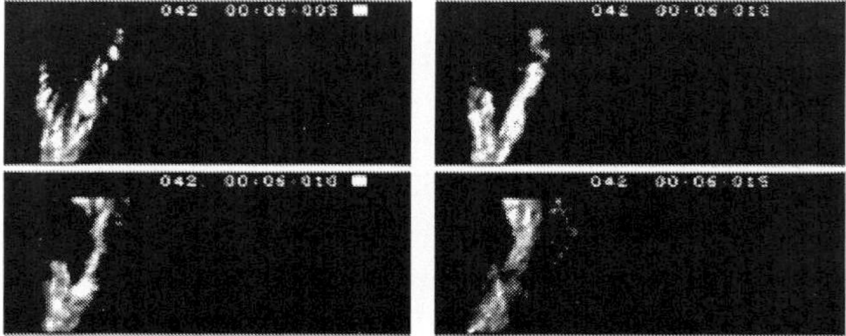

Fig. 7. Sequence showing two braids rotating around each other. Continuous lighting. The horizontal field of view is 8 cm, and the time interval between frame is 2.5 ms.

5 Statistics of cavitating events

In this section we briefly describe how statistics of the pressure fluctuations can be deduced from measurements of the light scattering from the cavitating bubbles. The basic idea is that the air microbubbles are stable at the mean pressure and not detectable by the photo detector (or at least account for the noise level), but they explode when enter a region where the pressure is below the cavitation threshold p_c and light scattering from this cavitating bubbles is detected. The easiest way to relate the level of cavitation activity at a given hydrostatic pressure to the fluctuation pressure distribution is to assume the black-and-white scheme:

$$p(\mathbf{x}, t) \leq p_c \Rightarrow \text{ cavitation}$$
$$p(\mathbf{x}, t) \geq p_c \Rightarrow \text{ no cavitation.}$$

Synchronized measurements with the photo detector and the video system used in the previous section gave evidence that the measured light signal corresponds to the presence of cavitating bubbles.

We define the *cavitation on-fraction* as the fraction of the time that cavitation is occurring in the 3 cm^3 central region (see figure 1). Indeed, in the limit of small volume, this cavitation on-fraction at a given hydrostatic pressure p_0 is given by

$$\wp(p \leq p_c; p_0) \propto \int_{-\infty}^{p_c} f(p; p_0)\, dp, \tag{2}$$

where $f(p; p_0)$ is the pressure probability density function (pdf), satisfying $\langle p \rangle = p_0$. Thus, varying the hydrostatic pressure p_0, we adjust the distance to the cavitation threshold $p_0 - p_c$, and we are able to measure point by point this cumulative distribution function $\wp(p \leq p_c; p_0)$.

From the light collected by the photo detector we have computed the cavitation on-fraction (2), following a procedure described in La Porta *et al* [1]. These curves are well fit by exponential, meaning that the pressure pdf, given by (minus) the on-fraction derivative, can be written

$$f(p; p_0) \sim e^{-(p-p_0)/\alpha}, \tag{3}$$

where the pressure constant α in $f(p; p_0)$ is the same as the one in $\wp(p \leq p_c; p_0)$. We obtain from best fit α =26.6, 35.4 and 48.1 mbar, for frequency rates f =7, 8 and 9 Hz, an increase reflecting the widening of the pressure pdf with increasing Reynolds number.

Other statistical measures, bridging from cavitation activity to pressure pdfs, are also discussed in the Ref. [1]. One is the cavitation waiting time between bursts of high activity level, and can be roughly seen as the complementary of the cavitation on-fraction, since the fraction of time that cavitation is *not* occurring is given by the average of the waiting time distribution. These methods lead to the same observation of an exponential distribution of pressure fluctuations.

In order to check the similarity of these pressure pdfs for different Reynolds numbers, we should rescale them by the pressure variance. However, this quantity is not measurable from cavitation, since the procedure described above is only supposed to work for extreme negative fluctuations. Assuming similarity, the pressure variance is expected to scale as $\rho U^2 \sim \mathrm{Re}^2$. The figure 8 collects the values of the pressure constant α for the three different Reynolds numbers. Although the range of Reynolds number available is too small to provide a strong

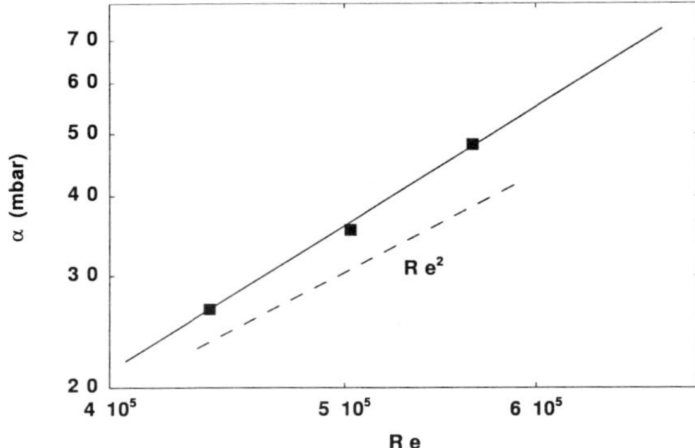

Fig. 8. Scaling of the pressure constant, as defined in (3), of the cavitation on-fraction with the Reynolds number. The dashed lines indicate the expected Re^2 scaling, and the straight line is a best fit by a power law, $\mathrm{Re}^{2.3}$.

test of the scaling, a linear fit to the log-log plot indicates that the pressure constant scales as $\mathrm{Re}^{2.3}$, slightly faster than the expected Re^2. However, the similarity scaling can not been ruled out, because self-seeding effects of cavitation may increase the cavitation activity level at higher Reynolds numbers, introducing a systematic bias for this measure.

6 Discussion and conclusion

We have shown that cavitation can be a useful tool to probe locally the turbulent fluctuating pressure in the bulk, allowing to visualize the structures resulting from a spatial cut in the pressure field at a well defined threshold close to the vapor pressure. Moreover, when the cavitation rate is low enough, this method allows to deduce statistics of the pressure fluctuations, usually not available from non invasive bulk measurement.

Our observations lead us to the conclusion that these structures can be seen as *large scale* objects:

- Their length, of the order of the disk diameter, and their orientation, along the vertical axis, suggest that these objects are essentially controlled by the large scales of the flow, even at very high Reynolds number.
- Their maximal azimuthal velocity is of the order of the disks velocity $U_d = 2\pi Rf$ (where R is the disk radius), and the corresponding depression $\sim \rho U_d^2$. Here again the filament characteristics are controlled by the large scale velocity U_d and *not* by the Kolmogorov velocity scale $\sim U_d(r/R)^{1/3}$.
- Both their life time and mean waiting time are of the order of the turnover time $1/f$. This timescale is related to their mechanism of formation (shear layer instability) and amplification (axial stretching due to the centrifugal pumping), which are properties of this specific stirring geometry.

These conclusions are in good agreement with other studies [8,22]: these low-pressure filaments appear to be the trace of primary instabilities of this flow, rather than turbulence itself. However, we have noted that visualizations of migrating bubbles at very low hydrostatic pressure revealed the existence of weaker filaments, whose orientation seems to be more randomly distributed. This suggests that weaker filaments are less sensitive to the large scales of the flow.

Additionally, statistics of scattering light from cavitation allow us to reconstruct the negative tail of the pressure pdf, in the central region of the flow where strongly anisotropic filaments are observed. These tails are found to decrease exponentially, with pressure constant scaling roughly as Re^2, in agreement with similarity hypothesis (discrepancy from this scaling may be due to the self-seeding effect of cavitation). Together with the previous observations, this means that the far tails of the exponential pressure distributions are controlled by the strongest, anisotropic, filaments, whereas the weaker ones contribute to an intermediate range of these distributions.

These observations raise the issue of the *universality* of pressure fluctuations in turbulence. Are our exponential distributions related to the ones observed in numerical simulations [5,23]? Pumir (1994) [5] notes that, in isotropic and homogeneous numerical turbulence, the lowest depressions are due to the worms [2,3], small scale vortices, whose radius is of the order of the Kolmogorov scale. Shall we conclude that, in an inhomogeneous and anisotropic case, depressions due to the large scale vortices hide the more *universal* properties of smaller scale pressure fluctuations?

The authors acknowledge P. Tabeling, H. Willaime and J.S. Andersen for fruitful discussions.

References

1. A. La Porta, G.A. Voth, F. Moisy and E. Bodenschatz: Using cavitation to measure statistics of low-pressure events in large-Reynolds-number turbulence, submitted to *Phys. Fluid* (1999). Preprint available on `http://milou.msc.cornell.edu/publications.html`.
2. F. Belin, J. Maurer, P. Tabeling, and H. Willaime: Observation of intense filaments in fully developed turbulence, *J. Phys II* **6**, 573–584 (1996).
3. J. Jimenez: Small scale intermittency in turbulence *Eur. J. Mech. B/Fluids* **17** (4), 405–419 (1998).
4. E.D. Siggia: Numerical study of the smale-scale intermittency in three-dimensional turbulence, *J. Fluid. Mech.* **107** (3), 375–406 (1981).
5. A. Pumir: A numerical study of pressure fluctuations in three-dimensional, incompressible, homogeneous, isotropic turbulence *Phys. Fluids* **6** (6), 2071–2083 (1994).
6. M.E. Brachet: The geometry of small-scale structures of the Taylor-Green vortex, *C.R.Acad.Sci.Paris* **311**, 775–780 (1990).
7. S. Douady, Y. Couder, and M.E. Brachet: Direct observation of the intermittency of intense vorticity filaments, *Phys. Rev. Lett* **67**, 983–986 (1991).
8. E. Villermaux and Y. Gagne: Intense vortical structures in grid-generated turbulence *Phys. Fluids* **7** (8), 2008–2013 (1995).

9. W.K. George, P.D. Beuther and R.E.A. Arndt: Pressure spectra in turbulent free shear flows *J. Fluid. Mech.* **148**, 155–191 (1984).
10. F. Moisy and P. Petitjeans: Pressure measurements in a stretched vortex, in this Volume.
11. P. Abry, S. Fauve, P. Flandrin, and C. Laroche : Analysis of pressure fluctuations in swirling turbulent flows *J. Phys. II France* **4** 725–733 (1994).
12. O. Cadot, S. Douady, and Y. Couder: Characterisation of the low pressure filaments in a 3D turbulent shear flow, *Phys. Fluids* **7**, 630–646 (1995).
13. R.E.A. Arndt: Cavitation in fluid machinery and hydraulic structures *Ann. Rev. Fluid. Mech.* **13**, 273–328 (1981).
14. T.J. O'Hern: An experimental investigation of turbulent shear flow cavitation, *J. Fluid. Mech.* **215**, 365–391 (1990).
15. B. Belahadji, J.P. Franc and J.M. Michel:Cavitation in the rotational structures of a turbulent wake *J. Fluid. Mech.* **287**, 383–403 (1995).
16. O. Boulon, M. Callenaere, J.P. Franc and J.M. Michel: An experimental insight into the effect of confinement on tip vortex cavitation of an elliptical hydrofoil, *J. Fluid. Mech.* **390**, 1–23 (1999).
17. B. Ran and J. Katz: The response of microscopic bubbles to dussen changes in the ambiant pressure *J. Fluid. Mech.* **224**, 91–115 (1991).
18. B. Ran and J. Katz: Pressure fluctuations and their effect on cavitation inception within water jets *J. Fluid. Mech.* **398**, 1–43 (1999).
19. G.A. Voth, K. Satyamarayan and E. Bodenschatz: Lagrangian Acceleration Measurements at Large Reynolds Numbers, *Phys. Fluids* **10** (9), 2268–2280 (1998).
20. M.S. Plesset and A. Prosperetti: Bubble dynamics and cavitation *Ann. Rev. Fluid. Mech.* **9**, 145–185 (1977).
21. S. Fauve, C. Laroche and B. Castaing: Pressure fluctuations in swirling turbulent flows *J. Phys. II France* **3**, 271–278 (1993).
22. S. Fauve, S. Aumaitre, P. Abry, J.-F. Pinton, and R. Labbe: Large scale fluctuations in swirling flows, Advances in Turbulence VII, Kluwer Ac. Publishers, Uriel Frisch Ed. (1998).
23. V. Prakash and P.K. Yeung: Similarity scaling of acceleration and pressure statistics in numerical simulations of isotropic turbulence, *Phys. Fluids* **11**, 1208–1220 (1999).

On the Feedback of Vorticity on Stretching

Caroline Nore[1], Bruno Andreotti[2], Stéphane Douady[2] and Malek Abid[3]

[1] Université de Paris-Sud, LIMSI, Bâtiment 508, F-91403 Orsay Cedex, France
[2] Laboratoire de Physique Statistique de l'Ecole Normale Supérieure associé au CNRS et aux Universités Paris 6 et 7, 24 rue Lhomond, 75005 Paris, France
[3] Institut de Recherche sur les Phénomènes Hors Equilibre, UMR CNRS et Université d'Aix-Marseille I, service 252, Centre St-Jérôme, 13397 Marseille Cedex 20, France

1 Introduction

It is known since pioneering work by Taylor [1] that vortex stretching, namely the fact that rotation is amplified by conservation of the angular momentum when the streamlines locally converge, is one of the most important mechanisms acting in vortex dynamics. The Burgers' vortex [2] is a well-known model of stretched vortex where the stretching γ is kept constant in space and time, the meridional velocity being given by:

$$v_r = -\frac{1}{2}\gamma r,$$
$$v_z = \gamma z.$$

The evolution of the vortex is governed by the competition between vortex stretching which tends to reduce the core radius R and viscosity which diffuses it, the circulation Γ remaining constant. In the stationary regime, the azimuthal velocity reads:

$$v_\theta = \frac{\Gamma}{2\pi r}\left(1 - \exp\left(-\frac{r^2}{R^2}\right)\right), \tag{1}$$

R being of the order of:

$$R \simeq \sqrt{\frac{\nu}{\gamma}}. \tag{2}$$

The particularity of this solution and more generally of all the Burgers-like models is that the stretching does not evolve: it is either permanent and spatially uniform [2–7] or spatially localised but imposed [8–10].

However, in real flows, vortices and more generally rotating regions are neither isolated nor submitted to a constant stretching field. What occurs if a vortex is subject to a stretching due to a localised structure and thus not permanent nor uniform? Our aim is to investigate both experimentally and numerically more complex situations where both the vortex and the stretching freely evolve.

2 Experimental Observations

The set up is formed of a cylindrical tank with a rotating bottom, a system well known to create a central vortex [11,12]. In our system (Fig. 1), an axial pumping is added on the vortex axis. When the pumping is switched on it generates an axial velocity gradient which stretches the vortex. Adjusting the flow rate, we control the velocity U in the pumping hole. The axial velocity u_z being null on the disc, we are able to impose the mean stretching $\bar{\gamma}$ on the axis:

$$\bar{\gamma} = \frac{U}{H} \tag{3}$$

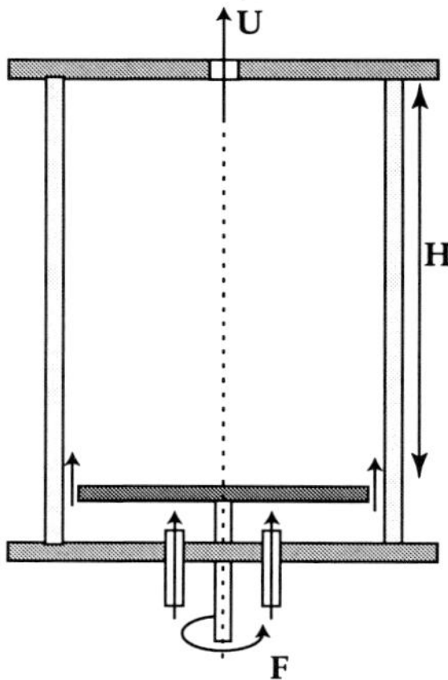

Fig. 1. Experimental set-up

We measure the axial and radial velocity components (u_z and u_r) in a diametral plane, using PIV method, during the transient formation of the vortex. Using the fact that the flow is nearly axisymmetric, the streamfunction Ψ is deduced from these measurements by integrating the equations:

$$u_r = -\frac{1}{r}\frac{\partial}{\partial z}\Psi, \tag{4}$$

$$u_z = \frac{1}{r}\frac{\partial}{\partial r}\Psi. \tag{5}$$

Several Ψ isocontours are shown in Fig. 2 for a pumping hole diameter of 24 mm, a velocity in the hole of $U = 0.2\ m\ s^{-1}$, a cell height $H = 0.15\ m$ and a $F = 10\ Hz$ disk frequency. The corresponding azimuthal vorticity has the same topology [13]. At short time ($t = 0.3\ s$), the effect of pumping is essentially localised around the pumping hole. Just below the hole, the streamlines abruptly converge toward the axis, indicating the presence of a localised and strong stretching. This stretching bubble then propagates towards the disc. It lets behind it a column by which the fluid is pumped, which also corresponds to the vortex core. After $t = 7\ s$, the stretching bubble reaches the disc and the last inhomogeneities of the column finally disappear at $t = 10\ s$.

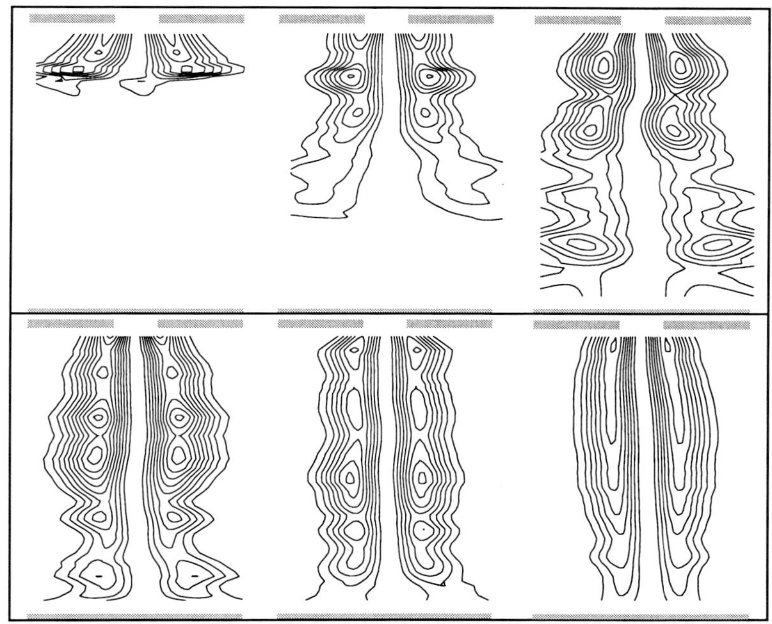

Fig. 2. Time evolution of the streamlines during the transient formation of the vortex. From left to right and from top to bottom, $t = 0.3\ s$, $t = 0.8\ s$, $t = 2.8\ s$, $t = 4\ s$, $t = 7\ s$ and $t = 10\ s$

In the stationary regime, the stretching γ (the axial velocity gradient) is of the order of $0.3\ s^{-1}$ in the flow bulk and increases to $4\ s^{-1}$ in the boundary layers. We observe a depletion of stretching: the mean gradient $\bar{\gamma}$ is more than four time larger than the stretching γ in the bulk for the set of control parameters above. This ratio can be even larger when the disk frequency is increased. As can be seen on Fig. 2, the stretching starts to be reduced (the nearly 2D column starts developing) as soon as the vortex rotates fast. In the stationary state, we furthermore observe that the core radius and the pumping column width are fixed by the pumping hole size. But the vortex core is also at viscous equilibrium which determines, contrarily to Burgers' models, not its radius (fixed by geometry) but the residual stretching γ:

$$\gamma \simeq \frac{\nu}{R^2}. \tag{6}$$

This shows that the stretching amplifies the vortex rotation but in turn, the vortex rotation retroacts on the stretching itself and reduces it. This is a kind of Lenz law which is related to the global tendency of rotating flows to bidimensionality.

3 Numerical Results

In order to investigate numerically the interaction between a vortex and a localised stretching field, we choose an initial condition which superimposes a Lamb's vortex (1), (Fig. 3 a) to a periodic array of alternate tori of wavelength λ (Fig. 3 b). The latter stretch the vortex, making the axial vorticity increase and the core size shrink at the centre (Fig. 4 a). In the figures presented here, the ratio between the vortex core radius and the array wavelength is $R/\lambda = 0.13$, the ratio between the vortex core radius and the Burgers viscous lengthscale is $R\sqrt{\gamma/\nu} = 19$, the vortex Reynolds number is $\Gamma/\nu = 1600$ and the ratio between the vortex and the tori typical velocities is around 3.7. At short time, the initial

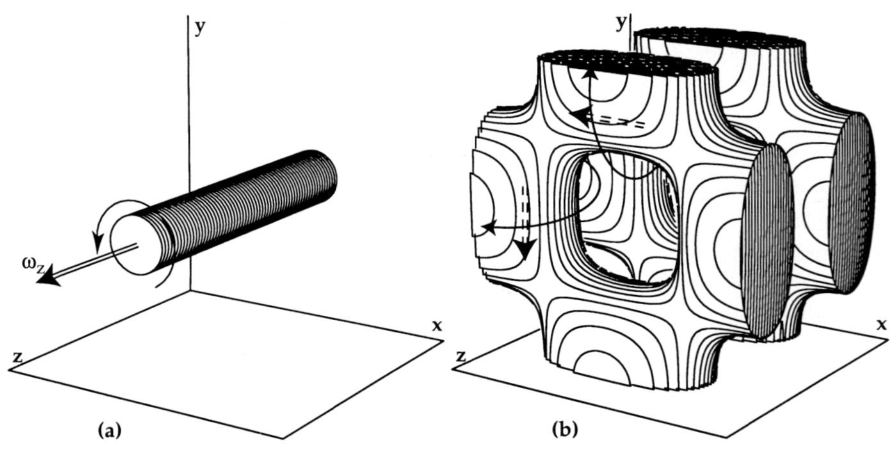

Fig. 3. Three-dimensional iso-countours of vorticity for the initial condition. (**a**) Axial vortex ω_z. (**b**) Perturbative azimuthal vorticity ω_θ

tori do not evolve significantly and the evolution is similar to that predicted by Burgers-like models. But the deformed vortex then retroacts on the stretching by inducing two new tori (Fig. 4 b,c). They turn in opposite way, their radius is smaller and they are more axisymmetric than the initial ones. They thus tend to straighten the vortex, the result being globally a counter-reaction (Fig. 4 b,c).

This negative feedback overshoots in its compression of the vortex. Again the vortex becomes non uniform along its axis and induces a negative feed-back against the inhomogeneous stretching to which it is submitted: some new tori appear inside the secondary ones (Fig. 4 d). This onion-like structuration of the flow continues until small scales are reached.

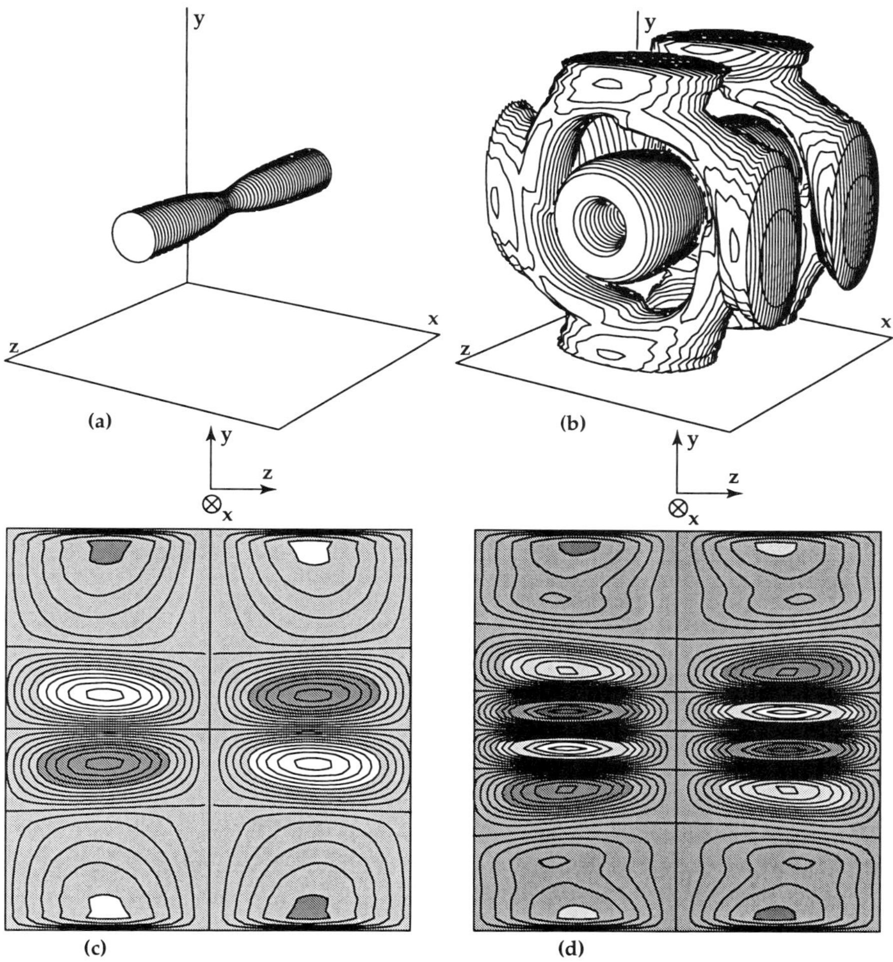

Fig. 4. Countours of vorticity during time evolution. (**a**) Deformed axial vortex ω_z. (**b**) Primary and secondary tori of azimuthal vorticity ω_θ. (**c**) Two-dimensional contours in the plane $x = 0$ at equally spaced levels of ω_θ corresponding to (**b**). The positive highest contour levels are shaded dark-gray while negative lowest contour levels are white. (**d**) Two-dimensional contours in the plane $x = 0$ at equally spaced levels showing new tertiary tori ω_θ at later time

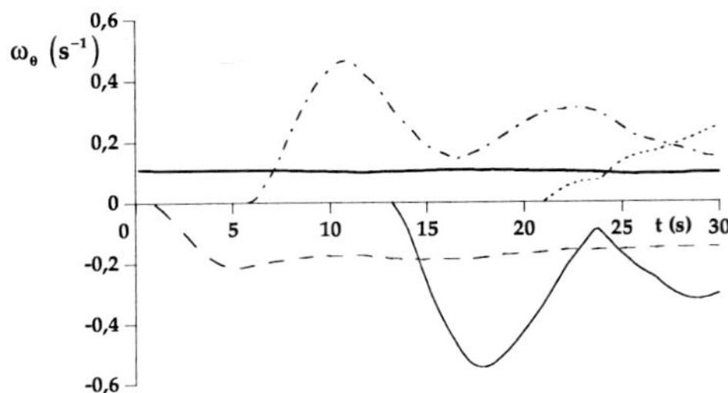

Fig. 5. Time evolution of the azimuthal vorticity of each torus: the constant initial torus (*continuous line*), the secondary torus (*long-dashed line*), the tertiary torus (*dot-dashed line*), the quaternary torus (*thin continuous line*), the next one (*dotted line*)

Figure 5 shows the azimuthal vorticity amplitude of the successive tori. The amplitude of initial rings remains nearly constant in time (around 0.1 s^{-1}, *continuous line*). Then appear the secondary tori inside the vortex (Fig. 4 b,c), which amplitude saturates around $-0.2\ s^{-1}$ (*long-dashed line*). The next generations of tori have amplitudes which slightly oscillate, but which never change sign, indicating a partial overlap of the induced structures. This confirms quantitatively that the feedback mechanism consists in the induction of new stretching structures rather than a global oscillation.

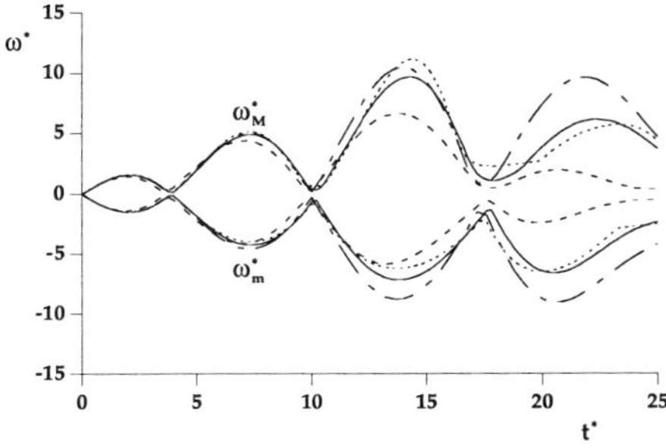

Fig. 6. Time evolution of the non-dimensionalised maximum axial vorticity ω^*_M and minimum axial vorticity ω^*_m versus the non-dimensionalised time $t^* = t\ (\omega\ R\ /\lambda)$ for different runs

Figure 6 shows the time evolution of the maximum and minimum axial vorticities normalised by $\gamma\lambda/R$ for various runs (with different vortex radii R and vorticities ω, and different initial stretchings γ [14]). The collapse, using the non-dimensional time $t^* = (\omega\ R\ /\lambda)\ t$, is qualitatively verified until $t^* = 17$.

Thus the oscillations frequency scales as:

$$f \propto \frac{\omega R}{\lambda}. \tag{7}$$

4 Discussion

In this paper, we have experimentally and numerically investigated the evolution of a vortex submitted to a non uniform stretching of typical amplitude γ. The vortex is deformed by this stretching on a timescale $\tau_{\gamma \to \omega} = 1/\gamma$. But it turns out that the vortex reacts by inducing secondary tori shaped structures which screen the effect of the primary stretching. This basically comes from the fact that there exists a differential rotation along the vortex axis which leads to a variation of the centrifugal force. The pressure gradient cannot balance this inhomogeneous centrifugal force anymore, the latter having a solenoidal part. As a consequence, a secondary toroidal flow is induced which stretches the vortex in the regions where it was slowly rotating and compresses it where it was rapidly rotating. Globally, the vortex tends to reduce the stretching to which it is submitted on a timescale $\tau_{\omega \to \gamma} = \lambda/\omega R$, where λ holds for the characteristic length over which stretching is inhomogeneous.

The ratio between the two timescales is then given by the non-dimensional number

$$B = \omega R/\gamma\lambda. \tag{8}$$

which characterises, by construction, the relative importance of the feedback mechanism and of the stretching process. For homogeneous stretching (Burgers' limit), λ tends to infinity and B tends to zero: there is no retroaction. On the contrary, if $B \gg 1$, $\tau_{\omega \to \gamma} \ll \tau_{\gamma \to \omega}$, the negative feedback of rotation on stretching plays a crucial role. Burgers-like models [2–10] thus correctly describe the vortex dynamics for small B, when $\tau_{\omega \to \gamma} \gg \tau_{\gamma \to \omega}$. This corresponds to the situation where the rotation speed ωR is much smaller than the characteristic velocity associated to the stretching $\gamma\lambda$, that is to say, when there is no vortex but a nearly pure stretching flow. As soon as the rotation becomes important when compared to the stretching (even in the numerical simulation presented here for which $B = 0.56$), the retroaction of rotation on stretching becomes an essential part of the dynamics.

References

1. G.I. Taylor: Proc. Roy. Soc. A **146**, 107 (1934)
2. J.M. Burgers: Proc. Acad. Sci. Amsterdam **43**, 2 (1940)
3. J.C. Neu: J. Fluid Mech. **143**, 253 (1984)
4. A.D.D. Craik and W.O. Criminale: Proc. R. Soc. Lond. **406**, 13 (1986)
5. O.S. Kerr and J.W. Dold: J. Fluid Mech. **276**, 307 (1994)
6. H.K. Moffatt, S. Kida and K. Ohkitani: J. Fluid Mech. **259**, 241 (1994)

7. S. Le Dizès, M. Rossi and H.K. Moffatt Phys. Fluids **8**, 2084 (1996)
8. C.D. Donaldson and R.D. Sullivan: Proc. Heat Transfer & Fluid Mech. Inst., Stanford University (1960)
9. M.V. Melander and F. Hussain: Phys. Fluids **5**, 1992 (1993)
10. R. Verzicco, J. Jimenez and P. Orlandi: J. Fluid Mech. **299**, 367 (1995)
11. J. Turner: J. Fluid Mech **25**, 377 (1966)
12. M.P. Escudier: Exp. Fluids **2**, 189 (1984)
13. B. Andreotti: PHD thesis (1999)
14. M. Abid, B. Andreotti, S. Douady and C. Nore: submitted to J. Fluid Mech (1999)

Characterisation of Vortex Flow Inside an Entrained Cavity

Alecsandra Rambert[1], Afif Elcafsi[1] and Pierre Gougat[1]

LIMSI-CNRS, B.P. 133, 91403 Orsay, France

Abstract. A number of studies have referred to the existence of a vortex cell within an urban street canyon when ambient winds aloft are perpendicular to the street. The understanding of vortex dynamics or vorticity distribution in a such configuration is of great interest. Vortex structures play an important role in the dynamics of pollutant dispersion. This configuration was simulated by the interaction between a boundary layer and a cavity. Experimental characterisation of the vortex structures evolution was developed by flow velocity measurements inside and out of the cavity. Classical methods like hot wire and Laser Doppler Velocimetry (LDV) display only local measurements. Particle Image Velocimetry (PIV) method based on the optical flow technique permitted global velocity measurements. This technique emphasis the vortex structures inside the cavity which present small scales as well as large scales related to the cavity geometry. Theses vortices are usually non-stationary.

1 Introduction

The pollutant dispersion prediction is an important factor in the urban pollution field and become essential to estimate qualitatively and quantitatively pollutant concentrations and their spatial and time evolution, strongly connected to the flow movements in this region. The majority of recent studies of circulation in street canyon have been performed by physical and numerical simulations. Physical simulation studies concerned pollutant levels in street canyons under various geometries show increases of concentration predicted consistent with a vortex flow pattern [1]. Much attention has been directed to the study of the various canyon flow regimes [2] since air flow is responsible for the transport of properties such as pollutants, heat and moisture. Canyon geometry is an important determinant of characteristic airflow observed within urban canyons. Previous studies used a k-e turbulence model of air flow to confirm the wind tunnels observations. The model was based on that developed by [3] to predict pressure experienced by walls of buildings within cities. It solves the Navier-Stokes equations for momentum and the equations for the transportation and dissipation of turbulent kinetic energy.

The aim of our work was to develop an experimental set-up and associated measurement techniques which permit to characterise the spatial and time dependant flow pattern in a boundary layer-cavity interaction. The numerical and experimental quantitative prediction of the vortex flow in this configuration remains still approximate so we decided to study a simplified configuration represented by the interaction between a laminar boundary layer and a cavity.

2 Material and Methods

The experimental set-up is presented in Fig. 1. The cavity characteristic dimensions are indicated in meters in Table 1 were H is the height, L the length and l the width of the cavity.

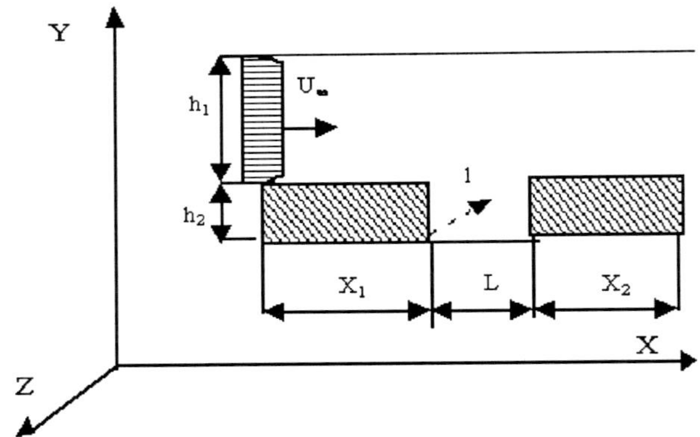

Fig. 1. Geometrical configuration

Table 1. Characteristic dimensions of the cavity.

h_1	h_2	l	L	X_1	X_2
0.07	0.05	0.02	0.1	0.2	0.3

Velocity profiles in the boundary layer were obtained by hot wire measurements. The interaction region between the boundary layer and the cavity and the vortex burst from the cavity were characterised by Laser Doppler Velocimetry measurements. The flow inside the cavity was characterised using Particle Image Velocimetry based on Optical Flow Classical flow visualisation is based on direct observation of tracer particles. Analysis of subsequent images searching for local displacements allows quantitative measurement of two-dimensional flow fields. The optical flow method offers a new approach for analysing flow images. It largely improves spatial accuracy and minimises the number of spurious vectors. Optical flow computation consists in extracting a dense velocity field from an image sequence assuming that the intensity is conserved during the displacement. Several techniques have been developed for the computation of optical flow. The technique that was chosen for PIV application was introduced by Quénot [4] as Orthogonal Dynamic Programming algorithm for optical flow detection from a pair of images. Compared with the classical PIV method, the optical flow has the following advantages: (i) it can be applied simultaneously to sequences of more than two images; (ii) it performs a global image match by enforcing continuity and regularity constrains on the flow field. This helps in ambiguous or low particle density regions; (iii) It provides dense velocity fields (iiii) local correlation is iteratively searched for in regions whose shape is modified by the flow, instead of being searched by fixed windows. This greatly improves the accuracy in regions with strong velocity gradients.

A sequence of 50 images was recorded by a digital camera of 30 images/s (Pulnix 9701). The time between two images is 1500 μ s. Lens magnification is 0.16. Lycopodium particles and smoke were used as seeding.

3 Results

At the entry of the cavity (U_∞ is equal to 1.27 m/s, Re_x=1.6×10^4), the boundary layer is laminar and it is matching with a Blasius profile (η versus U/U_∞,$\eta = \frac{U_\infty y}{Re_x^{0.5}}$). The mean velocity histogram obtained by Laser Doppler Velocimetry at $X = X_1$ and $h = h_1/2$ is presented on Fig.2a (U_∞ equal to 1.4m/s $Re_x = 1.6 \times 10^4$).The corresponding velocity profile obtained by hot wire measurements is presented in Fig. 2b.

A visualisation realised in this region (x=300 mm, y=50mm) point out the presence of vortex structures coming out from the cavity and their movement (Fig. 3).

Measurements made after the cavity showed that the initial laminar profile was modified. In Fig. 2c we presented the velocity profiles obtained after the cavity at x=310. A quantitative analysis of the velocity profiles after the cavity was made in terms of the boundary layer thickness δ , displacement thickness δ_1 and momentum thickness δ_2. The exp[erimental results were compared with the theoretical formula corresponding to a turbulent boundary layer (logarithmic profile) :

$\delta = \frac{0.38X}{Re_x^{0.2}}$ $\delta_1 = \frac{0.045X}{Re_x^{0.2}}$ $\delta_2 = \frac{0.037X}{Re_x^{0.2}}$

The results are presented in Table 2. One can see that the experimental profiles don't fit zith a logarithmic “standard ”profile. This behaviour can be explained by the vortex burst from the cavity which modify the initial laminar velocity profile.

Table 2. Experimental and theoritical turbulent boundary layers thickness.

U(m/s)	delta (m)		delta 1 (m)		delta 2 (m)	
	exp	turbulent	exp	turbulent	exp	turbulent
1.2	1.58 10^{-2}	1.73 10^{-2}	2.82 10^{-3}	2.17 10^{-3}	1.00 10^{-3}	1.69 10^{-3}
3	1.40 10^{-2}	1.45 10^{-2}	1.87 10^{-3}	1.82 10^{-3}	1.44 10^{-3}	1.42 10^{-3}
5	1.60 10^{-2}	1.32 10^{-2}	2.10 10^{-3}	1.66 10^{-3}	1.64 10^{-3}	1.29 10^{-3}

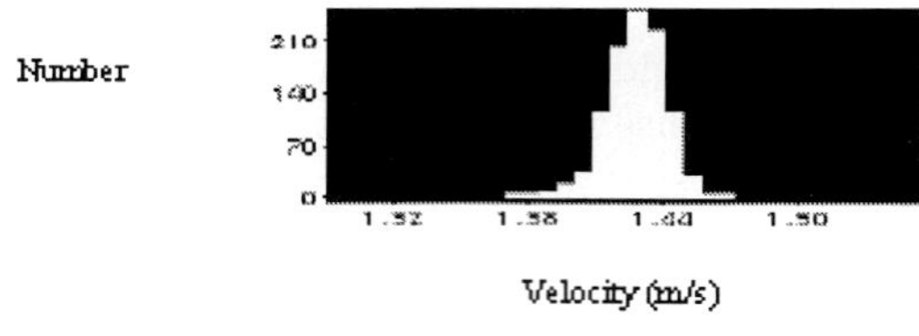

Fig. 2a

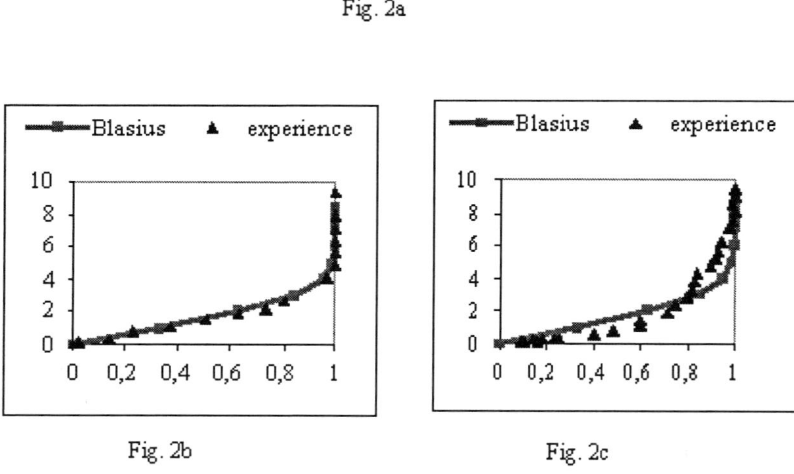

Fig. 2b

Fig. 2c

Fig. 2. Mean velocity and velocity profiles in boundary layer. (2a) Mean velocity histogramm at x=200 and $h = h_1/2$, (2b) before the cavity (x=200 mm), (2c) after the cavity (x=310 mm)

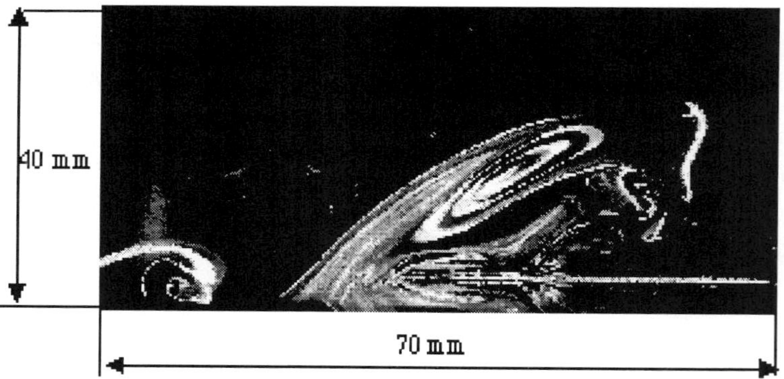

Fig. 3. Smoke visualisation of vortex release from the cavity

For the same mean velocities and at the same location the velocity profiles measured in the absence of the cavity fitted well with a Blasius profile.

After the cavity, the same measurements made by hot wire show that the velocity profile in the boundary layer is no longer laminar without becoming turbulent (Fig. 3). This behaviour can be explained by the vortex burst from the cavity which influence the laminar profile.

Knowing the time between images and the magnification one can estimate the mean translation velocity of the vortex structure which is sensibly the same as the mean velocity flow. At the same location, measurements performed by LDV furnish quantitative information about the vortex structures as mean velocity and frequency.

Figure 4 shows an example of velocity histogram and its time dependence obtained at x=310mm y=75 mm. The auto-correlation analysis of the signal permitted to identify a characteristic frequency about 17 Hz.

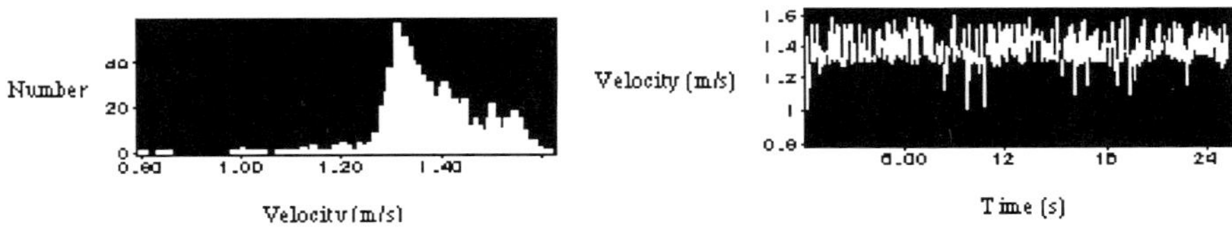

Fig. 4. Velocity histogram and its time dependence (x=310mm y=75mm)

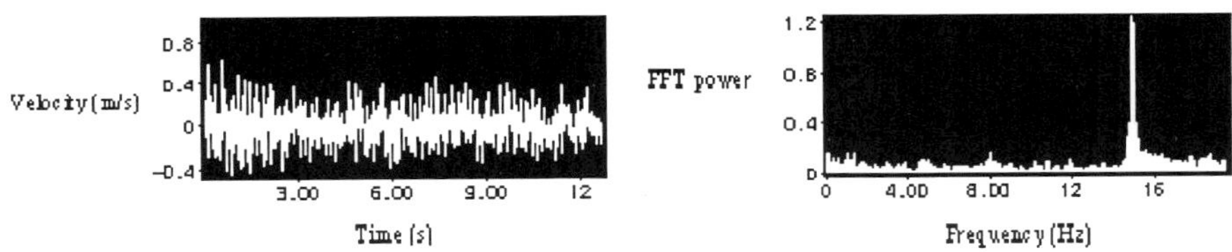

Fig. 5. Velocity autocorrelation signal and FFT power density (x=310 mm, y=75mm

Measurements were performed for different cavity shape ratio and for different mean velocities. The different shape ratio was obtained by (i) changing the height of the cavity h_2 and keeping the length L constant and (ii) by changing the length L of the cavity and keeping the height h_2 constant. The results of the vortex escape frequency corresponding to theses cases are presented in Figure 6a and Figure 6b as a function of the adimensional factors U/h_2 and U/L, respectively.

The vortex escape depends om the manner that the shape ratio was obtained, by changing the h_2 or the L cavity characteristic dimensions. For the shape ratio $h_2/L_{h_2=constant}$ and $h_2/L_{L=constant}$ equal to 0.5 and for the same velocities, there is no difference between the vortex escape frequencies. We notice that

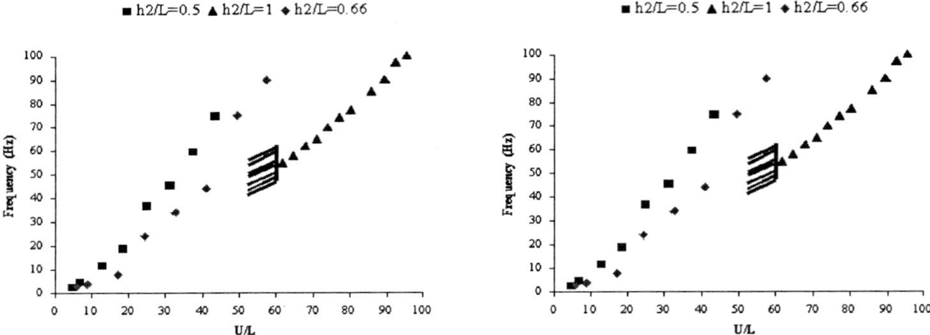

Fig. 6. (a) Vortex frequency as a function of U/h_2 for different cavities shape ratio $h_2/L_{L=constant}$, (b) Vortex frequency as a function of U/L for different cavities shape ratio $h_2/L_{h_2=constant}$

for the shape ratio $h_2/L_{h_2=constant}$ equal to 1 there is no vortex escape for a mean velocity less than 3m/s. The frequencies performed in this case are sensibly greater then those obtatined for the same shape factor $h_2/L_{L=constant}$. For the case of a shape ratio $h_2/L_{L=constant}$ equal to 2 there is no vortex escape for a mean velocity more than 4 m/s. Measurements made for $h_2/L_{h_2=constant}$ equal to 2 showed that there is not vortex escape at all for any mean velocity value.

Comparing the vortex escape frequencies for the same velocities for different $h_2/L_{L=constant}$ values we notice that this shape factor has not a significant influence on the vortex escape behaviour. On the contrary, analysing the results obtained for the $h_2/L_{h_2=constant}$ case we conclude that the length L of the cavity is a characteristic dimension which plays an important role in the vortex escape phemomenon.

For the same aerodynamic conditions and geometrical configuration a LES (Large Eddy Simulation) numerical code was developped [5]. At the entry of the cavity the numerical velocity profile matches with a Blasius profile. An analysis of the numerical velocity signal corresponding to x=310 mm and y=75 mm point out a characteristic frequency about 17 Hz which agrees well with the experimental results (Fig. 6). The numerical code will be tested for other different Reynolds numbers in order to compare the numerical results with the experimental ones.

To improve the characterisation of the vortex structures global velocity measurements were performed by PIV based on optical flow inside the cavity. An example of such a velocity field is presented in Fig. 6.

Two counter rotating vortex are observed with eddies of smaller dimensions turning around them. The vortex has a diameter equal to the height of the cavity. The velocity inside the cavity is about 0.2–0.3 m/s. In order to better characterise the movement of the main vortex, measurements were performed only in the right side of the cavity (Fig. 7).

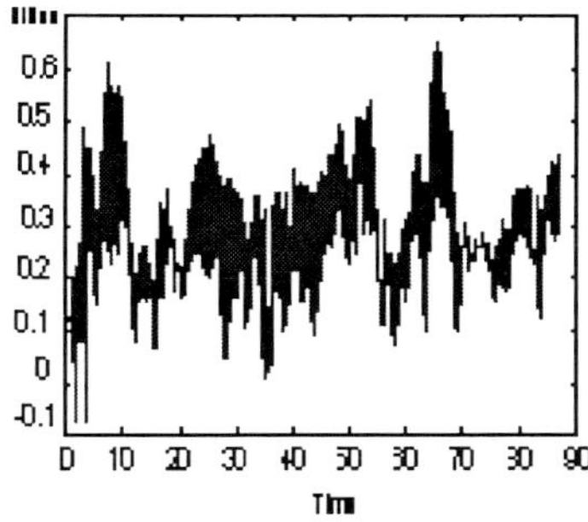

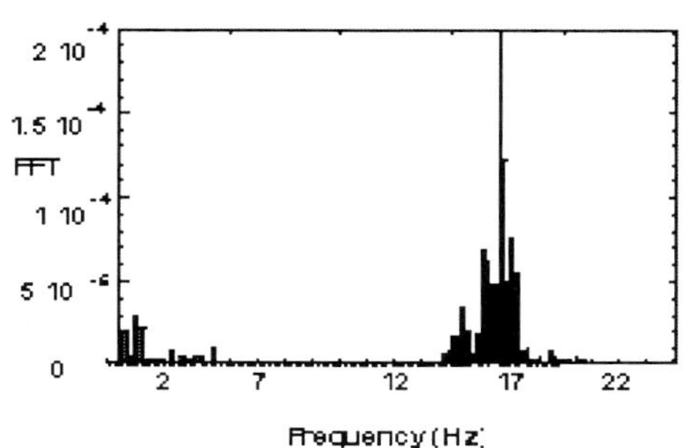

Fig. 7. Numerical velocity time dependence and FFT density power.

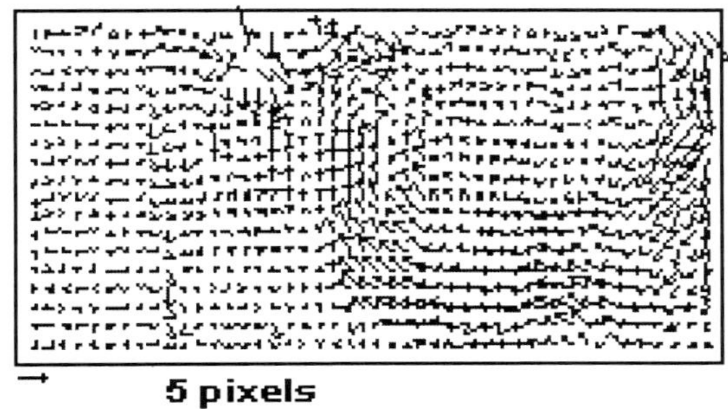

Fig. 8. Flow velocity fields obtained inside the cavity

Time between two velocity fields is 66ms. Due to the viscous effects this structure entrains the environmental fluid and its diameter growths. We can identify an ascension movement in order to burst out from the cavity.

4 Conclusion

The main application of this work is the study of the pollutant transport and dispersion in a canyon street. The study of the interaction between a boundary layer and a cavity is a first approach in characterisation of flow regimes within the urban canyon.

The movement dynamics was characterised by local (Laser Doppler Velocimetry) and global velocity measurements (Particle Image Velocimetry). Laser Doppler Velocimetry measurements furnish information about the time evolution of the fluid velocity. A vortex release frequency was identified.

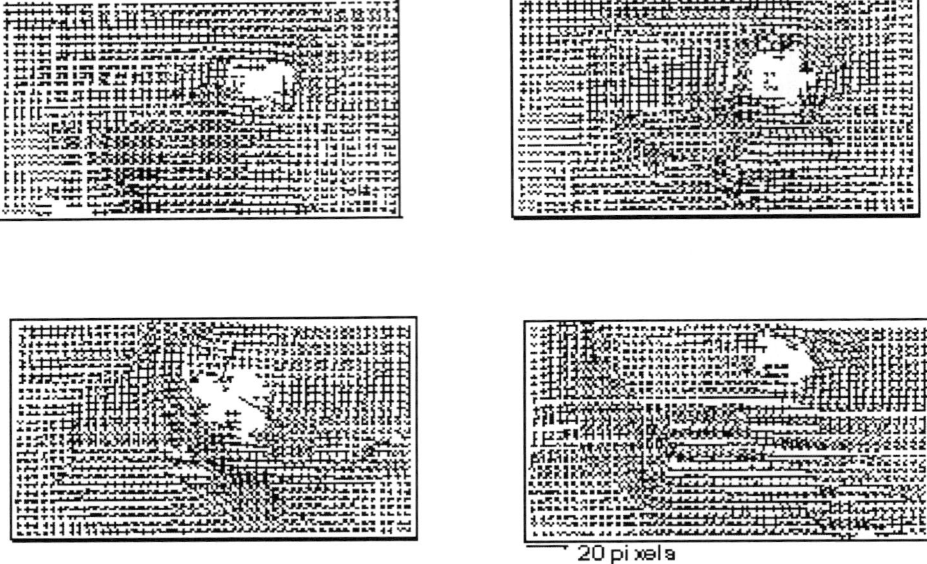

Fig. 9. Time evolution of instantaneous velocity fields (time between fields is 66 ms)

An Optical Flow technique based on the use of Dynamic Programming has been successfully applied to Particle Image Velocimetry yielding a significant increase in the accuracy and spatial resolution.

References

1. J.B. Wedding, D. Lombardi and J. Cermak: J. Air Pollut. Control Ass, **27**, 557-566 (1977)
2. F. T. DePaul and C.M. Sheih: Atmospheric Environment, **20**, 455-459 (1986)
3. D.A. Paterson, and C.J. Apelt: Bldng. Envir, **24**, 39-50 (1989)
4. G.M. Quénot: 'The Orthogonal Algorithm for Optical Flow detection using Dynamical Programming'In: Proc. IEE ICASSP, San Francisco, (1992), vol. 3, pp. 249-252
5. A. Rambert : 'Étude de l'interaction particule-écoulement : dans le cas d'une pollution urbaine'. Thèse Paris 11, 1998

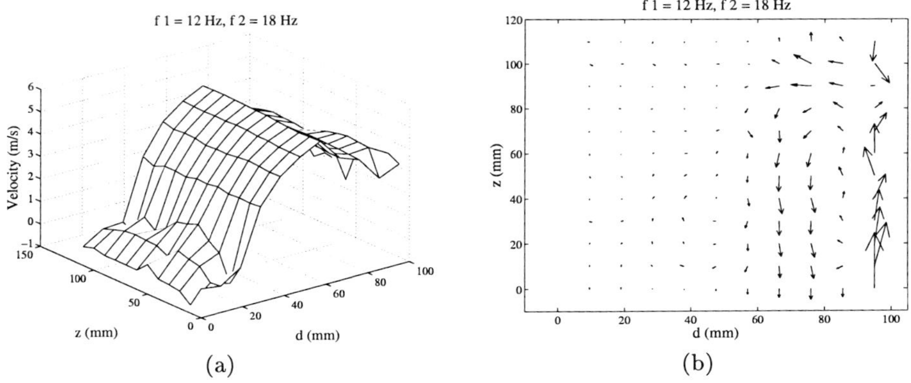

Fig. 3. $f_1 = 12$ Hz, $f_2 = 18$ Hz. (**a**) Orthoradial velocity profiles. (**b**) Meridian section of velocity

2 Solid Particle Motion

Information on the dynamics of the flow and about hydrodynamic forces can be obtained by observation of the motion of solid particles. We seeded the flow with one or several plastic beads, with a density lower than that of water (0.9) and diameter of 2 mm. A low density is chosen to favour the particle concentration in vortical regions [4]; however the beads do not act as lagrangian tracers due to their large size compared to that of the velocity gradients into the flow (e.g. Kolmogorov's length). Frames from a video recorded film of the motion of the particles are presented in Fig. 4.

We observe that in the (GR) case, the sphere rotates at a fixed distance from the axis ($d \approx 2$–3 cm), nearly constant in time. The mechanism which fixes this position for the sphere is not clearly understood, further studies are underway. In (DR), the sphere rotates less regularly, and has a more important axial motion. Information about its dynamics is revealed when we analyze short successive time intervals. Figure 5 presents four successive sequences extracted from the average sequence in Fig. 4 b, each one 2 seconds long.

One observes intermittent sudden concentration of the sphere on the axis, then a dispersion in the tank, and finally a return to an irregular movement of rotation.

These visualisations are consistent with other observations made with kalliroscope (tiny reflecting flakes) and air bubbles. They confirm the existence of two distinct regimes in this flow geometry: a stable swirling structure when the disks rotate at the equal speeds (GR), and an instable vortex dynamics at different rotation rates (DR). These feature are also in agreement with direct vorticity measurements using acoustic scattering techniques [5].

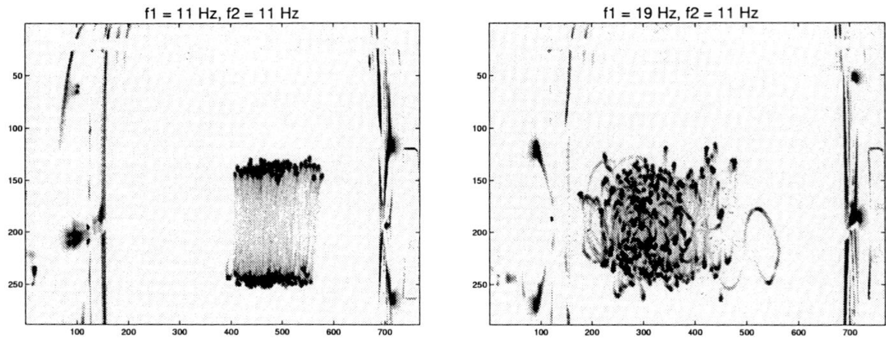

Fig. 4. Motion of a sphere, viewed from the superposition of 250 frames – 10 seconds, i.e. about 100 rotations. Diameter 2 mm and density 0.9 polypropylen sphere. (**a**) $f_1 = f_2 = 11$ Hz, (GR) regime; (**b**) $f_1 = 19$ Hz and $f_2 = 11$ Hz, (DR) regime. Gravity does not influence the axial motion (the rotation axis is horizontal) or the orthoradial velocity (high rotation rate)

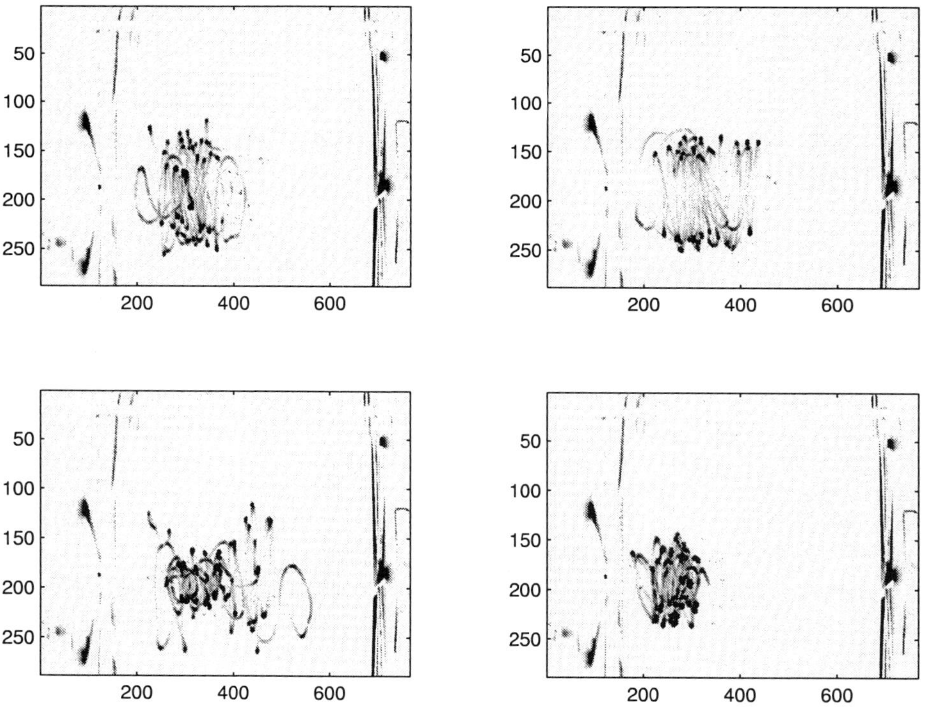

Fig. 5. Four successive sequences extracted from the average sequence in Fig. 4 b, each one 2 seconds long

3 Local Velocity Measurements Using a Hot-Film Probe in Air

While providing a topology of the flow field, the LDV measurements have a rather low resolution in time, so that they must be supplemented by a faster technique in order to describe the (small scales) fluctuations in this flow. We describe here hot-wire measurements, in air – the driving disks are in this case fitted with a set of vertical blades to increase entrainment. Their radius R is 9.85 cm, the distance H between them is 30 cm, and the tank radius is 11.7 cm.

Figures 6 and 7 show averaged velocity profiles of the azimuthal and meridian velocity component, measured at mid-distance between the disks. The profiles are obtained from single probe measurement, with the hot-wire aligned either parallel to the axis of rotation or perpendicular to it. It measures the azimuthal velocity $V_{d\theta} = \sqrt{u_d^2 + u_\theta^2}$ when it is set parallel to the axis, and the meridian velocity $V_{dz} = \sqrt{u_d^2 + u_z^2}$ in the other case.

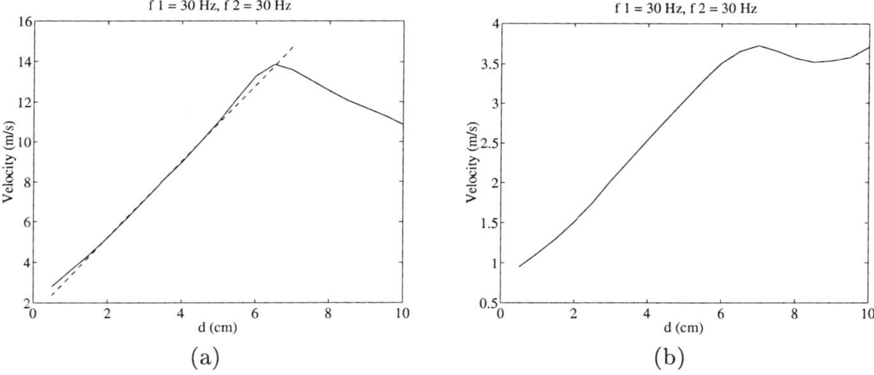

Fig. 6. $f_1 = f_2 = 30$ Hz. (**a**) Azimuthal velocity. (**b**) Meridian velocity component. In the first case the wire measures essentially the V_θ component of velocity for $d < 6$ cm

In the (GR) case ($f_1 = f_2 = 30$ Hz, Fig. 6) we observe that the meridian velocity components are much smaller than the azimuthal, even though not null: rotation is favoured and the azimuthal profile between $d = 2$ cm and $d = 5$ cm is linear in d. The slope computed by a best fit corresponds to a frequency equal to that of the disks, as in a solid body rotation.

This is not so in (DR) ($f_1 = 12$ Hz, $f_2 = 40$ Hz, see Fig. 7). Near the axis the pumping effect induced by the fastest disk causes the meridian velocity to be higher than the azimuthal, and essentially equal to V_z.

These observations are in qualitative agreement with the LDV measurements. However, we must report that introducing a probe into the flow may generate significant perturbations. Visualisations in water show that the vortex is deformed by the probe proximity: it bends to avoid the probe tip. Similar visualisations

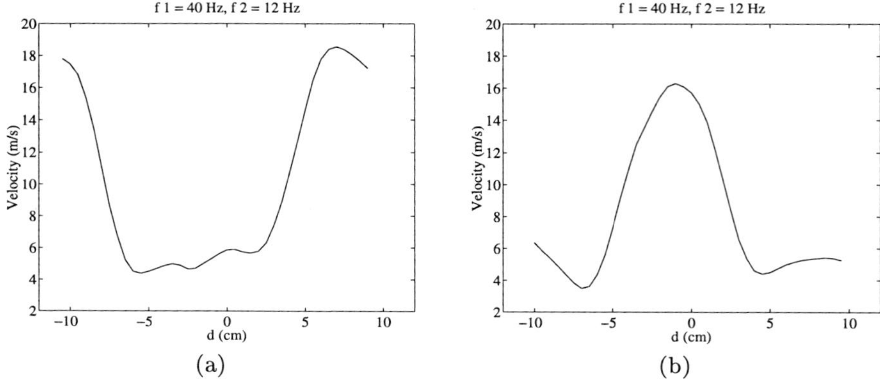

Fig. 7. $f_1 = 40$ Hz, $f_2 = 12$ Hz. (**a**) Azimuthal velocity. (**b**) Meridian velocity component. Measurement is obtained in the same way as in the preceeding figure, the probe being now translated over the entire diameter of the tank. Remind that the hot-wire measures the absolute value of velocity

using smoke tend to show that this effect is less important in air, but this could explain that $V_{d\theta}$ is not null in the center.

Finally, we consider the possible interactions between the vortex structure, its dynamics, and the velocity fluctuations at small scales. Several studies show that the statistical properties of turbulence are influenced by the proximity of a vortex [6–8].

We obtain the evolution for power spectra of azimuthal velocity as a fonction of distance to the rotation axis represented Fig. 8.

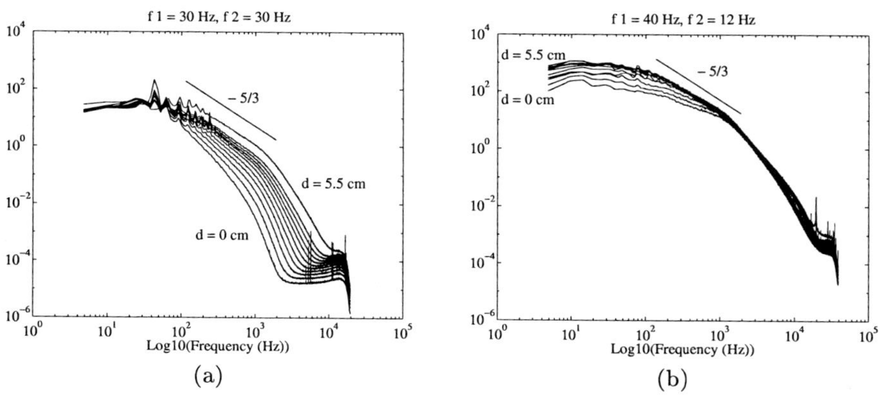

Fig. 8. Azimuthal velocity temporal fluctuations spectra; evolution in fonction of the distance d between the probe and the rotation axis. (**a**) $f_1 = f_2 = 30$ Hz. (**b**) $f_1 = 40$ Hz, $f_2 = 12$ Hz

Very strong differences in behavior are observed in different regions of the flow and for the two flow regimes. Indeed, when the rotation velocities of the disks are equal (GR), the spectra slope varies continuously from - 5/3 to nearly -3 when one translates the probe towards the rotation axis. On the other hand, for the (DR) regime, the spectral slopes vary from - 5/3 to about -1 when one brings the probe nearer to the rotation axis. We thus observe that the flow is strongly spatially inhomogeneous in regards to the intensity of velocity fluctuations and that it is influenced by the large scale dynamics.

This also appears in the energy transfers as estimated by the skewness of velocity increments $S_3(r) = \langle (u(x+r) - u(x))^3 \rangle$ (Fig. 9). Note that in this calculation, we have resampled the temporal data using a local Taylor hypothesis [9,10], in order to obtain spatial series.

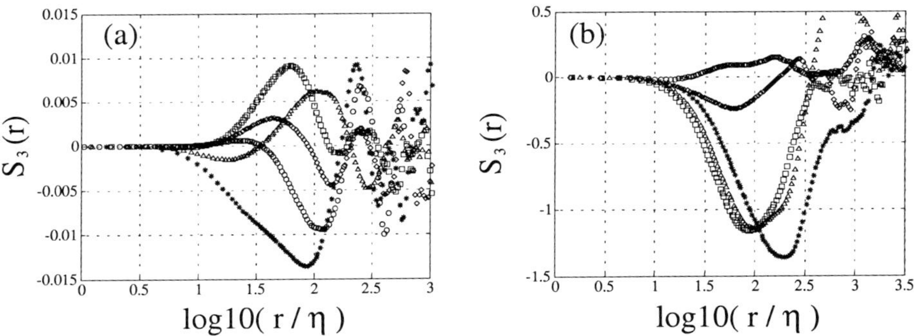

Fig. 9. Skewness of the increments of azimuthal velocity at distances d = 0.5(○); 1.5(◇); 2.5(⊔); 3.5(△) and 4.5(⋆) cm from the axis. (**a**) $f_1 = f_2 = 30$ Hz. (**b**) $f_1 = 40$ Hz, $f_2 = 12$ Hz

In the (GR) case, we observe that the skewness is very small and displays an unusual positive sign when measured close to the axis. It thus seems that the presence of stable vorticity and rotation reduces the activity of the turbulent cascade, with possible inverse transfers to the large scales. This matter certainly deserves further investigations.

On the contrary, in the DR case, we observe a direct energy cascade, with magnitude comparable to what is usually measured in homogeneous turbulence.

Finally, we emphasize that, in both regimes, the small scales velocity fluctuations are intermittent, in the sense that the probability density fonctions of velocity increments vary with the size of the increment (Fig. 10).

In the (GR) regime, both the ω^{-3} spectra behaviour in the inertial range and the motion of embedded particles point to a bidimensionalisation of the flow, but intermittency is observed: this is a sharp difference with 2D turbulence where no intermittency is detected [11].

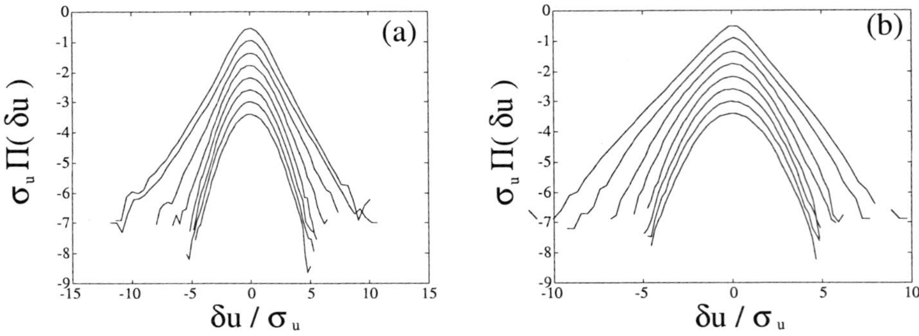

Fig. 10. Probability density fonctions of azimuthal velocity increments, at distance $d = 2.5$ cm from the axis. Values of the velocity have been normalised by the *rms* fluctuation level. For an easy reading, curves had been translate. (**a**) $f_1 = f_2 = 30$ Hz and $r/\eta = 1, 4, 19, 44, 104, 244, 571, 1336$ with $\eta = 165\mu$m. (**b**) $f_1 = 40$ Hz, $f_2 = 12$ Hz and $r/\eta = 1, 7, 31, 73, 172, 401, 941, 2199$ with $\eta = 49\mu$m

These preliminary studies show that the confined von Kármán flow is remarkably rich; it provides a variety of vortex geometries and dynamics. In each case, the large scale characteristics strongly influence the turbulent features, such as the distribution of fluctuations intensities along the scales (power spectra), the balance of energy transfers (third order structure function) and intermittency.

We would like to thank F. Daviaud and J. Burguete from CEA for their collaboration for the LDV measurements, and N. Mordant for many useful discussions.

References

1. P.J. Zandbergen, D. Dijkstra: Ann. Rev. Fluid Mech. **19**, 465 (1987)
2. R. Labbé, J.F. Pinton, S. Fauve: Phys. Fluids **8**, 914 (1996)
3. J.F. Pinton, F. Chillà, N. Mordant: Eur. J. Mech. B/Fluids **17**, 535 (1998)
4. R. Wunenburger, B. Andreotti, P. Petitjeans: Exp. Fluids **27**, 181 (1999)
5. N. Mordant, M. Baudry, *Undergraduated report*, (1997)
6. F. Chillà, J.F. Pinton, R. Labbé: Europhys. Lett. **34**, 271 (1996)
7. B. Andreotti, J. Maurer, Y. Couder, S. Douady: Eur. J. Mech. B/Fluids **17**, 451 (1998)
8. E. Gaudin, B. Protas, S. Goujon-Durand, J. Wojciechowski, J.E. Wesfreid: Phys. Rev. E **57**, R9 (1998)
9. J.F. Pinton, R. Labbé: J. Phys. II France **17**, 1461 (1994)
10. Y. Malécot: *PhD thesis*, Université de Grenoble 1, (1998)
11. J. Paret, P. Tabeling: Phys. Fluids **10**, 3126 (1998)

A Criterion of Centrifugal Instabilities in Rotating Systems

Denis Sipp and Laurent Jacquin

ONERA, 29 avenue de la Division Leclerc, BP 72, F-92322 Chatillon CEDEX, France

Abstract. This article deals with the stability of incompressible inviscid 2D planar flows in a rotating frame. We give a sufficient condition for such flows to undergo 3D short-wave centrifugal-type instabilities. This criterion states that a steady 2D planar flow subject to rotation Ω is unstable if there exists a streamline for which at each point $2(V/\mathcal{R}+\Omega)(\omega+2\Omega)<0$ where ω is the vorticity of the streamline, $\mathcal{R}$ is the local algebraic radius of curvature of the streamline and V is the local norm of the velocity. If this condition is satisfied then the flow is unstable to short-wavelength perturbations. When the streamlines are closed, it is further shown that a localized unstable normal mode can be constructed in the vicinity of a streamline.

1 Introduction

In this article, we give a simple criterion of centrifugal instability for inviscid, incompressible, non-axisymmetric, 2D planar flows in rotating systems. This criterion is a sufficient condition for instability or equivalently a necessary condition for stability. It is in accordance with the ones that already exist in some restrictive situations. For parallel shear flows in rotating systems, it amounts to the Pedley [1] or Bradshaw-Richardson [2] criterion. For axisymmetric flows in inertial frames, it yields the Rayleigh centrifugal criterion [3]. With axisymmetric flows in rotating frames, we obtain the criterion given by Kloosterziel & van Heijst [4] and Mutabazi *et al.* [5] And, with general 2D planar flows in inertial frames, we obtain Bayly's criterion [6]. The issue on curvature and rotation induced instabilities has also been addressed by Matsson & Alfredsson [7] in the context of channel flows.

The proof of linear instability will be given by means of two different approaches. A first proof is based on the geometrical optics method introduced by Lifschitz & Hameiri [8]. This method, roughly speaking, reduces to following a wave-packet along the flow trajectories, using a WKB approximation. This wave-packet is a localized shortwave perturbation characterized by a wavevector $\mathbf{k}$ and an amplitude vector $\mathbf{a}$. The flow is unstable if there exists a streamline on which the amplitude $\mathbf{a}(t)$ of a particular wave-packet grows unboundedly as $t\to\infty$. A second proof of instability will be given in terms of unstable normal modes. If the flow lies in the (x,y) plane and the rotation axis is parallel to the z axis, these normal modes are sought under the form $[\mathbf{u}',p'](x,y,z,t)=\exp(ikz+st)[\tilde{\mathbf{u}},\tilde{p}](x,y)$ where k is the vertical wavenumber and s the complex amplification rate. The flow is unstable if a normal mode exhibits a complex amplification rate s with a positive real part. It has been shown in the case of a Taylor-Green flow - which is an infinite 2D array of counter-rotating vortices [9] - subject to rotation, that these two approaches are consistent[10,11]. In particular, elliptic and centrifugal-type normal modes have been identified whose characteristics (spatial structure and eigenvalues) are in accordance with the results given by the geometrical optics method.

This article is an extension to rotating frames of a paper due to Bayly [6]. It is also a continuation of the works of Cambon *et al.* [12] and Leblanc & Cambon [13] in the search of a generalized criterion for the stability of two-dimensional flows in planes perpendicular to the rotation axis. Leblanc & Cambon [13] made a first attempt by suggesting that this criterion could be the local negativity of the second invariant of the inertial tensor somewhere in the flow. This effectively accounts for the special cases of Bradshaw-Richardson (or Pedley), Rayleigh and Kloosterziel & van Heijst, but it fails with Bayly's result. The reason lies in the fact that this criterion is only valid if the norm of the velocity V is constant along a streamline, which is not true in the general case.

Our criterion, which is not subject to these limitations, is used in Ref. [14] to study the centrifugal-type instabilities in the Stuart vortices [15], which is a family of exact solutions describing a row of periodic co-rotating eddies.

2 A sufficient criterion for centrifugal instability

We consider an inviscid incompressible flow in a frame rotating at a constant angular velocity $\Omega = \Omega\hat{\mathbf{z}}$, where $\hat{\mathbf{z}}$ is the unit vector in the positive z direction. We use the Euler equations referred to the rotating frame, so that a Coriolis term $2\Omega \times \mathbf{u}$ has to be considered. The relative motion of the basic flow $\mathbf{u}$ is supposed to be two-dimensional and to lie in a plane perpendicular to $\hat{\mathbf{z}}$, so that there exists a streamfunction $\psi(x,y)$ such as $\mathbf{u}(x,y) = \nabla \times [\psi(x,y)\hat{\mathbf{z}}]$ (see figure 1). The basic flow is a steady solution of the 'modified' Euler equations:

$$\mathbf{u} \cdot \nabla \mathbf{u} + 2\Omega \times \mathbf{u} = -\nabla p, \qquad \nabla \cdot \mathbf{u} = 0 \tag{1}$$

The relative vorticity, $\omega = \nabla \times \mathbf{u} = \omega\hat{\mathbf{z}}$ where $\omega = -\nabla^2\psi$, is therefore constant on each streamline ψ and is a function of ψ, $\omega = \omega(\psi)$. In the following, we will consider the local algebraic radius of curvature $\mathcal{R}$ at a given point of a streamline. This quantity is defined at all points of a particle trajectory $[\mathbf{x}(t) = x(t)\hat{\mathbf{x}} + y(t)\hat{\mathbf{y}}$ where $\mathbf{x}' = d\mathbf{x}/dt = \mathbf{u}]$ as:

$$\frac{V^3}{(\nabla\psi) \cdot [(\nabla \mathbf{u})\mathbf{u}]} \tag{2}$$

where $V(x,y) = |\mathbf{u}(x,y)|$ is the norm of the velocity field. Hence, $\mathcal{R} > 0$ if the flow is locally counter-clockwise and $\mathcal{R} < 0$ if the flow is locally clockwise.

Let us introduce the following quantities:

$$\delta(\mathbf{x}) = 2\left(\frac{V}{\mathcal{R}} + \Omega\right)(\omega + 2\Omega) \tag{3}$$

$$\Delta(\psi) = \max_{\psi} \delta(\mathbf{x}) \tag{4}$$

where $\max_\psi$ denotes the maximum over the streamline ψ. In the literature, $\delta(x)$ is referred to as Rayleigh's discriminant.

The flow is unstable if there exists ψ_0 such as:

$$\Delta(\psi_0) < 0 \tag{5}$$

It will be shown below that instability is achieved both with respect to the geometrical optics method (see section 4) and with respect to a classical normal mode analysis (see section 5). The perturbations $(\mathbf{u}', p')$ are 3D in both cases and are governed by the linearized 'modified' Euler equations:

$$\frac{\partial \mathbf{u}'}{\partial t} + \mathbf{u} \cdot \nabla \mathbf{u}' + \mathbf{u}' \cdot \nabla \mathbf{u} + 2\Omega \times \mathbf{u}' = -\nabla p', \qquad \nabla \cdot \mathbf{u}' = 0 \tag{6}$$

The criterion $\Delta(\psi) < 0$ compares the sign of the absolute angular velocity of the particle $V/\mathcal{R} + \Omega$ to the sign of its absolute vorticity $\omega + 2\Omega$. If these two quantities have opposite signs along a whole streamline ψ_0, then the flow is unstable. The quantity $\delta(\mathbf{x})$ is a polynom of the second order in Ω. The two roots of this polynom, $-V/\mathcal{R}$ and $-\omega/2$, have a special importance since if Ω is in the interval bounded by the two roots, then $\delta(\mathbf{x}) < 0$. The two roots are linked since:

$$\frac{V}{\mathcal{R}} = \frac{\omega}{2} + \mathcal{S} \tag{7}$$

where $\mathcal{S} = \mathbf{t} \cdot \left[1/2(\mathcal{L} + \mathcal{L}^T)\mathbf{n}\right]$. Here $\mathcal{L} = \nabla \mathbf{u}$ designates the velocity gradient tensor. The term $\mathcal{S}$ is therefore the non-diagonal term of the symmetric part of the velocity gradient tensor in the Serret-Frenet vector basis ($\mathbf{t} = \mathbf{u}/V, \mathbf{n} = \nabla\psi/V$). So, it can be referred to as the intrinsic shear. Equation (7) shows that the relative angular velocity of a fluid particle $V/\mathcal{R}$ is the sum of two terms. The first one, $\omega/2$, is due to the vorticity of the basic flow and the second one, $\mathcal{S}$, represents the angular velocity due to the shear. The criterion $\Delta(\psi) < 0$ can be satisfied only if the angular velocity Ω lies at each point of the streamline ψ between $-V/\mathcal{R}$ and $-\omega/2$. This requires that the intrinsic shear $\mathcal{S} = V/\mathcal{R} - \omega/2$ be either strictly positive or strictly negative along a whole streamline ψ. The physical mechanism of instability can therefore be traced back to the stretching of the vorticity of the perturbation by the intrinsic shear $\mathcal{S}$.

3 Special cases

In the particular case where the streamlines are circular, the criterion $\Delta(\psi) < 0$ is identical to the one given by Kloosterziel & van Heijst [4] and Mutabazi *et al.* [5]:

$$2\left(\frac{U(r)}{r} + \Omega\right)(\omega(r) + 2\Omega) < 0 \text{ for some radius } r_0 \tag{8}$$

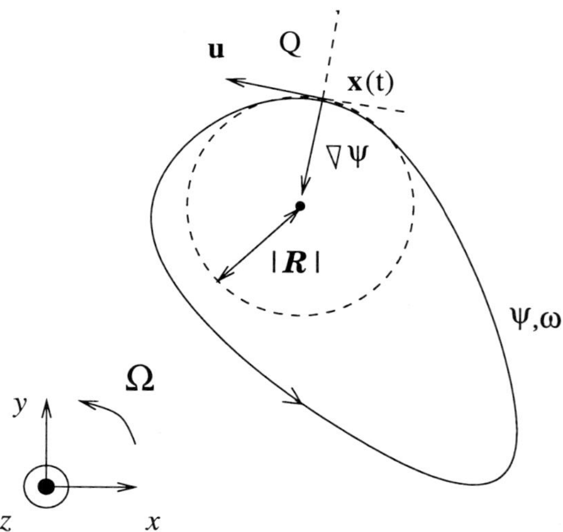

Fig. 1. Flow in the (x, y) plane. One streamline ψ is depicted. The flow is here assumed to be counterclockwise, so that $\nabla\psi$ and **u** have the disposition indicated. We have also shown the center of curvature, the radius of curvature $|\mathcal{R}|$, the quadrant **Q**.

where $U(r)$ is the orthoradial velocity of the basic flow. If in addition the background rotation is zero, then we obtain Rayleigh's [3] criterion:

$$\frac{U(r)}{r}\omega(r) < 0 \text{ for some radius } r_0 \tag{9}$$

If the radius of curvature $\mathcal{R}$ tends to infinity, we are led to the Pedley [1] or Bradshaw-Richardson [2] criterion, which gives sufficient conditions for a parallel shear flow $U(y)$ to be destabilized:

$$2\Omega\left(2\Omega - \frac{dU}{dy}\right) < 0 \text{ for some } y_0 \tag{10}$$

In an inertial frame, we also get Bayly's [6] result which applies to general non-axisymmetric basic flows with convex streamlines: "sufficient conditions for centrifugal instability are that streamlines be convex closed curves in some region of the flow, with the magnitude of the circulation decreasing outward". In our formalism, this can be restated in the following manner:

$$\frac{V}{\mathcal{R}}\omega < 0 \text{ on a whole streamline } \psi_0 \tag{11}$$

4 Proof of instability by means of the geometrical optics method

4.1 Presentation of the geometrical optics method

Following Lifschitz & Hameiri [8], the basic flow field $[\mathbf{u}(\mathbf{x}), p(\mathbf{x})]$ is perturbed by the following WKB-form velocity/pressure field:

$$\begin{bmatrix} \mathbf{u}' \\ p' \end{bmatrix} (\mathbf{x}, t) = \exp\left[i\phi(\mathbf{x}, t)/\epsilon\right] \left\{ \begin{bmatrix} \mathbf{a} \\ \pi \end{bmatrix} (\mathbf{x}, t) + \epsilon \begin{bmatrix} \mathbf{a}_\epsilon \\ \pi_\epsilon \end{bmatrix} (\mathbf{x}, t) \right\} + \epsilon \begin{bmatrix} \mathbf{u}_r \\ \pi_r \end{bmatrix} (\mathbf{x}, t) \quad (12)$$

where ϵ is a small parameter and ϕ is a phase field which is real. In the following, $\mathbf{k} = \nabla\phi$ is the wavevector of the perturbation field. This perturbation is substituted into the incompressible linearized Euler equations (6). The resulting equations contain terms of various orders in ϵ. Equating the lowest-order terms yields $\pi = 0$, the incompressibility condition $\mathbf{k} \cdot \mathbf{a} = 0$ and the eikonal equation:

$$(\partial_t + \mathbf{u} \cdot \nabla)\, \phi = 0 \quad (13)$$

i.e. the phase field ϕ is passively advected. The next-lowest-order terms yield the evolution equation for the velocity envelope function $\mathbf{a}$:

$$(\partial_t + \mathbf{u} \cdot \nabla)\, \mathbf{a} + \mathcal{L}(\mathbf{x})\mathbf{a} + 2\Omega \times \mathbf{a} = -i\mathbf{k}\pi_\epsilon \quad (14)$$

Projecting this equation in the plane perpendicular to $\mathbf{k}$, i.e. multiplying this equation by the operator $\mathcal{I} - (\mathbf{k} \otimes \mathbf{k})/|\mathbf{k}|^2$ where $\mathcal{I}$ is the identity tensor and $\otimes$ the tensor product, yields:

$$(\partial_t + \mathbf{u} \cdot \nabla)\, \mathbf{a} = \left(\frac{2\mathbf{k} \otimes \mathbf{k}}{|\mathbf{k}|^2} - \mathcal{I} \right) \mathcal{L}(\mathbf{x})\mathbf{a} + \left(\frac{\mathbf{k} \otimes \mathbf{k}}{|\mathbf{k}|^2} - \mathcal{I} \right) (2\Omega \times \mathbf{a}) \quad (15)$$

Equations (13) and (15) form a system evolving locally along particle trajectories. We can therefore write them as:

$$\frac{d\mathbf{x}}{dt} = \mathbf{u}(\mathbf{x}) \quad (16)$$

$$\frac{d\mathbf{k}}{dt} = -\mathcal{L}^T(\mathbf{x})\mathbf{k} \quad (17)$$

$$\frac{d\mathbf{a}}{dt} = \left(\frac{2\mathbf{k} \otimes \mathbf{k}}{|\mathbf{k}|^2} - \mathcal{I} \right) \mathcal{L}(\mathbf{x})\mathbf{a} + \left(\frac{\mathbf{k} \otimes \mathbf{k}}{|\mathbf{k}|^2} - \mathcal{I} \right) (2\Omega \times \mathbf{a}) \quad (18)$$

Here $\mathbf{x}(t)$ is the position at time t of a fluid particle and the superscript T is the transpose. Lifschitz & Hameiri [8] proved that a sufficient criterion for instability is that this system has at least one solution for which the amplitude $\mathbf{a}(t)$ increases unboundedly as $t \to \infty$.

4.2 Proof of instability

In this section, we prove that if there exists a streamline ψ_0 such as $\Delta(\psi_0) < 0$, then there exists a wave-packet for which the amplitude $\mathbf{a}(t)$ increases unboundedly as $t \to \infty$.

The condition $\Delta(\psi_0) < 0$ can be satisfied in the two following cases:

$$\Omega + \omega(\psi_0)/2 < 0 \text{ and } \Omega + \min_{\psi_0} V/\mathcal{R} > 0 \tag{19}$$

or

$$\Omega + \omega(\psi_0)/2 > 0 \text{ and } \Omega + \max_{\psi_0} V/\mathcal{R} < 0 \tag{20}$$

where $\min_{\psi_0}$ and $\max_{\psi_0}$ designate the minimum and maximum over the streamline ψ_0. Only the first case (19) will be treated here, the demonstration of the other case being similar. The proofs follow the treatment given by Bayly [6]. Here, we extend his formalism to the rotating case.

Let us consider spanwise perturbations where the wavevector is perpendicular to the flow field ($\mathbf{k} \parallel \hat{\mathbf{z}}$). Equation (18) therefore reduces to:

$$\frac{d\mathbf{a}}{dt} = -\mathcal{L}(\mathbf{x})\mathbf{a} - 2\Omega \times \mathbf{a} \tag{21}$$

This equation yields the following equation for the horizontal component of $\mathbf{a}$:

$$\frac{d}{dt}\begin{pmatrix} \mathbf{a}\cdot\mathbf{u} \\ \mathbf{a}\cdot\nabla\psi \end{pmatrix} = \begin{pmatrix} 0 & \omega + 2\Omega \\ -2(V/\mathcal{R} + \Omega) & 2V'/V \end{pmatrix}\begin{pmatrix} \mathbf{a}\cdot\mathbf{u} \\ \mathbf{a}\cdot\nabla\psi \end{pmatrix} \tag{22}$$

where $V' = dV/dt = \mathbf{u}\cdot\nabla V$. The key idea to obtain (22) is to use the relation $(\nabla\psi)\cdot(\mathcal{L}\mathbf{u}) = V^3/\mathcal{R}$ and the fact that the antisymmetric part of the velocity gradient tensor $(\mathcal{L} - \mathcal{L}^T)/2$ equals the operator $(\omega/2) \times ()$ where $\times$ stands for the cross product.

At each point $\mathbf{x}$ of the streamline ψ_0, the vectors $\mathbf{u}$ and $\nabla\psi$ divide the plane into four quadrants. The quadrant $\mathbf{Q}(\mathbf{x})$ consists of all vectors $\mathbf{v}$ satisfying $\mathbf{v}\cdot\mathbf{u} > 0$ and $\mathbf{v}\cdot\nabla\psi < 0$ (see figure 1). We first prove that if $\mathbf{a}(t=0)$ lies in the quadrant $\mathbf{Q}(t=0)$, then $\mathbf{a}(t)$ remains in the quadrant $\mathbf{Q}(t)$ for all $t > 0$. In other words, we show that if $\mathbf{a}\cdot\mathbf{u} > 0$ and $\mathbf{a}\cdot\nabla\psi < 0$ hold at $t = 0$ then they continue to hold for all $t > 0$. To prove this, suppose that this were to be violated. So there must be some time $t > 0$ when either

$$\mathbf{a}\cdot\nabla\psi = 0 \text{ with } \mathbf{a}\cdot\mathbf{u} > 0 \text{ and } (d/dt)(\mathbf{a}\cdot\nabla\psi) \geq 0 \tag{23}$$

or

$$\mathbf{a}\cdot\mathbf{u} = 0 \text{ with } \mathbf{a}\cdot\nabla\psi < 0 \text{ and } (d/dt)(\mathbf{a}\cdot\mathbf{u}) \leq 0 \tag{24}$$

must hold. We show that neither of these two assumptions can occur.

First, suppose that $\mathbf{a} \cdot \nabla\psi = 0$ occurred with $\mathbf{a} \cdot \mathbf{u} > 0$ for some time t. Therefore $\mathbf{a} = c\mathbf{u}$ for some positive number c. Thus, from (22):

$$\frac{d}{dt}(\mathbf{a} \cdot \nabla\psi) = -2cV^2 \left(V/\mathcal{R} + \Omega\right) \tag{25}$$

Now, assumption (19) yields $V/\mathcal{R} + \Omega > 0$ along ψ_0, so that $(d/dt)(\mathbf{a} \cdot \nabla\psi) < 0$ which contradicts (23).

Secondly, suppose that $\mathbf{a} \cdot \mathbf{u} = 0$ with $\mathbf{a} \cdot \nabla\psi < 0$ for some time t. Now, (22) yields:

$$\frac{d}{dt}(\mathbf{a} \cdot \mathbf{u}) = (\omega + 2\Omega)(\mathbf{a} \cdot \nabla\psi) \tag{26}$$

so that $(d/dt)(\mathbf{a} \cdot \mathbf{u}) > 0$, since $\omega + 2\Omega < 0$ (assumption (19)). This contradicts (24).

Now, using the fact that the amplitude $\mathbf{a}(t)$ remains at all times in the quadrant $\mathbf{Q}(t)$, it can be shown that it grows exponentially. Consider the angle $\theta(t)$ between $\mathbf{u}(\mathbf{x}(t))$ and $\mathbf{a}(t)$. Since (21) is linear in $\mathbf{a}$, θ satisfies a first-order ordinary differential equation depending only on θ itself and $\mathbf{x}(t)$. Equation (25) shows that $d\theta/dt(\theta = 0)$ $[= 2\left(V/\mathcal{R} + \Omega\right)]$ is separated by a finite amount from zero. If the streamline ψ_0 is closed, it can be deduced that there is a finite angle $\theta_c > 0$ such as $d\theta/dt(\theta = \theta_c)$ is greater than or equal to zero on the whole streamline ψ_0. This fact remains true in the case of parallel shear flows where $\mathcal{R} \to \infty$ and $V = cst$. So if $\theta(t = 0) \geq \theta_c$, then $\theta(t) \geq \theta_c$. This implies, using (26), that:

$$\mathbf{a}(t) \cdot \mathbf{u}(\mathbf{x}(t)) \geq \exp\left[-t\ (\omega + 2\Omega) \tan\theta_c\right] \left[\mathbf{a}(t = 0) \cdot \mathbf{u}(\mathbf{x}(t = 0))\right] \tag{27}$$

In the case of a closed streamline ψ_0, $\mathbf{u}(\mathbf{x}(t))$ is periodic with a period $T(\psi_0)$, so that $\mathbf{a}\left(nT(\psi_0)\right)$ grows exponentially and monotically as $t \to \infty$. The same conclusion, i.e. the exponential growth of $\mathbf{a}(t)$ as $t \to \infty$, is reached in the case of parallel shear flows. The flow is therefore unstable with respect to the geometrical optics method.

The fact that $\mathbf{a}(t)$ always remains in the quadrant $\mathbf{Q}(t)$ for all times means that the perturbation exchanges fluid from the region $\psi < \psi_0$ with fluid from the region $\psi > \psi_0$. We thus retrieve one of the basic features of centrifugal instability.

5 Proof of instability by means of a normal mode analysis

We consider closed streamlines ψ with $\Delta(\psi) < 0$. We show that a localized normal mode can be constructed in the neighborhood of a streamline. We follow Bayly's formalism[6] and extend it to the rotating case which is considered here. Normal modes are sought in the usual way by considering a vertical wavelength k and a complex amplification rate s:

$$\begin{bmatrix} \mathbf{u}' \\ p' \end{bmatrix} (x, y, z, t) = \exp(ikz + st) \begin{bmatrix} \tilde{\mathbf{u}} \\ \tilde{p} \end{bmatrix} (x, y) \tag{28}$$

The main idea is to use a particular vector field basis $\mathbf{f}_i$ for the representation of the eigenmode $\tilde{\mathbf{u}}(x,y)$:

$$\tilde{\mathbf{u}}(x,y) = \tilde{u}(x,y)\mathbf{f}_1(x,y) + \tilde{v}(x,y)\mathbf{f}_2(x,y) + \tilde{w}(x,y)\mathbf{f}_3(x,y) \tag{29}$$

$\mathbf{f}_i$ is constructed using the eigenvalues m_i and the eigenvectors $\mathbf{e}_i$ of the fundamental Floquet matrix associated to the differential equation (21). This matrix is $\mathcal{A}(T(\psi))$ where $T(\psi)$ refers to the time-period on the streamline ψ and $\mathcal{A}(t)$ is obtained from:

$$\frac{d\mathcal{A}}{dt} = -\mathcal{L}(\mathbf{x})\mathcal{A} - 2\Omega \times \mathcal{A}, \qquad \mathcal{A}(0) = \mathcal{I} \tag{30}$$

Since the basic flow lies in the (x,y) plane, the eigenvalues/eigenvectors $(m_i, \mathbf{e}_i)$ can be taken as follows: $(m_3 = 1, \mathbf{e}_3 = \hat{\mathbf{z}})$ and $(\mathbf{e}_1, \mathbf{e}_2)$ lies in the (x,y) plane. ¿From the incompressibility of the basic flow, the two corresponding complex eigenvalues (m_1, m_2) where $|m_1| \geq |m_2|$ must multiply to $1 : m_1 m_2 = 1$. As $\Delta(\psi) < 0$, there exists in equation (21) a perturbation whose amplitude $\mathbf{a}(t)$ exponentially increases as $t \to \infty$ (see section 4.2). This can be restated here by saying that m_1 and m_2 are real and reciprocals, so that $|m_1| > 1$ and $0 < |m_2| < 1$. The Floquet exponents, which are the natural logarithms of $|m_i|$ divided by the period $T(\psi)$ are therefore as follows: $s_1 = \sigma$, $s_2 = -\sigma$ and $s_3 = 0$ where $\sigma > 0$. The vector fields $\mathbf{f}_i$ are then defined in the following way: $\mathbf{f}_i(\mathbf{x}) = \exp\left[-s_i t\right] \mathcal{A}(t)\mathbf{e}_i$ where $\mathbf{x}(t)$ is a particle trajectory. From section 4.2, it follows that $\mathbf{f}_1$ lies in the quadrant $\mathbf{Q}(\mathbf{x})$. Considering σ as a function of ψ, we shall further assume that $\sigma(\psi)$ takes a quadratic maximum on a given streamline ψ_0, i.e. $\sigma'(\psi_0) = 0$ and $-\sigma''(\psi_0) > 0$. This streamline, where $\sigma(\psi_0) > 0$, turns out to be the one in whose neighborhood we can construct a localized instability.

In the limit $k \to \infty$, the eigenmodes are sought with the following asymptotic behavior in k:

$$\tilde{u} = \mathcal{U}, \; \tilde{v} = k^{-1}\mathcal{V}, \; \tilde{w} = k^{-1/2}\mathcal{W}, \tilde{p} = k^{-3/2}\mathcal{P} \text{ and } s = \sigma(\psi_0) - \mu/k \tag{31}$$

where $\tilde{u}$, $\tilde{v}$ and $\tilde{w}$ have been introduced in (29) and μ is a constant to be determined. Hence, as $k \to \infty$, the amplification rate s of the constructed eigenmode (28) converges towards the predicted maximum value of the geometrical optics method $\sigma(\psi_0)$. The solution (31) is localized within a region of width $O(k^{-1/2})$ around the streamline ψ_0. The following rescaled streamfunction coordinate is therefore considered: $\eta = k^{1/2}(\psi - \psi_0)$, so that $\mathcal{U}, \mathcal{V}, \mathcal{W}, \mathcal{P}$ are now functions of η. Introducing these expansions in the linearized Euler equations, we are led to the quantum harmonic oscillator[13,16]:

$$\frac{d^2\mathcal{U}}{d\eta^2} + \left(\frac{\mu}{C(\psi_0)} - \lambda^2\eta^2\right)\mathcal{U} = 0 \text{ with } \mathcal{U}(\pm\infty) = 0 \tag{32}$$

where:

$$C(\psi_0) = \frac{1}{T(\psi_0)} \int_0^{T(\psi_0)} (\mathbf{f}_1^{\dagger} \cdot \nabla\psi) \left[\sigma(\psi_0) + \frac{d}{dt}\right] (\mathbf{f}_1 \cdot \nabla\psi) \, \mathrm{dt} \tag{33}$$

$$\lambda^2 = -\frac{\sigma''(\psi_0)}{2C(\psi_0)} \tag{34}$$

In these equations, $\mathbf{f}_i^\dagger(x, y)$ is the adjoint vector field corresponding to $\mathbf{f}_i(x, y)$: $\mathbf{f}_i^\dagger(x, y) \cdot \mathbf{f}_j(x, y) = \delta_{ij}$.

Now, the condition $\Delta(\psi_0) < 0$ implies that $C(\psi_0) > 0$. This can be shown by combining the proof given in Bayly's paper [6] and the modifications introduced in section 4.2 to cope with the rotating frame. At last, $-\sigma''(\psi_0)$ is positive by assumption, so that $\lambda^2 > 0$. Equation (32) has therefore the following solution:

$$\mathcal{U}(\eta) = \exp\left(-\frac{\lambda\eta^2}{2}\right) \tag{35}$$

with $\mu = \lambda C(\psi_0)$. This eigenmode is exponentially concentrated in the neighborhood of the streamline ψ_0 on a characteristic length scale $1/\sqrt{\lambda}$. It's amplification rate is $s = \sigma(\psi_0) - \mu/k$ with $\sigma(\psi_0) > 0$. The flow is therefore unstable with respect to a classical normal mode analysis.

¿From a physical point of view, the constructed unstable normal mode again reflects the basic physics of the centrifugal instability. The localized modes take the form of highly elongated eddies. The fluid motion in these instabilities is predominantly in the horizontal plane along the unstable $\mathbf{f}_1$ direction, which belongs to the quadrant $\mathbf{Q}(\mathbf{x})$. This again shows that an exchange of fluid occurs between the region $\psi < \psi_0$ and the region $\psi > \psi_0$. This exchange is modulated in the z direction by the term $\exp(ikz)$. In order to preserve exact incompressibility, a pressure field and a z velocity component are also present in the perturbation, but at higher order in k^{-1}. This reflects the pressure-less nature of the centrifugal instability.

6 Conclusion

In this paper, we have presented a new criterion which gives a sufficient condition for centrifugal-type instabilities to occur in a general inviscid steady two-dimensional flow. This criterion is a generalization of the Rayleigh, Kloosterziel & van Heijst, Mutabazi *et al.*, Pedley, Bradshaw-Richardson and Bayly's criteria. The proof of instability has been given using both the geometrical optics approach and a classical normal mode analysis. This criterion shows that a 2D planar basic flow may undergo centrifugal instability only if its intrinsic shear is either positive or negative along a whole streamline. The related perturbations are shown to induce inward/outward exchanges of fluid in a vortex, the fluid motion lying at first order in the horizontal plane.

References

1. T.J. Pedley, "On the stability of viscous flow in a rapidly rotating pipe," J. Fluid Mech. **35**, 97 (1969).
2. P. Bradshaw, "The analogy between streamline curvature and buoyancy in turbulent shear flow," J. Fluid Mech. **36**, 177 (1969).
3. J.W.S. Rayleigh, "On the dynamics of revolving flows," Proc. R. Soc. London Ser. A **93**, 148 (1916).

4. R.C. Kloosterziel, G.J.F. van Heijst, "An experimental study of unstable barotropic vortices in a rotating fluid," J. Fluid Mech. **223**, 1 (1991).
5. I. Mutabazi, C. Normand, J.E. Wesfreid, "Gap size effects on centrifugally and rotationally driven instabilities," Phys. Fluids A **4**(6), 1199 (1992).
6. B.J. Bayly, "Three-dimensional centrifugal-type instabilities in inviscid two-dimensional flows," Phys. Fluids **31**(1), 56 (1988).
7. O.J.E. Matsson, P.H. Alfredsson, "Curvature- and rotation-induced instabilities in channel flow," J. Fluid Mech. **210**, 537 (1990).
8. A. Lifschitz, E. Hameiri, "Local stability conditions in fluid dynamics," Phys. Fluids A **3** (11), 2644 (1991).
9. G.I. Taylor, A.E. Green, "Mechanism of the production of small eddies from large ones," Proc. Roy. Soc. Lond. A **158**, 499 (1937).
10. D. Sipp, L. Jacquin, "Elliptic instability in two-dimensional flattened Taylor-Green vortices," Phys. Fluids **10**(4), 839 (1998).
11. D. Sipp, E. Lauga, L. Jacquin, "Rotating vortices: centrifugal, elliptic and hyperbolic-type instabilities," Phys. Fluids **11**(12), 3716 (1999).
12. C. Cambon, J.-P. Benoit, L. Shao, L. Jacquin, "Stability analysis and large eddy simulation of rotating turbulence with organized eddies," J. Fluid Mech. **278**, 175 (1994).
13. S. Leblanc, C. Cambon, "On the three-dimensional instabilities of 2D planar flows subjected to Coriolis force," Phys. Fluids, **9** (5), 1307 (1997).
14. D. Sipp, L. Jacquin, "Three-dimensional centrifugal-type instabilities of two-dimensional flows in rotating systems", Submitted to Phys. Fluids (1999).
15. J.T. Stuart, "On finite amplitude oscillations in laminar mixing layers," J. Fluid Mech. **29**, 417 (1967).
16. C.M. Bender, S.A. Orszag, *Advanced mathematical methods for scientists and engineers* (McGraw-Hill, New-York, 1978).

Large Scale Power Fluctuations and Dynamic Properties of Isolated Vortices in a Turbulent Flow

Jean Hugues Titon and Olivier Cadot

Université du Havre, Laboratoire de Mécanique, 25 rue Philippe Lebon, BP 540 76 058 Le Havre Cedex, France

1 Introduction

We present preliminary results concerning the correlation between fluctuations of power injected at large scales in a turbulent flow and the dynamic properties of low pressure filaments . Those two physical variables have already been studied, one at a time, in similar experimental geometries. Douady et al. (1991) [1] and Cadot et al. (1995) [2] have shown that low pressure filament form structures are in fact intense vortical intermittent structures whose size correspond to the integral length and whose dynamic vortex breakdown properties suggest a rapid transfer of energy towards smaller scales (dissipative scales). Labbé et al.(1996) [3] have shown that the power injected fluctuates with a significant degree and that this could be associated to a phenomenon of power intermittency at large scale.

2 Experimental set-up and measurments

The experimental set-up (Fig. 1) is similar to the one used by Cadot et al. (1995) [2] but of different dimensions. The fluid used is water seeded with bubbles that make possible the visualisation of high and low pressure regions [1] [2] . The experimental cell has a closed geometry : it stands as a vertical cylinder with an internal diameter of 20 cm and a height of 40 cm, corresponding to a volume of 12,5 litres. Two coaxial counter-rotating discs of 17,5 cm diameter, each equipped with eight regularly disposed blades are used as stirring device to produce a swirling turbulent flow. The stirrers are forced to rotate at constant angular velocity by two 500 watts PARVEX RS 520 G type servomotors. For this operating mode, the regulation board of the motors delivers an instantaneous measurement of the torque of each motor in the form of a voltage signal. These values of torque are recorded on a PC using LABVIEW? with a sampling frequency of 1 kHz. Simultaneously, the volume between the two stirrers is filmed using a digital video-camera with a sampling frequency of 50 Hz. The experimental cell is illuminated from behind and regions of high bubble concentrations appear as dark zones on a bright background. The visualisation field is adjusted on the central portion of the experimental cell with an optimum zoom setting in order to capture the maximum of events for the present operating mode. The present results are obtained for an acquisition duration of 3 minutes, for a shutter speed of 1/125 s and for a high density of bubbles that allows the visualisation of pressure events occurring near the walls of the cell essentially.

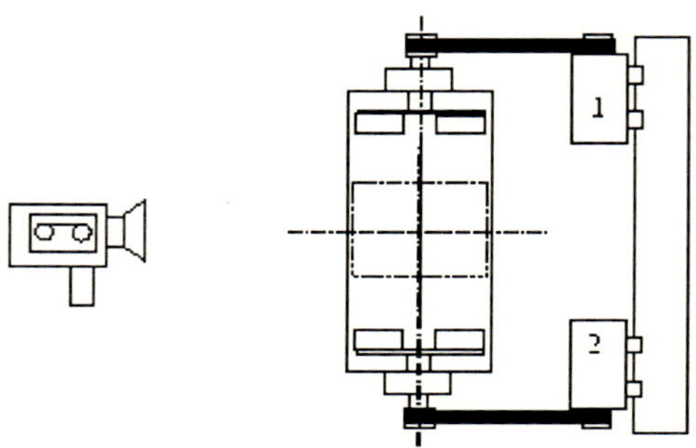

Fig. 1. Experimental set-up ; the visualisation field is represented in broken lines

3 Data processing

3.1 Power injected

Data acquired is processed with LABVIEW.

The values of the frequencies of rotation (Ω_1 and Ω_2) are fixed at Ω_1 = - Ω_2 = $\Omega = 5$ Hz and we verify that these two quantities fluctuate less than 1*percent* by analysing their time series. The Reynolds number for the present geometry is given by the following expression :

$$Re = 2\pi \Omega R^2/\nu,$$

where ν is the kinemetic viscosity of the fluid used (water), and R is the radius of the stirrers : Re is worth 240000 for the chosen frequency of rotation. The time-series of torque ($\Gamma_1(t)$ and $\Gamma_2(t)$) are measured on both servomotors, and Ω_1, Ω_2, $\Gamma_1(t)$ and $\Gamma_2(t)$ are used to build the time-series of power injected by each motor ($P_1(t)$ and $P_2(t)$) represented on Fig. 2 :

$$P_1(t) = \Omega_1.\Gamma_1(t); P_2(t) = \Omega_2.\Gamma_2(t).$$

The instantaneous values of the total power injected $P_T(t)$, in the experimental set-up by both servomotors are obtained by adding $P_1(t)$ and $P_2(t)$:

$$P_T(t) = P_1(t) + P_2(t).$$

Both signals fluctuate at a low frequency that corresponds to the turn-over time $T = 2\pi/\Omega$. The fluctuations of power injected are analysed and presented in the

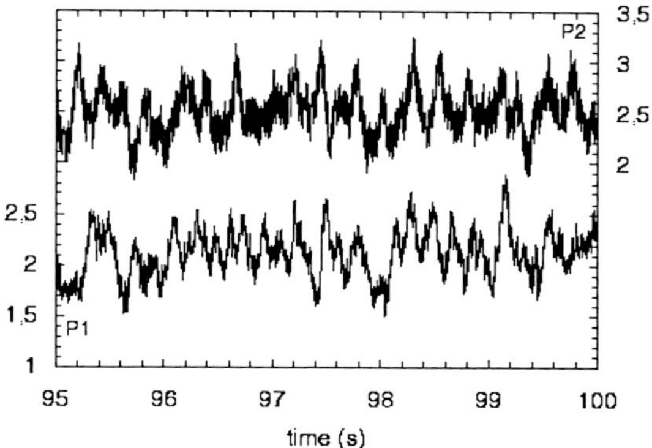

Fig. 2. 5 seconds portion of time-series of power injected by each motor, $P_1(t)$ and $P_2(t)$

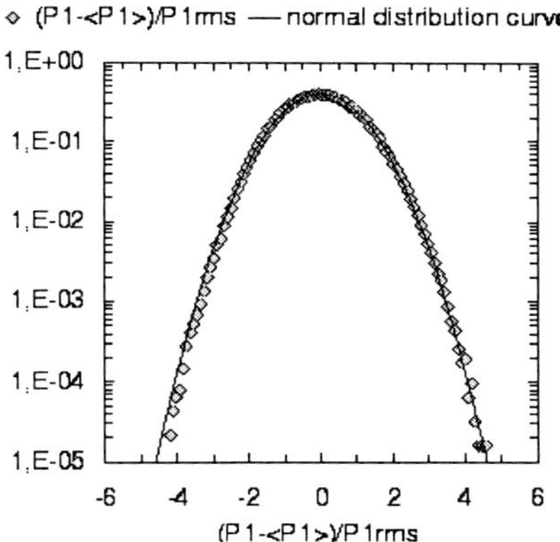

Fig. 3. Probability Density Function of power injected by motor 1, $P_1(t)$ and superposition of a normal distribution curve

form of probability density functions (PDF's). The PDF of power injected by motor 1 is represented on Fig. 3. We observe that $P_1(t)$, as well as $P_2(t)$, follows a normal distribution since motor 1 and motor 2 have similar properties and are mounted in a symmetrical way. Consequently, the total power injected $P_T(t)$ is also quasi-gaussian.

What is particularly noticeable is the simultaneity of appearance of similar events on signals $P_1(t)$ and $P_2(t)$, over the same turn-over time (see Fig. 2). The fluctuations of power $\delta P_1(t)$ and $\delta P_2(t)$ injected by motors 1 and 2 respectively, are analysed through the computation of their cross-correlation. Fig. 4, where we have displayed the values of $< \delta P_1(t).\delta P_2(t+\tau) >$ for $\tau = 0$, shows that these two variables are strongly correlated for a time lapse smaller than one turn-over time ($t/T < 1$). We deduce from the positive sign of the correlation that both servomotors witness either simultaneous drag increases, or simultaneous drag decreases. Such a behaviour prefigures the influence of similar and simultaneous events on both stirrers : visualising the turbulent flow seeded with bubbles to track corresponding power and image events stands as the necessary next step.

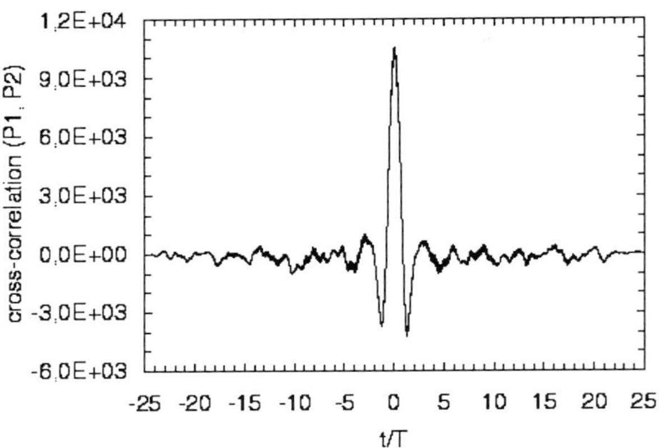

Fig. 4. Cross-correlation of $P_1(t)$ and $P_2(t)$. Note the strong correlation over one turn-over time

3.2 Image analysis

The technique of visualisation used is the same as that designed by Douady et al. (1991) [1] : low pressure coherent structures result from the migration of bubbles to form filamentary dark zones. The proportion of air bubbles in the fluid is optimised so that a maximum of events is recorded. The film is analysed, one image at a time, to build a time-series that is representative of the filamentary activity. The temporal variations of the global luminous intensity for the whole sequence of images recorded are studied and for each image, the corresponding histogram of 255 grey levels is extracted. Fig. 5 (a) is an example of histogram associated to an image where no filament appeared. The probability density function corresponds to a random distribution of bubbles convoluted to a non-homogeneous background intensity. The formation of a low pressure filament (Fig. 5 (b)) is accompanied by a significant decrease of the global luminous intensity of the corresponding images. The associated histograms show an increase of asymmetry, translated by the growth of a tail towards the less intense (or darker) grey levels. The filaments are not stable, and disappear following a dynamic process similar to a vortex breakdown. The initial form of the histogram is restored after the dissipation of the filament.

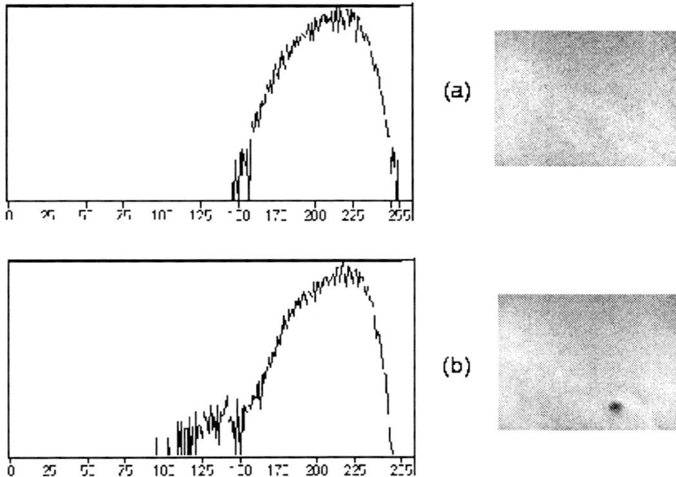

Fig. 5. Histogram of spatial luminous intensity (or levels of grey) and the corresponding image : (a) without a filament, (b) with a filament formed alongside the wall of the experimental cell

The above observations have carried us to study the coefficient of asymmetry or skewness, S, calculated for each histogram of luminous intensity for the whole sequence of images recorded. A time-series is built from the instantaneous values of S and is representative of the filamentary activity (Fig. 6 (b)). From the formation to the total dissipation of a dark coherent structure, the skewness signal displays a large deviation (for the filaments captured in the optimum visualisation field), towards more negative values (t = 126,6 s ; Fig. 6 (b)).

4 Power fluctuations and low pressure filaments

We have devised a series of tools, (a) that allow us to measure the power injected in a closed experimental geometry and appreciate how this physical quantity fluctuates, and, (b) that allow us to detect the presence of and visualise the intermittent coherent structures in a swirling turbulent flow. Our aim is at length, to describe the dynamic properties of the large scale intermittency filaments by studying the fluctuations of power injected at integral scale. The same power is afterwards dissipated at smaller scales of the turbulent flow.

Fig. 6 (a) and (b) shows the preceding 5 seconds portion of skewness signal and the corresponding portion of total power injected signal. We observe that the striking event on the skewness signal coincides with a noticeable decrease in power injected. Such cases of observable correlation are exceptional but there seems to be a repeated tendency for skewness events to be correlated with impor-

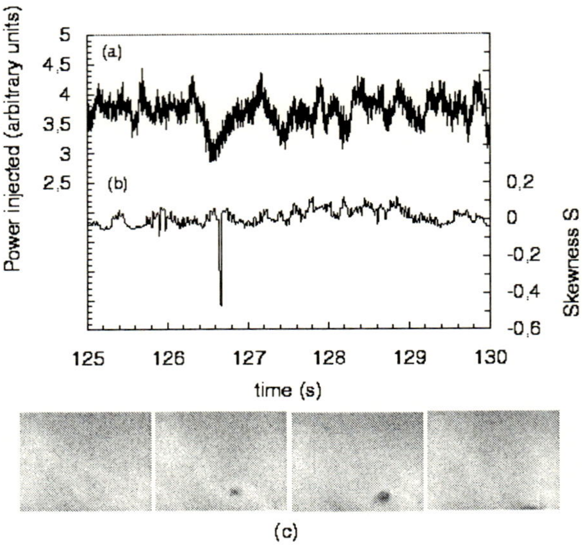

Fig. 6. (a) 5 seconds window of the time series of power injected at large scale, (b) coinciding event on the corresponding window of the time series of skewness of intensity and (c) images associated to the event of skewness of intensity that occurred at 126,5 s, the time lapse between two images is 0,04 s

tant drops of $P_T(t)$. However, the large decrease in the power injected recorded at 127,4 s on Fig. 6 (a) seems to have no repercussion on the skewness signal. Parallel to this case, we have observed on other portions of signals that some large events of skewness have no correspondence with large events of reduction of the total power injected. Signal analysis alone is not enough to establish a clear correlation.

The images corresponding to the event at 126,6 s were extracted from the film (Fig. 6 (c)) : we can observe the formation of a very intense filament close to the wall of the experimental cell. The filament leaves the field captured by the camera on the last image proposed before its total dissipation. We deduce that a certain number of coherent structures, among those formed alongside the experimental wall only (it is very difficult to record volume and wall events at the same time with the same camera), are missed by the camera, since the turbulent flow cannot be filmed integrally. This first argument brings us to criticise the skewness signal : not all events are represented on it. Furthermore the luminous intensity of an event recorded will depend on its position relative to the focal plane of the camera. The small fluctuations on the skewness signal may in fact correspond to very intense filaments. Another difficulty lies in the fact that the measurement of power is integrated over all the large scales of the turbulent flow. Hence, an event associated to a reduction of total power injected recorded by the camera could have been erased on the time-series of power injected by another event, this time associated to an increase of power injected, but formed beyond the field taken into account by the camera. This time, the presence of a filament is taken into account on the time-series of skewness but no correspondence is found on the time-series $P_T(t)$.

5 Conclusion and perspectives

The spatial analysis of the images yields interesting information concerning the distribution and dynamic properties of regions of high and low pressure and the presence of intermittent structures, minimising the importance of the weak points of the method. This preliminary study of the correlation of a suitable variable representative of the filamentary activity with the total power injected in the turbulent flow seems to associate the presence of low pressure filaments to a reduction of the global power injected. This can be interpreted by visualising the formation of these filaments as a storage of energy at large scale. The transfer of this energy to the smaller scales is hence delayed. However when the filaments burst, the energy is redistributed to the smaller dissipative scales. This scenario suggests a mechanism of intermittency of power injected at the large scales of the turbulent flow.

Further flow-seeding conditions and new settings of the video-camera are actually tested to find out the optimum parameters that will allow us to obtain the best visualisation and extract the time-series the most representative of the filamentary activity. Different frequencies of rotation are also imposed to find the ideal counter-rotating regime for which no power event is masked. The characterisation of the energetic properties of intermittent phenomena such as apparition of low pressure filaments, through the study of their correlation with global measurements of the power injected in the turbulent flow, remains our principal aim.

References

1. S. Douady, Y. Couder, M.-E. Brachet: Phys. Rev. Lett. **67**, 983 (1991)
2. O. Cadot, S. Douady, Y. Couder: Phys. Fluids A **7**, 630 (1995)
3. R. Labbé, J.-F. Pinton, S. Fauve: J. Phys. II **6**, 1099 (1996)

Index

Lecture Notes in Physics

For information about Vols. 1–519
please contact your bookseller or Springer-Verlag

Vol. 520: J. A. van Paradijs, J. A. M. Bleeker (Eds.), X-Ray Spectroscopy in Astrophysics. EADN School X. Proceedings, 1997. XV, 530 pages. 1999.

Vol. 521: L. Mathelitsch, W. Plessas (Eds.), Broken Symmetries. Proceedings, 1998. VII, 299 pages. 1999.

Vol. 522: J. W. Clark, T. Lindenau, M. L. Ristig (Eds.), Scientific Applications of Neural Nets. Proceedings, 1998. XIII, 288 pages. 1999.

Vol. 523: B. Wolf, O. Stahl, A. W. Fullerton (Eds.), Variable and Non-spherical Stellar Winds in Luminous Hot Stars. Proceedings, 1998. XX, 424 pages. 1999.

Vol. 524: J. Wess, E. A. Ivanov (Eds.), Supersymmetries and Quantum Symmetries. Proceedings, 1997. XX, 442 pages. 1999.

Vol. 525: A. Ceresole, C. Kounnas, D. Lüst, S. Theisen (Eds.), Quantum Aspects of Gauge Theories, Supersymmetry and Unification. Proceedings, 1998. X, 511 pages. 1999.

Vol. 526: H.-P. Breuer, F. Petruccione (Eds.), Open Systems and Measurement in Relativistic Quantum Theory. Proceedings, 1998. VIII, 240 pages. 1999.

Vol. 527: D. Reguera, J. M. G. Vilar, J. M. Rubí (Eds.), Statistical Mechanics of Biocomplexity. Proceedings, 1998. XI, 318 pages. 1999.

Vol. 528: I. Peschel, X. Wang, M. Kaulke, K. Hallberg (Eds.), Density-Matrix Renormalization. Proceedings, 1998. XVI, 355 pages. 1999.

Vol. 529: S. Biringen, H. Örs, A. Tezel, J.H. Ferziger (Eds.), Industrial and Environmental Applications of Direct and Large-Eddy Simulation. Proceedings, 1998. XVI, 301 pages. 1999.

Vol. 530: H.-J. Röser, K. Meisenheimer (Eds.), The Radio Galaxy Messier 87. Proceedings, 1997. XIII, 342 pages. 1999.

Vol. 531: H. Benisty, J.-M. Gérard, R. Houdré, J. Rarity, C. Weisbuch (Eds.), Confined Photon Systems. Proceedings, 1998. X, 496 pages. 1999.

Vol. 532: S. C. Müller, J. Parisi, W. Zimmermann (Eds.), Transport and Structure. Their Competitive Roles in Biophysics and Chemistry. XII, 400 pages. 1999.

Vol. 533: K. Hutter, Y. Wang, H. Beer (Eds.), Advances in Cold-Region Thermal Engineering and Sciences. Proceedings, 1999. XIV, 608 pages. 1999.

Vol. 534: F. Moreno, F. González (Eds.), Light Scattering from Microstructures. Proceedings, 1998. XII, 300 pages. 2000

Vol. 535: H. Dreyssé (Ed.), Electronic Structure and Physical Properties of Solids: The Uses of the LMTO Method. Proceedings, 1998. XIV, 458 pages. 2000.

Vol. 536: T. Passot, P.-L. Sulem (Eds.), Nonlinear MHD Waves and Turbulence. Proceedings, 1998. X, 385 pages. 1999.

Vol. 537: S. Cotsakis, G. W. Gibbons (Eds.), Mathematical and Quantum Aspects of Relativity and Cosmology. Proceedings, 1998. XII, 251 pages. 1999.

Vol. 538: Ph. Blanchard, D. Giulini, E. Joos, C. Kiefer, I.-O. Stamatescu (Eds.), Decoherence: Theoretical, Experimental, and Conceptual Problems. Proceedings, 1998. XII, 345 pages. 2000.

Vol. 539: A. Borowiec, W. Cegła, B. Jancewicz, W. Karwowski (Eds.), Theoretical Physics. Fin de Siècle. Proceedings, 1998. XX, 319 pages. 2000.

Vol. 540: B. G. Schmidt (Ed.), Einstein's Field Equations and Their Physical Implications. Selected Essays. 1999. XIII, 429 pages. 2000

Vol. 541: J. Kowalski-Glikman (Ed.), Towards Quantum Gravity. Proceedings, 1999. XII, 376 pages. 2000.

Vol. 542: P. L. Christiansen, M. P. Sørensen, A. C. Scott (Eds.), Nonlinear Science at the Dawn of the 21st Century. Proceedings, 1998. XXVI, 458 pages. 2000.

Vol. 543: H. Gausterer, H. Grosse, L. Pittner (Eds.), Geometry and Quantum Physics. Proceedings, 1999. VIII, 408 pages. 2000.

Vol. 544: T. Brandes (Ed.), Low-Dimensional Systems. Interactions and Transport Properties. Proceedings, 1999. VIII, 219 pages. 2000

Vol. 545: J. Klamut, B. W. Veal, B. M. Dabrowski, P. W. Klamut, M. Kazimierski (Eds.), New Developments in High-Temperature Superconductivity. Proceedings, 1998. VIII, 275 pages. 2000.

Vol. 546: G. Grindhammer, B. A. Kniehl, G. Kramer (Eds.), New Trends in HERA Physics 1999. Proceedings, 1999. XIV, 460 pages. 2000.

Vol. 547: D. Reguera, G. Platero, L.L. Bonilla, J.M. Rubí(Eds.), Statistical and Dynamical Aspects of Mesoscopic Systems. Proceedings, 1999. XII, 357 pages. 2000.

Vol. 548: D. Lemke, M. Stickel, K. Wilke (Eds.), ISO Surveys of a Dusty Universe. Proceedings, 1999. XIV, 432 pages. 2000.

Vol. 549: C. Egbers, G. Pfister (Eds.), Physics of Rotating Fluids. Proceedings, 1999. XIV, 439 pages. 2000.

Vol. 550: M. Planat (Ed.), Noise, Oscillators and Algebraic Randomness. Proceedings, 1999. VIII, 417 pages. 2000.

Vol. 551: B. Brogliato (Ed.), Impacts in Mechanical Systems. Analysis and Modelling. Lectures, 1999. IX, 273 pages. 2000.

Vol. 552: Z. Chen, R. E. Ewing, Z.-C. Shi (Eds.), Numerical Treatment of Multiphase Flows in Porous Media. Proceedings, 1999. XXI, 445 pages. 2000.

Vol. 553: J.-P. Rozelot, L. Klein, J.-C. Vial Eds.), Transport of Energy Conversion in the Heliosphere. Proceedings, 1998. IX, 215 pages. 2000.

Vol. 554: K. R. Mecke, D. Stoyan (Eds.), Statistical Physics and Spatial Statistics. The Art of Analyzing and Modeling Spatial Structures and Pattern Formation. Proceedings, 1999. XII, 415 pages. 2000.

Vol. 555: A. Maurel, P. Petitjeans (Eds.), Vortex Structure and Dynamics. Proceedings, 1999. XII, 319 pages. 2000.

Monographs

For information about Vols. 1–20
please contact your bookseller or Springer-Verlag

Vol. m 21: G. P. Berman, E. N. Bulgakov, D. D. Holm, Crossover-Time in Quantum Boson and Spin Systems. XI, 268 pages. 1994.

Vol. m 22: M.-O. Hongler, Chaotic and Stochastic Behaviour in Automatic Production Lines. V, 85 pages. 1994.

Vol. m 23: V. S. Viswanath, G. Müller, The Recursion Method. X, 259 pages. 1994.

Vol. m 24: A. Ern, V. Giovangigli, Multicomponent Transport Algorithms. XIV, 427 pages. 1994.

Vol. m 25: A. V. Bogdanov, G. V. Dubrovskiy, M. P. Krutikov, D. V. Kulginov, V. M. Strelchenya, Interaction of Gases with Surfaces. XIV, 132 pages. 1995.

Vol. m 26: M. Dineykhan, G. V. Efimov, G. Ganbold, S. N. Nedelko, Oscillator Representation in Quantum Physics. IX, 279 pages. 1995.

Vol. m 27: J. T. Ottesen, Infinite Dimensional Groups and Algebras in Quantum Physics. IX, 218 pages. 1995.

Vol. m 28: O. Piguet, S. P. Sorella, Algebraic Renormalization. IX, 134 pages. 1995.

Vol. m 29: C. Bendjaballah, Introduction to Photon Communication. VII, 193 pages. 1995.

Vol. m 30: A. J. Greer, W. J. Kossler, Low Magnetic Fields in Anisotropic Superconductors. VII, 161 pages. 1995.

Vol. m 31 (Corr. Second Printing): P. Busch, M. Grabowski, P.J. Lahti, Operational Quantum Physics. XII, 230 pages. 1997.

Vol. m 32: L. de Broglie, Diverses questions de mécanique et de thermodynamique classiques et relativistes. XII, 198 pages. 1995.

Vol. m 33: R. Alkofer, H. Reinhardt, Chiral Quark Dynamics. VIII, 115 pages. 1995.

Vol. m 34: R. Jost, Das Märchen vom Elfenbeinernen Turm. VIII, 286 pages. 1995.

Vol. m 35: E. Elizalde, Ten Physical Applications of Spectral Zeta Functions. XIV, 224 pages. 1995.

Vol. m 36: G. Dunne, Self-Dual Chern-Simons Theories. X, 217 pages. 1995.

Vol. m 37: S. Childress, A.D. Gilbert, Stretch, Twist, Fold: The Fast Dynamo. XI, 406 pages. 1995.

Vol. m 38: J. González, M. A. Martín-Delgado, G. Sierra, A. H. Vozmediano, Quantum Electron Liquids and High-Tc Superconductivity. X, 299 pages. 1995.

Vol. m 39: L. Pittner, Algebraic Foundations of Non-Com-mutative Differential Geometry and Quantum Groups. XII, 469 pages. 1996.

Vol. m 40: H.-J. Borchers, Translation Group and Particle Representations in Quantum Field Theory. VII, 131 pages. 1996.

Vol. m 41: B. K. Chakrabarti, A. Dutta, P. Sen, Quantum Ising Phases and Transitions in Transverse Ising Models. X, 204 pages. 1996.

Vol. m 42: P. Bouwknegt, J. McCarthy, K. Pilch, The W3 Algebra. Modules, Semi-infinite Cohomology and BV Algebras. XI, 204 pages. 1996.

Vol. m 43: M. Schottenloher, A Mathematical Introduction to Conformal Field Theory. VIII, 142 pages. 1997.

Vol. m 44: A. Bach, Indistinguishable Classical Particles. VIII, 157 pages. 1997.

Vol. m 45: M. Ferrari, V. T. Granik, A. Imam, J. C. Nadeau (Eds.), Advances in Doublet Mechanics. XVI, 214 pages. 1997.

Vol. m 46: M. Camenzind, Les noyaux actifs de galaxies. XVIII, 218 pages. 1997.

Vol. m 47: L. M. Zubov, Nonlinear Theory of Dislocations and Disclinations in Elastic Body. VI, 205 pages. 1997.

Vol. m 48: P. Kopietz, Bosonization of Interacting Fermions in Arbitrary Dimensions. XII, 259 pages. 1997.

Vol. m 49: M. Zak, J. B. Zbilut, R. E. Meyers, From Instability to Intelligence. Complexity and Predictability in Nonlinear Dynamics. XIV, 552 pages. 1997.

Vol. m 50: J. Ambjørn, M. Carfora, A. Marzuoli, The Geometry of Dynamical Triangulations. VI, 197 pages. 1997.

Vol. m 51: G. Landi, An Introduction to Noncommutative Spaces and Their Geometries. XI, 200 pages. 1997.

Vol. m 52: M. Hénon, Generating Families in the Restricted Three-Body Problem. XI, 278 pages. 1997.

Vol. m 53: M. Gad-el-Hak, A. Pollard, J.-P. Bonnet (Eds.), Flow Control. Fundamentals and Practices. XII, 527 pages. 1998.

Vol. m 54: Y. Suzuki, K. Varga, Stochastic Variational Approach to Quantum-Mechanical Few-Body Problems. XIV, 324 pages. 1998.

Vol. m 55: F. Busse, S. C. Müller, Evolution of Spontaneous Structures in Dissipative Continuous Systems. X, 559 pages. 1998.

Vol. m 56: R. Haussmann, Self-consistent Quantum Field Theory and Bosonization for Strongly Correlated Electron Systems. VIII, 173 pages. 1999.

Vol. m 57: G. Cicogna, G. Gaeta, Symmetry and Perturbation Theory in Nonlinear Dynamics. XI, 208 pages. 1999.

Vol. m 58: J. Daillant, A. Gibaud (Eds.), X-Ray and Neutron Reflectivity: Principles and Applications. XVIII, 331 pages. 1999.

Vol. m 59: M. Kriele, Spacetime. Foundations of General Relativity and Differential Geometry. XV, 432 pages. 1999.

Vol. m 60: J. T. Londergan, J. P. Carini, D. P. Murdock, Binding and Scattering in Two-Dimensional Systems. Applications to Quantum Wires, Waveguides and Photonic Crystals. X, 222 pages. 1999.

Vol. m 61: V. Perlick, Ray Optics, Fermat's Principle, and Applications to General Relativity. X, 220 pages. 2000.

Vol. m 63: R. J. Szabo, Ray Optics, Equivariant Cohomology and Localization of Path Integrals. XII, 315 pages. 2000.

Vol. m 64: I. G. Avramidi, Heat Kernel and Quantum Gravity. X, 143 pages. 2000.